Logica Universalis

Towards a General Theory of Logic

Second Edition

Jean-Yves Beziau

Editor

Birkhäuser
Basel · Boston · Berlin

Editor:

Jean-Yves Beziau
Swiss National Science Foundation
University of Neuchâtel
Espace Louis Agassiz 1
2000 Neuchâtel
Switzerland
e-mail: jean-yves.beziau@unine.ch

First Edition 2005

2000 Mathematics Subject Classification: Primary 03B22; Secondary 03B50, 03B47, 03C95, 03B53, 03G10, 03B20, 18A15, 03F03, 54H10

Library of Congress Control Number: 2007923780

Bibliographic information published by Die Deutsche Bibliothek:
Die Deutsche Bibliothek lists this publication in the Deutsche Nationalbibliographie; detailed bibliographic data is available in the internet at <http://dnb.ddb.de>

ISBN 978-3-7643-8353-4 Birkhäuser Verlag AG, Basel · Boston · Berlin

© 2007 Birkhäuser Verlag AG
Basel · Boston · Berlin
P.O. Box 133, CH-4010 Basel, Switzerland
Part of Springer Science+Business Media
Printed on acid-free paper produced from chlorine-free pulp. TCF ∞
Printed in Germany

ISBN 978-3-7643-8353-4   e-ISBN 978-3-7643-8354-1
9 8 7 6 5 4 3 2 1   www.birkhauser.ch

Contents

Part III. Tools and Concepts for Universal Logic

Preface

Logica Universalis (or *Universal Logic*, *Logique Universelle*, *Universelle Logik*, in vernacular languages) is not a new logic, but a general theory of logics, considered as mathematical structures. The name was introduced about ten years ago, but the subject is as old as the beginning of modern logic: Alfred Tarski and other Polish logicians such as Adolf Lindenbaum developed a general theory of logics at the end of the 1920s based on consequence operations and logical matrices. Talking about the papers of Tarski dealing with this topic, John Etchemendy says: "What is most striking about these early papers, especially against their historical backdrop, is the extraordinary generality and abstractness of the perspective adopted" [4]. After the second world war, this line of work was pursued mainly in Poland and became a bit of an esoteric subject. Jerzy Łoś's fundamental monograph on logical matrices was never translated in English and the work of Roman Suszko on abstract logics remained unknown outside of Poland during many years.

Things started to change during the 1980s. Logic, which had been dominated during many years by some problems related to the foundations of mathematics or other metaphysical questions, was back to reality. Under the impulsion of artificial intelligence, computer science and cognitive sciences, new logical systems were created to give an account to the variety of reasonings of everyday life and to build machines, robots, programs that can act efficiently in difficult situations, for example that can smoothly process inconsistent and incomplete information. John McCarthy launched non-monotonic logic, few years later Jean-Yves Girard gave birth to linear logic. Logics were proliferating: each day a new logic was born. By the mid eighties, there were more logics on earth than atoms in the universe. People began to develop general tools for a systematic study of this huge amount of logics, trying to put some order in this chaotic multiplicity. Old tools such as consequence operations, logical matrices, sequent calculus, Kripke structures, were revived and reshaped to meet this new goal. For example sequent calculus was the unifying instrument for substructural logics. New powerful tools were also activated, such as labelled deductive systems by Dov Gabbay.

Amazingly, many different people in many different places around the world, quite independently, started to work in this new perspective of a general theory of logics, writing different monographs, each one presenting his own way to treat the problem: Norman Martin's emphasis was on Hilbert systems [9], Richard Epstein's,

This research was supported by a grant of the Swiss National Science Foundation.

on semantical tools, in particular relational structures and logical matrices [5], Newton da Costa's, on non truth-functional bivalent semantics [7], John Cleave's, on consequence and algebra [3], Arnold Koslow's, on Hertz's abstract deductive systems [8]. This was also the time when was published a monograph by Ryszard Wójcicki on consequence operations making available for the first time to a wide public the main concepts and results of Polish logic [10], and the time when Dov Gabbay edited a book entitled *What is a logical system?* gathering a collection of papers trying to answer this question in many different ways [6]. Through all these publications, the generality and abstractness of Tarski's early work was being recovered. It is surrounded by this atmosphere that I was doing my PhD [2] and that I coined in the middle of a winter in Poland the expression "universal logic" [1], by analogy to the expression "universal algebra".

The present book contains recent works on universal logic by first-class researchers from all around the world. The book is full of new and challenging ideas that will guide the future of this exciting subject. It will be of interest for people who want to better understand what logic is. It will help those who are lost in the jungle of heterogeneous logical systems to find a way. Tools and concepts are provided here for those who want to study classes of already existing logics or want to design and build new ones.

In Part I, different frameworks for a general theory of logics are presented. Algebra, topology, category theory are involved. The first paper, written by myself, is a historical overview of the different logical structures and methods which were proposed during the XXth century: Tarski's consequence operator and its variants in particular Suszko's abstract logic, structures arising from Hertz and Gentzen's deductive systems, da Costa's theory of valuation, etc. This survey paper presents and explains many concepts that are used in other papers of the book. The following paper, by Marta García-Matos and Jouko Väänänen, gives a hint of how *abstract model theory* can be used for developing universal logic. Although abstract logic and abstract model theory are expressions which look similar, they refer to two different traditions. Abstract logic has been developed by Suszko in the context of the Polish tradition focusing on a general theory of zero-order logics (i.e. propositional logics). On the other hand, the aim of abstract model theory has been the study of classes of higher order logics. The combination of abstract model theory with abstract logic is surely an important step towards the development of universal logic. It is also something more than natural if we think that both theories have their origins in the work of Alfred Tarksi. Steffen Lewitzka's approach is also model-theoretical, but based on topology. He defines in a topological way logic-homomorphims between abstract logics, which are mappings that preserve structural properties of logics. And he shows that those model-theoretical abstract logics together with a strong form of logic-homomorphisms give rise to the notion of institution. Then comes the work of Ramon Jansana which is a typical example of what is nowadays called *abstract algebraic logic*, the study of algebraization of logics, a speciality of the Barcelona logic group. Within this framework, abstract logics are considered as generalized matrices and are used as models for logics. Finally, Pierre Ageron's paper deals with logics for which the law of *self-deductibility* does not hold. According to this law, a formula is always a consequence of itself, it was one of the basic axioms of Tarski's consequence operator. Ageron shows here how to develop logical structures without this law using tools from category theory.

The papers of Part II deal with a central problem of universal logic: the question of identity between logical structures. A logic, like classical logic, is not a given structure, but a class of structures that can be identified with the help of a given criterion. According to this criterion, we say that structures of a given class are equivalent, congruent or simply identical. Although this question may at first look trivial, it is in fact a very difficult question which is strongly connected to the question of what a logical structure is. In other words, it is not possible to try to explain how to identify different logical structures without investigating at the same time the very nature of logical structures. This is what makes the subject deep and fascinating. Three papers and seven authors are tackling here the problem, using different strategies. Caleiro and Gonçalves's work is based on concepts from category theory and they say that two logics are the same, *equipollent* in their terminology, when there exist uniform translations between the two logical languages that induce an isomorphism on the corresponding theory spaces. They gave several significant illustrations of equipollent and non equipollent logics. Mossakowski, Goguen, Diaconescu and Tarlecki also use category theory, more specifically their work is based on the notion of *institution*. They argue that every plausible notion of equivalence of logics can be formalized using this notion. Lutz Straßburger's paper is proof-theoretically oriented, he defines identity of proofs via *proof nets* and identity of logics via pre-orders.

In part III, different tools and concepts are presented that can be useful for the study of logics. The papers by Arnon Avron and by Carlos Caleiro et al. both deal with a concept very popular in the Polish tradition, the concept of logical matrices, the basic tool for many-valued logics. In his paper Avron studies the notion of *non-deterministic matrices* which allows to easily construct semantics for proof systems and can be used to prove decidability. This tool can be applied to a wide range of logics, in particular to logics with a formal consistency operator. Caleiro, Carnielli, Coniglio and Marcos discuss Suszko's thesis, according to which any logic is bivalent, and present some techniques which permit to construct in a effective way a *bivalent semantics*, generally not truth-functional, from a many-valued matrix. Their paper is illustrated by some interesting examples, including Belnap's four-valued logic. Then comes a paper by David Makinson, one of the main responsible for the revival of Tarski's consequence operator at the beginning of the 1980s. He used it at the main tool, on the one hand for the development together with Carlos Alchourrón and Peter Gärdenfors, of theory change (universally known today under the acronym AGM), on the other hand as a basis for a general theory of non monotonic logics. In both cases, Makinson's use of Tarski's theory was creative, he kept the original elegant abstract spirit, but widened and extended the basic underlying concepts. Here again he is innovative defining within classical propositional logic two new concepts, *logical friendliness and sympathy*, which lead to some consequence relations with non standard properties. The paper by Lloyd Humberstone is no less original and brilliant, he studies the very interesting phenomenon of *logical discrimination*. The question he examines is in which circumstances, discrimination, i.e. distinction between formulas, is correlated with the strength of a logic. The work of Humberstone is a very good example of the philosophical import of universal logic. By a careful examination of a phenomenon like discrimination, that requires a precise mathematical framework, one can see to which extent a statement with philosophical flavor saying that discrimination is inversely proportional to strength is true or not.

References

[1] J.-Y. Beziau, *Universal Logic*. Proceedings of the 8th International Colloquium - Logica'94, T. Childers and O. Majer (eds), Czech Academy of Sciences, Prague, 2004, pp.73-93.

[2] J.-Y. Beziau, *Recherches sur la Logique Universelle*. PhD Thesis, Department of Mathematics, University Denis Diderot, Paris, 1995.

[3] J.P. Cleave, *A study of logics*. Clarendon, Oxford, 1991.

[4] J.Etchemendy, *Tarski on truth and logical consequence*. The Journal of Symbolic Logic, **58** (1988), 51-79.

[5] R.L. Epstein, *The semantic foundations of logic*. Kluwer, Dordrecht, 1990.

[6] D. Gabbay (ed), *What is a logical system?* Clarendon, Oxford, 1994.

[7] N. Grana, *Sulla teoria delle valutazioni di N.C.A. da Costa*. Liguori, Naples, 1990.

[8] A. Koslow, *A structuralist theory of logic*. Cambridge University Press, New York, 1992.

[9] N.M. Martin, *Systems of logic*. Cambridge University Press, New York, 1989.

[10] R. Wójcicki, *Theory of logical calculi – Basic theory of consequence operations*. Kluwer, Dordrecht, 1988.

Jean-Yves Beziau
Neuchâtel,
January 31th, 2005.

Preface to the Second Edition

The first edition of this book was a success and was sold out in a short time. In this second edition you will find the same authors and table of contents, but most of the papers have been extended and improved.

Universal logic is making its way. A new journal entitled *Logica Universalis* has been created and soon a book series dedicated to the subject will also be launched by Birkhäuser: *Studies in Universal Logic*.

Jean-Yves Beziau
Neuchâtel, Switzerland
February, 2007.

Part I
Universal Logic: Frameworks and Structures

J.-Y. Beziau (Ed.), *Logica Universalis*, *2nd edition*, 3–17

From Consequence Operator to Universal Logic: A Survey of General Abstract Logic

Jean-Yves Beziau

Abstract. We present an overview of the different frameworks and structures that have been proposed during the last century in order to develop a general theory of logics. This includes Tarski's consequence operator, logical matrices, Hertz's Satzsysteme, Gentzen's sequent calculus, Suszko's abstract logic, algebraic logic, da Costa's theory of valuation and universal logic itself.

Mathematics Subject Classification (2000). Primary 03B22 ; Secondary 03B50, 03B47, 03B53, 03G10.

Keywords. Universal logic, consequence operator, abstract logic, substructural logic, algebraic logic, many-valued logics, truth-functionality.

1. Introduction

During the XXth century, numerous logics have been created: intuitionistic logic, quantum logic, modal logic, many-valued logic, relevant logic, paraconsistent logic, erotetic logic, polar logic, linear logic, non-monotonic logic, dynamic logic, free logic, fuzzy logic, paracomplete logic, etc. And the future will see the birth of many other logics that one can hardly imagine at the present time.

Facing this incredible multiplicity, one can wonder if there is not a way to find common features which allow one to unify the study of all these particular systems into a science called *logic*.

In what follows we describe various attempts that have been made during the XXth century to develop a general theory of logics.

This research was supported by a grant of the Swiss National Science Foundation. I would like to thank Newton da Costa and Alexandre Costa-Leite for useful comments.

2. Tarski's consequence operator

2.1. Tarski's three axioms

Undoubtedly, Tarski has, among many other things, to be considered as the initiator of a general theory of logics.

At the end of the twenties, he launched the theory of consequence operator [43]. This theory is about an "operator", a function Cn defined on the power set of a given set $\mathbb{S}$. Following the philosophical ideas of his master, Leśniewski, Tarski calls these objects "meaningful sentences". But in fact, the name does not matter, the important thing is that here Tarski is considering a very general theory, because the *nature* of these objects is not specified. For Tarski, these sentences can be sentences of any kind of scientific languages, since his work is concerned with the *methodology of deductive sciences*, and not only with *metamathematics*. The function Cn obeys three basic axioms, for any *theories* (i.e. sets of sentences) T and U:

[TAR1] $T \subseteq CnT$

[TAR2] if $T \subseteq U$ then $CnT \subseteq CnU$

[TAR3] $CnCnT \subseteq CnT$

Hereafter, a structure $\langle \mathbb{S}; Cn \rangle$ where Cn obeys the three above axioms will be called a *Tarski structure*. [1]

2.2. Axiomatizing axiomatic proof systems

Why these axioms? Tarski wanted to give a general characterization of the notion of *deduction*. At this time, the standard notion of deduction was the one given by what is called nowadays *Hilbert-type proof systems*, or *axiomatic proof systems*. It is easy to check that any notion of deduction defined with the help of this kind of systems obeys the three above axioms.

One could say that Tarski was in this sense axiomatizing axiomatic proof systems. It is very important however to understand the difference between the two occurrences of the word *axiom* here. Tarski's axioms are not axioms of a proof system, although they can be considered as such, at a more complex level. One should rather consider these axioms model-theoretically, as defining a certain class of structures.

One can wonder if these three axioms characterize exactly the notion of deduction in the sense that any Tarski structure $\langle \mathbb{S}; Cn \rangle$ verifying these axioms can be defined in a Hilbertian proof-theoretically way.

---

[1] In this paper we will do a bit of taxonomy, fixing names to different logical structures.

2.3. Semantical consequence and completeness

Before examining this question, let us note that these axioms axiomatize also the *semantical* or *model-theoretical notion of consequence* given later by Tarski [45]. We do not know if Tarski had also this notion in the back of his mind when he proposed the consequence operator.

Anyway, there is an interesting manner to connect these two notions which forms the heart of a general completeness theorem. If we consider a Tarski structure $\langle \mathbb{S}; Cn \rangle$ and the class of *closed theories* of this structure, i.e., theories such that $CnT = T$, this class forms a sound and complete semantics for this structure, in the sense that the semantical notion of consequence $\vDash$ defined by:

$$T \vDash a \text{ iff for any closed theory } U \text{ such that } T \subseteq U, a \in U$$

coincides with Cn.

Now back to the previous point, it is clear that Tarski's three axioms do not characterize the notion of Hilbertian proof-theoretical deduction, since for example there are some structures $\langle \mathbb{S}; Cn \rangle$ like second-order logic that obey these three axioms but cannot be defined in this proof-theoretical sense. This is because second-order logic is not compact, or more precisely, finite.

2.4. Compactness and finiteness

Tarski had also an *axiom of finiteness*:

[FIN] $CnT = \bigcup CnF \ (F \subseteq T, F \text{ finite})$

In classical logic, this axiom is equivalent to the *axiom of compactness*:

[COM]   if for every a, $a \in CnT$, then there is a finite $F \subseteq T$ such that
         for every a, $a \in CnF$

But in general they are not. Clearly it is the axiom of finiteness which characterizes the Hilbertian proof-theoretical notion of deduction.

Once one has this axiom, one has also a more interesting semantical notion of consequence. Let us call *maximal* a theory T, such that $CnU = \mathbb{S}$, for every strict extension U of T and *relatively maximal* a theory T such that there is a a such that $a \notin T$ and for every strict extension U of T, $a \in CnU$. The class of relatively maximal theories characterizes any finite consequence operator (i.e. a consequence operator obeying the finiteness axiom together with Tarski's three axioms). In the case of an *absolute* Tarski structure, i.e. a structure where all relatively maximal theories are also maximal [2], maximal theories characterize finite consequence operators, but it is not true in general [9].

---

[2] The terminology "absolute" was suggested by David Makinson.

3. Hertz and Gentzen's proof systems

3.1. Hertz's Satzsysteme

It is difficult to know exactly the origin of the work of Paul Hertz about *Satzsysteme* developed during the 1920s [31]. But something is clear, it emerged within the Hilbertian stream and it is proof-theoretically oriented, although it is a very abstract approach ; like in the case of Tarski's consequence operator, the nature of the basic objects is not specified.

What Hertz calls a *Satz* is something of the form $u_1 \ldots u_n \to v$. One could interpret this as the sentence "u_1 and...and u_n implies v", considering $\to$ as material implication, and Hertz himself suggests this, saying that $\to$ is taken from Whitehead and Russell's *Principia Mathematica*.

However, we shall interpret here Hertz's *Satz* $u_1 \ldots u_n \to v$ in the perspective of Gentzen's work considering it as the prototype of a Gentzen's sequent and we shall just call such a *Satz*, a protosequent. "u" and "v" are called "elements" in Hertz's terminology. Following again Gentzen, we will just consider that they are sentences of a possible unspecified language. Hertz uses the word "complex" to denote a finite site Γ of "elements", we will just call such a set a finite theory.

Hertz's notion of *Satzsysteme* is based on a notion of proof which is similar to the Hilbert's one, except that elements of the proof are protosequents. Therefore rules in *Satzsysteme* have as premises, protosequents, and as a conclusion a protosequent. Axioms are protosequents. There are only one kind of axiom and two rules in Hertz's system:

[HER1] $\Gamma \to \alpha \ (\alpha \in \Gamma)$

[HER2] $\dfrac{\Gamma \to \alpha}{\Gamma\Delta \to \alpha}$

[HER3] $\dfrac{\Gamma \to \alpha \qquad \Delta\alpha \to \beta}{\Gamma\Delta \to \beta}$

If we consider, something that Hertz didn't, the structure $\langle \mathbb{S}; \vdash \rangle$ generated by such a system in the following way: $T \vdash a$ iff there exists a finite subtheory Γ of T such that there is a proof of the protosequent $\Gamma \to a$ in the system, we have a structure which is equivalent, modulo trivial exchanges between $\vdash$ and Cn, to a Tarski structure.

As we have seen, Tarski's motivations are clear and one can perceive the interest of his proposal. In the case of Hertz's work, it is not clear at all. One can see a step towards a kind of generalization of Hilbert-type proof-theoretical concepts. But at first, Hertz's notion of *Satz* is quite strange. In the light of Gentzen's work, we are now conscious of the incredible power of this notion, but many people think that without Gentzen's work, Hertz's work would have been completely ignored.

In fact Hertz's work is generally not well-known. It is therefore important to stress that Gentzen started his researches, which will lead him to his famous sequent calculus, by studying Hertz's work, probably on a suggestion of Paul Bernays. Gentzen's first paper [29] is entirely devoted to Hertz's system and among several results he proves that this system is sound and complete with respect to the semantics of closed theories (although he doesn't use such a language). Due to his paper, the model-theoretical notion of consequence and this general related completeness theorem can be credited to Gentzen as well as to Tarski. Apparently his work was carried out in total independence to the work of Tarski.

3.2. Gentzen's sequent calculus

Gentzen's sequent calculus differs in several points to Hertz's system. Instead of Hertz's *S*atz, Gentzen considers "sequents", i.e, objects of the form $u_1, \ldots, u_n \to v_1, \ldots, v_n$ where $u_1, \ldots, u_n$ and $v_1, \ldots, v_n$ are sequences of sentences (mind the comma!), hence the name "sequent calculus". Gentzen's rules are divided in two categories: structural rules and logical rules. Logical rules are rules concerning logical operators. Such rules appear here because Gentzen is not only interested to work at the "abstract" level but also with specific logics, mainly classical and intuitionistic logics. Gentzen's structural rules are the following:

[GEN1] $\Sigma \to \alpha \ (\alpha \in \Sigma)$

[GEN2l] $\dfrac{\Sigma \to \Xi}{\alpha, \Sigma \to \Xi}$ [GEN2r] $\dfrac{\Sigma \to \Xi}{\Sigma \to \Xi, \alpha}$

[GEN3] $\dfrac{\Sigma \to \Xi, \alpha \qquad \alpha, \Omega \to \Pi}{\Sigma, \Omega \to \Xi, \Pi}$

[GEN4l] $\dfrac{\Sigma(\alpha, \alpha) \to \Xi}{\Sigma(\alpha) \to \Xi}$ [GEN4r] $\dfrac{\Sigma \to \Xi(\alpha, \alpha)}{\Sigma \to \Xi(\alpha)}$

[GEN5l] $\dfrac{\Sigma(\alpha, \beta) \to \Xi}{\Sigma(\beta, \alpha) \to \Xi}$ [GEN5r] $\dfrac{\Sigma \to \Xi(\alpha, \beta)}{\Sigma \to \Xi(\beta, \alpha)}$

where Σ, Ω, Π and Ξ are sequences and something like $\Sigma(\alpha, \beta)$ means that α and β are occurrences of formulas appearing in the sequence Σ in that order.

[GEN1], [GEN2] and [GEN3] are adaptations of [HER1], [HER2] and [HER3] to the sequent context. The rules [GEN4] and [GEN5] were implicit in the case of Hertz's system, but if one considers sequents instead of finite sets of sentences, these rules are necessary.

Later on, people started to work with Tait's version of Gentzen's sequent calculus. It is a Hertzianization of Gentzen's system where finite sets are considered instead of sequences, and where therefore there are no contraction [GEN4] and permutation [GEN5] rules.

In view of Tait's version, one may think that Gentzen's system is a useless *détour*. But it is not, as the recent development of substructural logics has dramatically shown.

3.3. Scott structure

An important point is that Gentzen is considering a multiplicity of sentences on the right. So even if one considers Tait's versions of Gentzen's structural rules, we have here something different from Hertz's rules, in particular we must have two thinning rules, [GEN2l] and [GEN2r].

The multiplicity of the sentences on the right is very important in Gentzen's original system, since, as it is known, if one reduces this multiplicity to unity on the right, one goes from classical logic to intuitionistic logic.

However, even if we stay at the abstract level, the multiplicity is an important thing that permits to work with more symmetry. The structure generated by an abstract (only structural rules) Tait's version of Gentzen's sequent calculus is a structure of type $\langle \mathbb{S}; \bowtie \rangle$ where $\bowtie$ is a relation on $\mathcal{P}(\mathbb{S}) X \mathcal{P}(\mathbb{S})$.

This relation obeys a straightforward generalization of the three Tarskian axioms. We will call such a structure, a Scott structure, since Dana Scott made important contribution working with this kind of structures, generalizing for example Lindenbaum theorem for them (see [38]). This kind of approach is usually known under the banner "multiple-conclusion logic" (see [39]).

3.4. Substructural structure

A substructural structure $\langle G; \bowtie \rangle$ is a Scott structure where a magma $G = \langle \mathbb{S}; * \rangle$ is considered instead of the naked set $\mathbb{S}$. A magma is just a set with a binary operation $*$.[3] Some specific axioms can be added for the operation $*$. Gentzen's notion of sequents can be designed in this way, and therefore substructural structures are a refinement of Gentzen's idea. Gentzen's notion of sequents is quite precise but for example associativity is an implicit supposition of it. Considering a magma, one can turn this hypothesis explicit, with an axiom of associativity for $*$, or withdraw it and work with non associativity. In a substructural structure in general there are also no specific axioms for $\bowtie$.

In the last twenty years the amazing development of linear logic [30] and non monotonic logics has shown the fundamental role of substructural structures (see [28], [37]).

3.5. Turning style

In the context of Gentzen's sequent calculus, the Hertz-Gentzenian symbol $\rightarrow$ is very often replaced by the turnstile $\vdash$ (in particular due to the fact that people now use $\rightarrow$ for material implication instead of the old $\supset$).

This change of symbol seems harmless, but in fact one has to be very careful, because it leads to a confusion between a proof-theoretical system with axioms and rules with the structure generated by this system. People like Scott and those working with substructural logics generally do not make this difference, and there is a tendency to use the same name for Gentzen's structural rules and for axioms applying to the relation $\bowtie$.

---

[3]The terminology "magma" is due to Charles Ehresmann.

For example the axiom stating the transitivity of $\bowtie$ is sometimes called cut. This can lead to serious misunderstandings of the cut-elimination theorem. Gentzen's system for classical logic with cut is equivalent to his system without cut, but the cut rule is not derivable in the cut-free system, although this system generates a transitive relation $\bowtie$, since the two systems are equivalent (see [8]).

Another example is the confusion between thinning rules [GEN2] and the axiom of monotonicity [TAR2]. A proof system can have no thinning rules and be monotonic.

4. Matrix theory and abstract logic

4.1. Łukasiewicz and Tarski's concept of logical matrices

Influenced by Tarski's theory of consequence operator, Polish logicians have developed since the 1930s a general theory of zero-order logics (i.e. propositional or sentential logics). This kind of stuff is generally known under the name Polish logic (see [12]). A central concept of Polish logic is the notion of matrix, or logical matrix. In fact one could say that Polish logic is the fruit of the wedding between the concepts of consequence operator and logical matrix. Polish logicians have not developed the theory of consequence operator by itself, at the abstract level, maybe because they thought it was sterile. Anyway they have shown that its combination with matrix theory is highly fruitful.

The concept of logical matrix was introduced in Poland by Łukasiewicz, through the creation of many-valued logic. However it is Tarski who saw the possibility of using this theory as a basic tool for a systematic study of logics. It is clear that matrix theory does not reduce to many-valued logics, as shown by its use for the proof of independence of axioms of the two-valued propositional logics. Matrices are models of zero-order non classical logics. In fact the consideration of *models of zero-order non classical logics* led Tarski to *classical first-order model theory* (see [38]).

4.2. Lindenbaum's matrix theorem

The first important general result about matrices is due to Lindenbaum. A matrix $\mathcal{M}$ is an algebra $\mathcal{A} = \langle \mathbb{A}; f \rangle$ together with a subset $\mathbb{D}$ of $\mathbb{A}$, whose elements are called designated values. When one uses logical matrices, one considers the set $\mathbb{S}$ of sentences of a logical structure as an algebra, an absolutely free algebra (explicit consideration of this fact is also credited to Lindenbaum). Operators of this algebra represent zero-order connectives.

Let us call a *Lindenbaum structure*, a structure $\langle \mathcal{S}; \mathbb{T} \rangle$ where $\mathcal{S}$ is an absolutely free algebra of domain $\mathbb{S}$ and $\mathbb{T}$ is a subset of $\mathbb{S}$. One can wonder if it is possible to find a logical matrix $\mathcal{M}$ which characterizes this structure in the sense that any homomorphism η from $\mathcal{S}$ to the algebra $\mathcal{A}$ of the matrix is such that:

$$\alpha \in \mathbb{T} \text{ iff } \eta(\alpha) \in \mathbb{D}$$

There are many logics, such as some modal logics like S5 or intuitionistic logic, that cannot be characterized by finite matrices (i.e. matrices where the domain of the algebra is finite). However Lindenbaum has shown that every Lindenbaum structure stable under substitution can be characterized by a matrix of cardinality superior or equal to the cardinality of the language (i.e. the domain of the structure).

Lindenbaum was killed during the second world war, but just after the war his work was disseminated in Poland through the monograph of Jerzy Łoś entirely devoted to logical matrices (see [32]). Lindenbaum's theorem was generalized by Wójcicki for the case of Łoś structures. A Łoś structure is a structure of type $\langle S; Cn \rangle$ where S is an absolutely free algebra and Cn a structural consequence operator, i.e. a consequence operator obeying the three basic Tarski's axioms and the following condition:

$$\text{for every endomorphism } \epsilon \text{ of } S, \ \epsilon CnT \subseteq Cn\epsilon T$$

In other words, this means that Cn is stable under substitution. This crucial notion was introduced in [33].

Matrix theory was also applied to Scott structures by Zygmunt (see [48]).

4.3. Suszko's abstract logic

Suszko and his collaborators have shown that all known logics are structural. Later on, Suszko developed a general study of logics that he called "abstract logic" considering as basic structure a Suszko structure, i.e. a structure of type $\langle \mathcal{A}; Cn \rangle$ where $\mathcal{A}$ is an abstract algebra and Cn a consequence operator obeying the three basic Tarski's axioms (see [18]).

Abstract logic in this sense is very close to universal algebra. Concepts of category theory and model theory can also be fruitfully applied for its development. With Suszko's abstract logic, the general theory of logics reached the level of mathematical maturity, turning really into a mathematical theory in the modern sense of the word.

One could say: "very well, this is mathematics, but this is not *about* mathematics!", since an abstract logic is a model for a propositional logic and we all know that such a logic, be it classical, intuitionistic or whatever, is not rich enough to fully represent mathematical reasoning.

I have proposed to generalize the notion of abstract logic considering structures of type $\langle \mathcal{A}; Cn \rangle$ where $\mathcal{A}$ can be an infinitary algebra, in order to represent logics of order superior to zero, taking in account the fact that higher order languages can be described by infinitary algebras (see [3]).

5. Algebraic logic

5.1. Logic and algebra

Algebraic logic is an ambiguous expression which can mean several things. One could think that it is crystal clear and that algebraic logic means the study of logic from an algebraic point of view. But this is itself ambiguous, because this in turn means two things:

(1) The study of logic using algebraic tools

(2) Logics considered as algebraic structures.

(2) implies (1) but not necessarily the converse. It is clear that when one considers a logic as a Lindenbaum structure or a Loś structure and considers the problem of characteristic matrices, this involves mainly algebraic concepts. One can even say that these structures as well as Suszko's abstract logics are algebraic structures. Roughly speaking this is right. But if one wants to be more precise, it is important to emphasize two points ; this will be the subject of the two next subsections.

5.2. Cross structures

First these structures are not exactly algebras, according to *Birkhoff's standard definition of algebra*. To call these structures algebras leads to a general confusion according to which any mathematical structure is called an algebra. In fact an abstract logic in the sense of Suszko is a mixture of topological concepts and algebraic concepts. Algebraic concepts are related to the structure of the language - algebraic operators representing logical operators - and topological concepts are related to the consequence operator Cn. In fact Tarski was probably influenced by topology when he developed the theory of consequence operator since topology was very popular at this time in Poland and Tarski himself was collaborating with Kuratowski. Loś and Suszko structures are in fact cross structures according to Bourbaki's terminology, they are the result of crossing two fundamental mother structures: topological and algebraic structures.

To call "algebraic logic" a general theory of logics involving algebraic concepts is misleading. Polish logic, which is such a theory, is often assimilated with algebraic logic, by opposition to a more traditional approach to logic based on intuitive concepts related to linguistics. But when the people, following this second approach, say that Polish logic is algebraic logic, they simply identify algebra with mathematics, or in the best case algebraic structures with mathematical structures.

5.3. Lindenbaum-Tarski algebras

It is not rare to hear that classical propositional logic is a boolean algebra. Tarski at the beginning of the 1930s showed how to reduce classical logic to a boolean algebra by factorizing the structure (cf. [44]). The factorized structure is called a *Lindenbaum-Tarski algebra*, LT-algebra for short. The concept of LT-algebra was then extended to other logics. Algebraic logic in this sense is the study of logics, via their LT-algebras, and more generally the study of algebras which can be considered as LT-algebras of some logics. This means in general that people are considering algebraic structures of type $\langle \mathcal{A}; \leqslant \rangle$ where $\leqslant$ is an order relation and $\mathcal{A}$ is an algebra whose operators have intended logical meaning: conjunction, disjunction, implication and negation. This is for example the case of the famous

Birkhoff-von Neumann's quantum logic which is in fact an algebraic structure of this kind. These structures are tightly linked with lattices. A general study of these structures has been developed by H.B.Curry (see [26]).

If one considers algebraic logic as the study of this kind of structures and this is probably the only rigorous way to use this terminology, it seems then that algebraic logic is too restricted for developing a general theory of logics. Firstly because the notion of logical consequence cannot be properly represented by an order relation $\leqslant$, one has to consider at least a consequence relation or a binary Scott-type relation; secondly because there are logics which cannot properly be handle through LT-algebras. This is the case of simple logics, logics that have no non-trivial congruence relations and which cannot be factorized, like da Costa's paraconsistent logic C1 (see [5]).

However these last twenty years work in algebraic logic has made important advances through the introduction of several new concepts such as protoalgebraization and the correlated refinement of LT-algebra (see [27], [17]).

6. Da Costa's theory of valuation

6.1. Every logic is two-valued

As we have seen, closed theories form a sound and complete semantics for any Tarski structure and relatively maximal theories form a sound and complete semantics for Tarski structures obeying the finiteness axiom. Now instead of considering theories, one can consider the characteristic functions of these theories, these are *bivaluations*.

The above results can therefore be reinterpreted as saying that every Tarski structure has a bivalent semantics, and they justify, at least if we restrict ourselves to such structures, a general theory of logics based on the concept of bivaluations. Newton da Costa's theory of valuation is such a theory.

The advantage of such a theory is that it is based on the semantical intuitive ideas of true and false and that it can be seen as a natural generalization of the bivalent semantics for classical propositional logic, which can be applied to non-classical logics and high-order logics.

6.2. Bivalency and truth-functionality

This generalization is however not so natural in the sense that one central feature of the semantics of classical propositional logic is lost in most of the cases: truth-functionality. Therefore the theory of valuation is mainly a theory of non truth-functional bivalent semantics. An interesting result due to da Costa shows that truth-functional bivalent semantics determine only logics which are sublogics of classical logic (see [23]).

Semantics of bivaluations can be developed for many-valued logics, such as Łukasiewicz's three-valued logic. But they are more interesting for logics which are not truth-functional, in the sense that they cannot be characterized by a finite matrix. In fact, originally da Costa built semantics of valuation for his paraconsistent logic C1, which is a non truth-functional logic [22].

6.3. Bivaluations, truth-tables and sequent calculus

Despite of the non truth-functionality of such semantics, it is possible to construct truth-tables which are quite similar to the classical ones, and which provide decision methods.

I have linked the theory of valuations with sequent calculus showing how it is possible to translate conditions defining bivaluations into sequent rules and vice-versa. Combining action of valuations upon sequent rules, in the spirit of Gentzen's 1932 proof [29], with Lindenbaum-Asser theorem, I have given a general version of the completeness theorem, from which it is possible to derive instantaneously many specific completeness theorems (see [2], [11]),

7. Universal logic

7.1. Universality and trivialization

Wójcicki said once that his objective was to trivialize the completeness theorem. What does this mean? It means finding a general formulation of this theorem from which particular theorems appear as trivial corollaries.

In a proof of a completeness theorem for a given logic, one may distinguish the elements of the proof that depend on the specificity of this logic and the elements that do not depend on this peculiarity, that we can call *universal.*

This distinction is important from a methodological, philosophical and mathematical point of view. The first proofs of completeness for propositional classical logic give the idea that this theorem is depending very much on classical features. Even one still gets this impression with recent proofs where the theorem is presented using the concept of maximal consistent set which seems to depend on classical negation. In fact this idea is totally wrong and one can present the completeness theorem for classical propositional logic in such a way that the specific part of the proof is trivial, i.e. one can trivialize the completeness theorem.

One central aim of a general theory of logics is to get some universal results that can be applied more or less directly to specific logics, this is one reason to call such a theory *universal logic.*

Some people may have the impression that such general universal results are trivial. This impression is generally due to the fact that these people have a concrete-oriented mind, and that something which is not specified has no meaning for them, and therefore universal logic appears as logical abstract nonsense. They are like someone who understands perfectly what is Felix, his cat, but for whom the concept of cat is a meaningless abstraction. This psychological limitation is in fact a strong defect because, as we have pointed through the example of the completeness theorem, what is trivial is generally the specific part, not the universal one which requires what is the fundamental capacity of human thought: abstraction.

7.2. Universal logic and universal algebra

Originally I introduced the terminology "universal logic" to denote a general theory of logics, by analogy with the expression "universal algebra" (cf. [1]).

What is universal algebra? During the XIXth century, lots of algebraic structures appeared and then some people started to turn this heterogeneous variety into a unified theory. In 1898, Whitehead wrote a book entitled *A treatise on universal algebra* (cf. [46]), but it is Garrett Birkhoff who is considered as the real founder of universal algebra [4]. Birkhoff was the first to give a very general definition of abstract algebra, as a set with a family of operators. He introduced further general concepts and proved several important universal results (see [14], [16]).

The idea beyond universal logic is to develop a general theory of logics in a similar way. This means that logics are considered as mathematical structures, general concepts are introduced and universal results are proved.

One central question is to know which kind of structures are logical structures. One may think that these structures are algebraic structures and that therefore universal logic is just a part of universal algebra, this was more or less the idea of Suszko. But as we have pointed out, it seems inappropriate to base essentially a general theory of logics on the notion of algebraic structures. Other types of structures are required.

7.3. Universal logic and the theory of structures

The idea I proposed about ten years ago is that logical structures must be considered as fundamental mother structures in the sense of Bourbaki, together with algebraic, topological and order structures. This was also the idea of a former student of de Possel, Jean Porte, 40 years ago (see [36]). In his work, Porte proposed several types of logical structures.

My idea was to focus on a logical structure of type $\langle \mathbb{S}; \vdash \rangle$ where $\vdash$ is a relation on $\mathcal{P}(\mathbb{S})X\mathbb{S}$. The important thing is that the structure of $\mathbb{S}$ is not specified, in fact, further on, any kind of structure can be put on $\mathbb{S}$, not only an algebraic structure. We are back therefore to something very close to Tarski's original theory of consequence operator. One important difference is that in this new definition of logical structure, no axioms are stated for the consequence relation $\vdash$, in the same way that no axioms are stated for the operators in Birkhoff's definition of abstract algebra.

Universal logic, like universal algebra, is just a part of the general theory of structures, logical abstract nonsense is a subfield of general abstract nonsense. If, as we have suggested, abstraction is the important thing, one could argue that what is really interesting is a general theory of structures, like category theory, and not a theory of specific structures like universal logic. The fact is that abstraction is really

---

[4]L. Corry erroneously says, in his otherwise excellent book [20], that the expression 'universal algebra" is due to Whitehead, this expression is due in fact to J.J. Sylvester, see [42].

a nonsense if it is considered only by itself. Abstraction is abstraction of something and when applied back it gives another view of this thing. Moreover there must be a continuous interplay between the specific and the general. Universal logic is an interesting material for the general theory of structures. For example, a central point in universal logic is to try to define properly a relation between logics which permits to compare them and to identify them (cf. [6], [13]). To solve this problem, new concepts and tools have to be introduced at the level of a general theory of structures, which can later on be applied to other fields of mathematics.

References

[1] J.-Y. Beziau, *Universal Logic*. Proceedings of the 8th International Colloquium - Logica'94, T. Childers and O. Majer (eds), Czech Academy of Sciences, Prague, 2004, pp.73-93.

[2] J.-Y. Beziau, *Recherches sur la Logique Universelle*. PhD Thesis, Department of Mathematics, University Denis Diderot, Paris, 1995.

[3] J.-Y. Beziau, *Sobre a verdade lógica*. PhD Thesis, Department of Philosophy, University of São Paulo, São Paulo, 1996.

[4] J.-Y. Beziau, *What is many-valued logic?* Proceedings of the 27th International Symposium on multiple-valued logic, IEEE Computer Society, Los Alamitos, 1997, pp.117-121.

[5] J.-Y. Beziau, *Logic may be simple (logic, congruence and algebra)*. Logic and Logical Philosophy, **5** (1997), 129-147.

[6] J.-Y. Beziau, *Classical negation can be expressed by one of its halves*. Logic Journal of the Interest Group in Pure and Applied Logics, **7** (1999), 145-151.

[7] J.-Y. Beziau, *A sequent calculus for Łukasiewicz's three-valued logic based on Suszko's bivalent semantics*. Bulletin of the Section of Logic, **28** (1999), 89-97.

[8] J.-Y. Beziau, *Rules, derived rules, permissible rules and the various types of systems of deduction*. in Proof, types and categories, PUC, Rio de Janeiro, 1999, pp. 159-184.

[9] J.-Y. Beziau, *La véritable portée du théorème de Lindenbaum-Asser*. Logique et Analyse, **167-168** (1999), 341-359.

[10] J.-Y. Beziau, *From paraconsistent logic to universal logic*. Sorites, **12** (2001), 5-32.

[11] J.-Y. Beziau, *Sequents and bivaluations*. Logique et Analyse, **176** (2001), 373-394.

[12] J.-Y. Beziau, *The philosophical import of Polish logic*. Logic, Methodology and Philosophy of Science at Warsaw University, M.Talasiewicz (ed), Warsaw, 2002, pp.109-124.

[13] J.-Y. Beziau, R.P. de Freitas and J.P. Viana, *What is classical propositional logic? (A study in universal logic)*. Logical Studies, **7** (2001).

[14] G. Birkhoff, *Universal algebra*. Comptes Rendus du Premier Congrès Canadien de Mathématiques, University of Toronto Press, Toronto, 1946, pp.310-326.

[15] G. Birkhoff, *The rise of modern algebra to 1936* and *The rise of modern algebra, 1936 to 1950*, in Men and institutions in American mathematics, Graduate Studies, Texas Technical Studies, **13** (1976), pp.65-85 and pp.41-63.

[16] G. Birkhoff, *Universal algebra*. Selected papers on algebra and topology by Garrett Birkhoff, G.-C. Rota and J.S. Oliveira (eds), Birkhäuser, Basel, 1987.

[17] W.J. Blok and D. Pigozzi, *Algebraizable logics*. Mem. Amer. Math. Soc. **77**(1989).

[18] S.L. Bloom, D.J. Brown and R. Suszko. Dissertationes Mathematicae, **102** (1973).

[19] N. Bourbaki, *The architecture of mathematics*. American Mathematical Monthly, **57** (1950), 221-232.

[20] L. Corry, *Modern algebra and the rise of mathematical structures*. Birkhäuser, Basel, 1996.

[21] N.C.A. da Costa, *Calculs propositionnels pour les systèmes formels inconsistants*. Comptes Rendus d l'Académie des Sciences de Paris, **257** 1963, 3790-3793.

[22] N.C.A. da Costa and E.H. Alves, *A semantical analysis of the calculi C_n*. Notre Dame Journal of Formal Logic, **18** (1977), 621-630.

[23] N.C.A. da Costa and J.-Y. Beziau, *Théorie de la valuation*. Logique et Analyse, **145-146** (1994), 95-117.

[24] N.C.A. da Costa and J.-Y. Beziau , *La théorie de la valuation en question*. Proceedings of the XI Latin American Symposium on Mathematical Logic (Part 2), Universidad Nacional del Sur, Bahia Blanca, 1994 pp.95-104.

[25] N.C.A. da Costa , J.-Y. Beziau and O.A.S. Bueno, *Malinowski and Suszko on many-valued logics: on the reduction of many-valuedness to two-valuedness*. Modern Logic, **6**(1996), pp.272-299.

[26] H.B. Curry, *Leçons de logique algébrique*. Gauthier-Villars, Paris/Nauwelaerts, Louvain, 1952.

[27] J. Czelakowski, *Protoalgebraic logcis*, Kluwer, Dordrecht, 2001.

[28] K. Dosen and P. Schröder-Heister, *Substructural logics*. Clarendon, Oxford, 1993.

[29] G. Gentzen, *Über die Existenz unabhängiger Axiomensysteme zu unendlichen Satzsystemen*, Mathematische Annalen, **107** (1932), 329-350.

[30] J.-Y. Girard, *Linear logic*. Theoretical Computer Science, **50** (1987), 1-112.

[31] P. Hertz, *Über Axiomensysteme für beliebige Satzsysteme*, Mathematische Annalen, **101** (1929), 457-514.

[32] J. Łoś, *O matrycach logicznych*. Travaux de la Société des Sciences et des Lettres de Wrocław, Série B, **19** (1949).

[33] J. Łoś and R. Suszko, *Remarks on sentential logics*. Indigationes Mathematicae, **20** (1958), 177-183.

[34] J. Łukasiewicz, *O logice trójwartosciowej*. Ruch Filozoficzny, **5** (1920), 170-171.

[35] G. Malinowski, *Many-valued logics*. Clarendon, Oxford, 1993.

[36] J. Porte, *Recherches sur la théorie générale des systèmes formels et sur les systèmes connectifs*. Gauthier-Villars, Paris / Nauwelaerts, Louvain, 1965.

[37] F. Paoli, *Substructural logics: a primer*. Kluwer, Dordrecht, 2002.

[38] D.S. Scott, *Completeness and axiomatizability in many-valued logic*. Proceedings of the Tarski Symposium, L. Henkin (ed.), American Mathematical Society, Providence, 1974, pp.411-435.

[39] D.J. Shoesmith and T.J. Smiley, *Multiple-conclusion logic*. Cambridge University Press, Cambridge, 1978.

[40] S.J. Surma, *Alternatives to the consequence-theoretic approach to metalogic*. Proceedings of the IX Latin American Symposium on Mathematical Logic. (Part 2), Universidad Nacional del Sur, Bahia Blanca, 1994, pp.1-30.

[41] R. Suszko, *Remarks on Łukasiewicz's three-valued logic*. Bulletin of the Section of Logic, **4** (1975), 377-380.

[42] J.J. Sylvester, *Lectures on the principles of universal algebra*. American Journal of Mathematics, **6** (1884), 270-286.

[43] A. Tarski, *Remarques sur les notions fondamentales de la méthodologie des mathématiques*. Annales de la Société Polonaise de Mathématiques, **6** (1928), 270-271.

[44] A. Tarski, *Grundzüge des Systemankalkül, Erster Teil*. Fundamenta mathematicae, **25** (1935), 503-526.

[45] A. Tarski, *O pojciu wynikania logicznego*. Przeglad Folozoficzny, **39** (1936), 58-68.

[46] A.N. Whitehead, *A treatise on universal algebra*. Cambridge University Press, Cambridge, 1898.

[47] R. Wójcicki, *Theory of logical calculi - Basic theory of consequence operations*. Kluwer, Dordrecht, 1988.

[48] J. Zygmunt, *An essay in matrix semantics for consequence relations*. Wydawnictwo Uniwersytetu Wroclawskiego, Wrocław, 1984.

Jean-Yves Beziau
Institute of Logic
University of Neuchâtel
Espace Louis Agassiz 1
2000 Neuchâtel
Switzerland
e-mail: jean-yves.beziau@unine.ch
www.unine.ch/unilog

J.-Y. Beziau (Ed.), *Logica Universalis*, 2nd edition, 19–33

Abstract Model Theory as a Framework for Universal Logic

Marta García-Matos and Jouko Väänänen

Abstract. We suggest abstract model theory as a framework for universal logic. For this end we present basic concepts of abstract model theory in a general form which covers both classical and non-classical logics. This approach aims at unifying model-theoretic results covering as large a variety of examples as possible, in harmony with the general aim of universal logic.

Mathematics Subject Classification (2000). Primary 03C95; Secondary 03B99.

Keywords. Abstract logic, model theoretic logic.

1. Introduction

Universal logic is a general theory of logical structures as they appear in classical logic, intuitionistic logic, modal logic, many-valued logic, relevant logic, paraconsistent logic, non-monotonic logic, topological logic, etc. The aim is to give general formulations of possible theorems and to determine the domain of validity of important theorems like the Completeness Theorem. Developing universal logic in a coherent uniform framework constitutes quite a challenge. The approach of this paper is to use a semantic approach as a unifying framework.

Virtually all logics considered by logicians permit a semantic approach. This is of course most obvious in classical logic and modal logic. In some cases, such as intuitionistic logic, there are philosophical reasons to prefer one semantics over another but the fact remains that a mathematical theory of meaning leads to new insights and clarifications.

Researchers may disagree about the merits of a semantic approach: whether it is merely illuminating or indeed primary and above everything else. It is quite reasonable to take the concept of a finite proof as the most fundamental concept in logic. From this predominantly philosophically motivated point of view semantics comes second and merely as a theoretical tool. It is also possible to take the mathematical concept of a structure as a starting point and use various logics merely as tools for the study of structures. From this predominantly mathematically motivated view, prevalent also in computer science logic, formal proofs are like certificates that we may acquire at will according to our needs. Finally, we may take the intermediate approach that formal languages are fundamental in logic, and there are many different ones depending on the purpose they are used for, but common to all is a mathematical theory of meaning. This is the approach of this paper.

Abstract model theory is the general study of model theoretic properties of extensions of first order logic (see [3]). The most famous result, and the starting point of the whole field, was Lindström's Theorem [22] (see Theorem 3.4 below) which characterized first order logic in terms of the Downward Löwenheim-Skolem Theorem and the Compactness Theorem. Subsequently many other characterizations of first order logic emerged. Although several characterizations of first order logic tailored for structures of a special form (e.g. topological [39]) were found, no similar characterizations were found for extensions of first order logic in the setting of ordinary structures. However, in [34] a family of new infinitary languages is introduced and these infinitary languages permit a purely model theoretic characterization in very much the spirit of the original Lindström's Theorem.

In Section 2 we present an approach to abstract model theory which is general enough to cover many non-classical logics. Even if few results exist in this generality, the approach suggests questions for further study.

Intertwined with the study of general properties of extensions of first order logic is naturally the study of the model theory of particular extensions. In this respect logics with generalized quantifiers and infinitary languages have been the main examples, but recently also fragments of first order logic such as the guarded fragment and finite variable fragments have been extensively studied. Chang [9] gives an early sketch of modal model theory, modern model theory of modal logic emphasizes the role of bisimulation (e.g. [26]).

It was noticed early on that particular properties of extensions of first order logic depended on set theoretical principles such as CH or $\Diamond$. A famous open problem is whether $L(Q_2)$ is axiomatizable (Q_2 is the quantifier "there exist at least $\aleph_2$ many"). In [35] the concept of a *logic frame* is introduced to overcome this dependence on metatheory. It becomes possible to prove results such as, if a logic of a particular form is axiomatizable, then it necessarily also satisfies the Compactness Theorem. We discuss logic frames in Section 5 and point out their particular suitability for the study of universal logic.

2. Abstract Model Theory

The basic concept of abstract model theory is that of an abstract logic. This concept is an abstraction of the concept of truth as a relation between structures of some sort or another and sentences of some sort or another. In its barest formulation, void of everything extra, when we abstract out all information about what kind of structures we have in mind, and also all information about what kind of sentences we have in mind, what is left is just the concept of a binary relation between two classes[1].

Definition 2.1. An *abstract logic* is a triple $L = (S, F, \models)$ where $\models \subseteq S \times F$. Elements of the class S are called the structures of L, elements of F are called the sentences of L, and the relation $\models$ is called the satisfaction relation of L.

Figure 1 lists some building blocks from which many examples of abstract logics can be constructed.

Some obvious conditions immediately suggest themselves, such as closure under conjunction (see Definition 2.5 below) and closure under permutation of symbols. However, the spectrum of different logics is so rich that it seems reasonable to start with as generous a definition as possible. Still, even this definition involves a commitment to truth as a central concept which limits its applicability as an abstraction of e.g. fuzzy logic.

The value of a general concept like the above depends upon whether we can actually say anything on this level of generality. Surprisingly, already this very primitive definition allows us to formulate such a central concept as compactness and prove some fundamental facts about compactness.

We say that a subset T of F *has a model* if there is $A \in S$ such that $A \models T$ i.e. $\forall \phi \in T (A \models \phi)$. An abstract logic $L = (S, F, \models)$ is said to satisfy the *Compactness Theorem* if every subset of F, every finite subset of which has a model, has itself a model. We can now demonstrate even in this quite general setup that compactness is inherited by sublogics, a technique frequently used in logic. First we define the sublogic-relation. The definition below would be clearer if we applied it only to logics which have the same structures. This seems an unnecessary limitation and we are quite naturally lead to allowing a translation of structures, too.

Definition 2.2. An abstract logic $L = (S, F, \models)$ is a *sublogic* of another abstract logic $L' = (S', F', \models')$, in symbols

$$L \leq L',$$

if there are a sentence $\theta \in S'$ and functions $\pi : S' \to S$ and $f : F \to F'$ such that

1. $\forall A \in S \exists A' \in S' (\pi(A') = A$ and $A' \models' \theta)$
2. $\forall \phi \in F \forall A' \in S' (A' \models \theta \to (A' \models' f(\phi) \iff \pi(A') \models \phi)).$

The idea is that the structures in S' are richer than the structures in S, and therefore we need the projection π. The role of θ is to cut out structures in S' that are meaningless from the point of view of L. Typical projections are in Figure 2.

Lemma 2.3. *If $L \leq L'$ and L' satisfies the Compactness Theorem, then so does L.*

---

[1] In lattice theory such relations are called Birkhoff polarities [5]. A recent study is [14]. We are indebted to Lauri Hella for pointing this out.

Structures	Sentences
Valuations -double valuations -three-valued	Propositional logic Relevance logic Paraconsistent logic
Relational structures -monadic -ordered -finite -pseudofinite -topological -Banach space -Borel -recursive -ω-models	Predicate logic -with two variables -guarded fragment -with infinitary operations -with generalized quantifiers -higher order -positive bounded, etc -logic of measure and category
Kripke structures -transitive reflexive -equivalence relation -etc	Intuitionistic logic Modal logic -S4,S5, etc
Many-valued structures	Many-valued logic, fuzzy logic
Games	Linear logic

FIGURE 1. Building blocks of abstract logics

Proof. We give the easy details in order to illustrate how the different components of Definition 2.2 come into play. Suppose $T \subseteq F$ and every finite subset of T has a model. Let $T' = \{\theta\} \cup \{f(\phi) : \phi \in T\}$. Suppose $T'_0 = \{\theta\} \cup \{f(\phi) : \phi \in T_0\} \subseteq T'$ is finite. There is $A \in S$ such that $A \models T_0$. Let $A' \in S'$ such that $\pi(A') = A$ and $A' \models' \theta$. Then $A' \models T'_0$. By the Compactness Theorem of L' there is $A' \in S'$ such that $A' \models' T'$. Thus $\pi(A') \models T$.   □

Another property that is inherited by sublogics is decidability (and axiomatizability). The formulation of these properties in abstract terms requires sentences of the abstract logic to be encoded in such a way that concepts of effectiveness apply. It is not relevant which concept of effectiveness one uses.

Definition 2.4. An abstract logic $L = (S, F, \models)$ is said to be *recursive* if F is effectively given. A recursive abstract logic $L = (S, F, \models)$ is an *effective sublogic* of another recursive abstract logic $L' = (S', F', \models')$, in symbols $L \leq_{\text{eff}} L'$, if $L \leq L'$ via θ, π and f such that $f(\phi)$ can be effectively computed from ϕ. A recursive abstract logic $L = (S, F, \models)$ is said to be *decidable* if the set $\{\phi \in F : \phi$ has a model$\}$ can be effectively decided inside F. It is *co-r.e. (or r.e.) for satisfiability* if $\{\phi \in F : \phi$ has a model$\}$ is co-r.e. (respectively, r.e.).

The above concept of co-r.e. for satisfiability could be appropriately called *effective axiomatisability* for logics closed under negation (see Definition 3.3 below).

Structure A	Projection $\pi(A)$	θ expresses
Ordered structure $(M, <)$	Structure M	Axioms of order
Zero-place relations	Valuation	
Binary structure	Kripke structure	Transitivity (e.g.)
Structure with unary predicates	Many-sorted structure	

FIGURE 2. Typical projections

The Independence Friendly logic IF (see [16]) is not closed under negation, it is r.e. for satisfiability, but neither effectively axiomatizable nor co-r.e. for satisfiability.

Definition 2.5. An abstract logic $L = (S, F, \models)$ is said to be *closed under conjunction* if for every $\phi \in F$ and $\psi \in F$ there is $\phi \wedge \psi \in F$ such that $\forall A \in S (A \models \phi \wedge \psi \iff A \models \phi$ and $A \models \phi)$. We say that a recursive abstract logic L is *effectively closed under conjunction* if $\phi \wedge \psi$ can be found effectively from ϕ and ψ inside F.

Lemma 2.6. *Suppose $L \leq_{eff} L'$ and L' is effectively closed under conjunction. If L' is decidable (or co-r.e. for satisfiability), then so is L.*

Proof. If $\phi \in F$ has a model A in S, then there is a model A' in S' such that $A' \models \theta$ and $\pi(A') = A$, whence $A' \models \theta \wedge f(\phi)$. Conversely, if $\theta \wedge f(\phi) \in F$ has a model A' in S', then ϕ has the model $\pi(A')$ in S. Since there is an effective algorithm for "$\theta \wedge f(\phi)$ *has a model in S'*", there is also one for "ϕ *has a model in S*". □

Example (Predicate logic). Let S be the set of all first order structures of various vocabularies, and F the set of all first order sentences (with identity) built from the atomic formulas and the usual logical symbols $\exists, \forall, \wedge, \vee, \neg, \rightarrow, \leftrightarrow$ and parentheses. The relation $A \models_L \phi$ is defined as usual for structures A and sentences ϕ of the same vocabulary. Predicate logic is a recursive abstract logic which satisfies the Compactness Theorem and is co-r.e. for satisfiability but not decidable. Predicate logic on finite structures does not satisfy Compactness Theorem. It is r.e. for satisfiability but not co-r.e. for satisfiability and hence not decidable. It is not a sublogic of predicate logic but it is an effective sublogic of the extension of predicate logic with the generalized quantifier Q_0 ("there exists infinitely many"). Predicate logic on ordered structures is important in computer science (especially on finite ordered structures). It is clearly an effective sublogic of predicate logic. Another variant of predicate logic is many-sorted logic [28]. It is an effective sublogic of the ordinary predicate logic.

Example (Two variable predicate logic). Let S be the set of all first order structures of various vocabularies, and F the set of all first order sentences (with identity) built from the atomic formulas with just the two variables x and y, and the usual logical symbols $\exists, \forall, \wedge, \vee, \neg, \rightarrow, \leftrightarrow$, and parentheses. The relation $A \models_L \phi$ is defined as usual for structures A and sentences ϕ of the same vocabulary. This is a decidable abstract logic [29], which is an effective sublogic of predicate logic.

Example (Guarded fragment of predicate logic [1]). Let S be the set of all first order structures of various vocabularies, and F the set of all guarded first order sentences i.e. first order formulas where all quantifiers are of the form $\exists \vec{x}(R(\vec{x}, \vec{y}) \wedge \phi(\vec{x}, \vec{y}))$ or $\forall \vec{x}(R(\vec{x}, \vec{y}) \rightarrow \phi(\vec{x}, \vec{y}))$, where $R(\vec{x}, \vec{y})$ is atomic. The relation $A \models_L \phi$ is defined as usual for structures A and sentences ϕ of the same vocabulary. This is a decidable abstract logic [1], which is an effective sublogic of predicate logic.

Example (Propositional logic). Let us fix a set P of propositional symbols $p_0, p_1, \ldots$. Let S be the set of all functions $v : P \rightarrow \{0, 1\}$, and F the set of all propositional sentences built from the symbols of P and the usual logical symbols $\wedge, \vee, \neg, \rightarrow, \leftrightarrow$ and parentheses. The relation $v \models_L \phi$ is defined to hold if $v(\phi) = 1$. Propositional logic is an effective sublogic of the (even two variable) predicate logic: We may treat p_n as a 0-place predicate symbol. A first order structure A gives rise to a valuation $\pi(A)$ which maps p_n to the truth value of P_n in A. Clearly, every valuation arises in this way from some structure.

Example (Modal logic). Let us fix a set P of propositional symbols $p_0, p_1, \ldots$. Let S be the set of all reflexive and transitive Kripke-structures, and F the set of all propositional modal sentences built from the symbols of P and the usual logical symbols of modal logic $\Box, \Diamond, \wedge, \vee, \neg, \rightarrow, \leftrightarrow$ and parentheses. The relation $\mathcal{K} \models_L \phi$ is defined as usual. Modal logic is an effective sublogic of predicate logic: A first order structure $A = (K, R, c, P_0, P_1, \ldots)$, where R is transitive and reflexive, gives rise to a Kripke-structure $\mathcal{K}$ in which (K, R) is the frame, c denotes the initial node and P_n indicates the nodes in which p_n is true. Clearly, every Kripke-structure arises in this way from some such structure A. Sentences are translated in the well-known way:

$$
\begin{aligned}
g(p_n, x) &= P_n(x) \\
g(\neg \phi, x) &= \neg g(\phi, x) \\
g(\phi \wedge \psi, x) &= g(\phi, x) \wedge g(\psi, x) \\
g(\phi \vee \psi, x) &= g(\phi, x) \vee g(\psi, x) \\
g(\Box \phi, x) &= \forall y (R(x, y) \rightarrow g(\phi, y)) \\
g(\Diamond \phi, x) &= \exists y (R(x, y) \wedge g(\phi, y)) \\
f(\phi) &= g(\phi, c)
\end{aligned}
$$

In view of Lemma 2.3 this sublogic relation gives immediately the Compactness Theorem, also for say, S4. In fact, basic modal logic is an effective sublogic of the two variable logic, and therefore by Lemma 2.6 decidable.

Example (Intuitionistic logic). We fix a set P of propositional symbols $p_0, p_1, \ldots$. Let S be the set of all transitive and reflexive Kripke-structures, and F the set of all propositional sentences built from the symbols of P and the usual logical symbols of intuitionistic logic $\supset, \wedge, \vee, \neg$ and parentheses. The relation $\mathcal{K} \models_L \phi$ is defined as usual. Intuitionistic logic is an effective sublogic of predicate logic, and satisfies therefore the Compactness Theorem.

3. Lindström Theorems

The most famous metatheorem about abstract logics is Lindström's Theorem characterizing first order logic among a large class of abstract logics [22]. This type of characterization results are generally called Lindström theorems even when the conditions may be quite different from the original result. The original Lindström's Theorem characterizes first order logic in a class abstract logics $L = (S, F, \models)$ satisfying a number of assumptions that we first review.

The most striking formulations of Lindström theorems assume negation. To discuss negation at any length we have to impose more structure onto the class S of structures of an abstract logic.

Definition 3.1. An *abstract logic with occurrence relation* is any quadruple $L = (S, F, \models, V)$, where $L = (S, F, \models)$ is an abstract logic and $V \subseteq S \times F$ is a relation (called *occurrence relation*) such that $\models \subseteq V$. An abstract logic $L = (S, F, \models, V)$ with occurrence relation is *classical* if S is a subclass of the class of all relational structures of various vocabularies, and L satisfies:

Isomorphism Axiom:	If $A \models \phi$ and $A \cong B \in S$ then $B \models \phi$.
Reduct Axiom:	If $V(B, \phi)$ and B is a reduct of $A \in S$, then $A \models \phi \iff B \models \phi$
Renaming Axiom:	Suppose every $A \in S$ is associated with $A' \in S$ obtained by renaming symbols in A. Then for every $\phi \in F$ there is $\phi' \in F$ such that for all $A \in S$, $V(A, \phi) \iff V(A', \phi')$ and $A \models \phi \iff A' \models \phi'$.

A *classical abstract logic with vocabulary function* is the special case of an abstract logic with occurrence relation where we have a *vocabulary function* τ mapping F into S and the occurrence relation is defined by $V(A, \phi) \iff A$ is a $\tau(\phi)$-structure. We denote such an abstract logic by $(S, F, \models, \tau)$.

Intuitively, $V(A, \phi)$ means that the non-logical symbols occurring in ϕ have an interpretation in A. In classical abstract logics this means the vocabulary of the structure A includes the vocabulary of ϕ, whence the concept of a vocabulary function.

We say that a classical $L = (S, F, \models, \tau)$ is a *classical sublogic* of another classical $L' = (S', F', \models', \tau')$, $L \leq_c L'$, if $L \leq L'$ via θ, π and f such that $\tau'(f(\phi)) = \tau(\phi)$ for all $\phi \in F$, and $\pi(A)$ and A have the same universe for all $A \in S'$. A classical abstract logic L satisfies the *Downward Löwenheim-Skolem Theorem* if whenever $\phi \in F$ has a model, ϕ has a model with a countable universe.

Lemma 3.2. *If $L \leq_c L'$, L' is closed under conjunction, and L' satisfies the Downward Löwenheim-Skolem Theorem, then so does L.*

Definition 3.3. An abstract logic $L = (S, F, \models, V)$ with occurrence relation is said to be *closed under negation* if for every $\phi \in F$ there is $\neg\phi \in F$ such that

$$\forall A(V(A, \phi) \iff V(A, \neg\phi) \text{ and } A \models \neg\phi \iff A \not\models \phi). \tag{3.1}$$

The Independence Friendly logic IF [16] is not closed under negation (one way to see this is Corollary 4.2) if we define $V(A, \phi)$ to mean that the non-logical symbols occurring in ϕ have an interpretation in A. However, if we use a different definition letting $V(A, \phi)$ mean that the semantic game of ϕ is determined on A, then IF *is* closed under negation. But then IF is not what we called *classical* above.

A classical abstract logic $(S, F, \models, \tau)$ with vocabulary function is *fully classical* if S is the *whole* class of all relational structures. Among fully classical abstract logics we assume the sublogic relation $L \leq L'$ satisfies always the natural assumptions that every model A satisfies $\pi(A) = A$ and $A \models \theta$. In such a case we say that the abstract logic L' *extends* L. Two fully classical abstract logics are *equivalent* if they are sublogics of each other. Now we are ready to state:

Theorem 3.4 (Lindström's Theorem [22]). *Suppose L is a fully classical abstract logic closed under conjunction and negation extending first order logic. Then L is equivalent to first order logic if and only if L satisfies the Compactness Theorem and the Downward Löwenheim-Skolem Theorem.*

This important result gives a purely model-theoretic syntax-free characterization of first order logic. It has lead to attempts to find similar characterizations for other logics, also for non-classical logics. Indeed, de Rijke [26] has obtained a characterization for basic modal logic in terms of a notion of "finite rank". Other characterizations can be found in Figure 3. Many of them are very close in spirit to Theorem 3.4.

Another result of Lindström [22] tells us that a recursive fully classical abstract logic satisfying the closure conditions of Theorem 3.4 (effectively), which satisfies the Downward Löwenheim-Skolem Theorem and is effectively axiomatizable, is an effective sublogic of first order logic. This result has an extra limitation on definability of the set F of formulas of the abstract logic, typically satisfied by, but not limited to, the extensions of first order logic by finitely many generalized quantifiers.

There are two traditions in abstract model theory. One based on back-and-forth systems, particularly suitable for infinitary logic and interpolation theorems. The work on bisimulation shows that it is a similarly suitable setup for modal logics. The other tradition is the method of identities associated with generalized quantifiers and compact logics [27, 32]. It seems difficult to combine these two traditions. This culminates in the open question whether there is an extension of first order logic satisfying both the Compactness Theorem and the *Interpolation Theorem*: If every model of ϕ is a model of ψ, then there is a sentence θ such that every model of ϕ is a model of θ, every model of θ is a model of ψ, and

Abstract logic L	Reference
Monadic logic	Tharp [37]
- with Q_1	Caicedo [7]
Predicate logic	Lindström [22]
- with a monotone generalized quantifier	Flum [10]
- on pseudofinite structures	Väänänen [38]
Infinitary logic	
- $L_{\kappa\omega}$	Barwise [2]
- L_κ^1	Shelah,Väänänen [34]
Modal logic	de Rijke [26]
Topological logic	Sgro [30]
Various topological and related logics	Ziegler [39]
Banach space logic	Iovino [18]

FIGURE 3. Some examples of Lindström theorems

θ contains (in the obvious sense) only non-logical symbols common to both ϕ and ψ. For logics closed under negation (see Definition 3.3) this is equivalent to the *Separation Theorem*: If ϕ and ψ have no models in common, then there is a sentence θ such that every model of ϕ is a model of θ, θ and ψ have no models in common, and θ contains only non-logical symbols common to both ϕ and ψ.

Where did the two traditions reach an impasse? With back-and-forth systems the problem arose that uncountable partially isomorphic structures need not be isomorphic. With identities and compact logics the problem is the existence of a fundamental function for the relevant identity, and that is a difficult partition theoretic question.

The difficulties of the study of extensions of first order logic raise (among others) the question, is there a logic of "many" that we could somehow understand, e.g. axiomatize. A recent result of Shelah [33] shows that it is consistent that $L(Q_1, Q_2)$ is non-compact. On the other hand, logics with cofinality quantifiers are axiomatizable and compact [32]. It seems that the cofinality quantifiers behave much better than the "many"-type quantifiers.

With the new infinitary languages of [34] one can express *"there is an uncountable sequence"* in a way which does not allow one to say *"there is an infinite sequence"*. This proves to be crucial. Note that the generalized quantifier *"there exists uncountably many"* is axiomatizable but the quantifier *"there exists infinitely many"* is not. The new infinitary logics of [34] transform this phenomenon from generalized quantifiers to infinitary logic. Thereby also a new Lindström theorem arises.

Structures	Sentences
Relational structures	Existential sentences
	Existential universal, FI logic [24]
	Transfinite game formulas [17]
	Existential second order, IF logic [16]
Banach space structures	Positive bounded formulas [18]
Kripke structures	Intuitionistic logic

FIGURE 4. Examples of lack of negation.

4. Abstract logic without negation

There are many examples of abstract logics $L = (S, F, \models, V)$ with a natural occurrence relation which are not closed under negation, see Figure 4. Several concepts of abstract model theory have definitions which are equivalent if we have negation but otherwise different. It is not immediately obvious which of these definitions are the most natural ones when we do not have negation. For the Interpolation Theorem it seems that the Separation Theorem is the right formulation in the absence of negation. This question is further studied in [11]. A formulation of Lindström's Theorem for logics not closed under negation states (as pointed out in [10]):

Theorem 4.1 (Lindström's Theorem without negation). *Suppose L is a fully classical abstract logic extending first order logic and closed under conjunction and disjunction, which satisfies the Compactness Theorem and the Downward Löwenheim-Skolem Theorem. If $\phi \in F$ and $\psi \in F$ have no models in common, then there is a first order sentence θ such that every model of ϕ is a model of θ but θ has no models in common with ψ.*

Corollary 4.2 (Total lack of negation). *Suppose L is a fully classical abstract logic extending first order logic and closed under conjunction and disjunction, which satisfies the Compactness Theorem and the Downward Löwenheim-Skolem Theorem. Then only the first order sentences in F have a negation (in the sense of (3.1).*

The Independence Friendly logic IF is an example of a logic which satisfies the assumptions of the above corollary [16]. Consequently, only the first order sentences in IF have a negation. As an illustration of an abstract logic with partial negation we consider the following example:

Definition 4.3 ([11]). Let us consider predicate logic as built from atomic and negated atomic formulas by means of $\forall, \exists, \wedge, \vee$. Let $L(m, n)$ be the extension of this predicate logic obtained by adding the generalized quantifiers

$$Q_m x \phi(x) \iff \text{there are at least } \aleph_m \text{ elements } x \text{ satisfying } \phi(x)$$

and

$$\check{Q}_n x \phi(x) \iff \text{all but fewer than } \aleph_n \text{ elements } x \text{ satisfy } \phi(x).$$

In other words, $L(m,n)$ is built from atomic and negated atomic formulas by means of $Q_m, \check{Q}_n, \forall, \exists, \wedge, \vee$.

Theorem 4.4 ([11]). *The abstract logic $L(m,n)$ satisfies the Compactness Theorem if $m < n$, and also (by [31]) if $m \geq n$ and $\aleph_m^\omega = \aleph_m$. It satisfies total lack of negation if $n \neq m$, and it satisfies the Separation Theorem if and only if $n < m$.*

We can define a kind of partial negation $\sim\!\phi$ in $L(m,n)$ as follows:

$$
\begin{aligned}
\sim\!\phi &= \neg\phi \text{ if } \phi \text{ atomic} \\
\sim\!\phi &= \phi \text{ if } \phi \text{ negated atomic} \\
\sim\!(\phi \wedge \psi) &= \sim\!\phi \vee \sim\!\psi \\
\sim\!(\phi \vee \psi) &= \sim\!\phi \wedge \sim\!\psi \\
\sim\!\exists x \phi &= \forall x \sim\!\phi \\
\sim\!\forall x \phi &= \exists x \sim\!\phi \\
\sim\!Q_m x \phi &= \check{Q}_n x \sim\!\phi \\
\sim\!\check{Q}_n x \phi &= Q_m x \sim\!\phi
\end{aligned}
$$

What can we say about $Q_m x \phi \wedge \sim\!Q_m x \phi$ for first order ϕ? The meaning of this sentence is that ϕ is satisfied by at least $\aleph_m$ elements but still all but fewer than $\aleph_n$ elements satisfy $\neg\phi$. If $m < n$ this is perfectly possible. So in this case $\sim\!\psi$ acts as a weak negation which is not even in contradiction with ψ unless ψ is first order. In a sense, $L(m,n)$ does not satisfy the Law of Contradiction if $\sim$ is interpreted as its negation. If $m > n$ then $Q_m x \phi \wedge \sim\!Q_m x \phi$ cannot hold but now $Q_m x \phi \vee \sim\!Q_m x \phi$ may fail. Thus in this case $\sim\!\psi$ acts as a strong negation which does not cover the complement of ψ unless ψ is first order. In a sense, $L(m,n)$ does not satisfy the Law of Excluded Middle if $\sim$ is interpreted as its negation. Finally, if $m = n$, $\sim$ acts as a perfect negation for $L(m,n)$, i.e. $A \models \sim\!\phi \iff A \not\models \phi$.

Open Question. Is there a model theoretic characterization of any abstract logic $L = (S, F, V)$ with occurrence relation which is not closed under negation?

5. Logic frames

Having totally neglected the aspect of syntax we now bring it back with the concept of a logic frame. Without specifying the axioms and rules of proof, for example, the question of the *Completeness Theorem* has to be reduced to the question whether the set of valid sentences is recursively enumerable. From such knowledge one could in principle devise axioms and rules that yield a Completeness Theorem. However, it is often relevant to know whether particular axioms and rules constitute a complete set. This is all the more important in the case of logics which are originally defined via axioms and rules, such as constructive logic. The below concept of a

Structures	Sentences	axioms and rules by
Valuations	Propositional logic	Post [25]
Relational structures	Predicate logic	Gödel [12]
-monadic		Löwenheim
-ω-models		Orey [23]
	-with two variables	Scott [29]
	-infinitary	Karp [19]
	-with Q_1	Keisler [20]
	-higher order	Henkin [15]
-topological	-invariant	Ziegler [39]
-analytic	-existential bounded	Iovino [18]
-Borel	-Borel logic	Friedman[36]
Kripke structures	Intuitionistic logic	Kripke [21]
	Modal logic, S4	Kripke [21]
Many-valued	Many-valued logic	Belluce-Chang [4]
structures	Fuzzy logic	Hajek [13]
Games	Linear logic, additive	Blass [6]

FIGURE 5. Examples of complete logic frames

logic frame captures in abstract form the combination of syntax, semantics and proof theory of a logic. An *axiom* of an abstract logic L is simply a sentence of L, i.e. an element of F. An *inference rule* is any collection of functions defined in the set F.

Definition 5.1 ([35]). A *logic frame* is a quadruple $L = (S, F, \models, A)$ where $(S, F, \models)$ is an abstract logic and A is a class of axioms and inference rules of L.

We write $T \vdash \phi$ if ϕ is derivable from T in the usual sense using the axioms and rules of L. A logic frame $L = (S, F, \models, A)$ satisfies: the *Soundness Theorem* if $T \vdash \phi$ implies every model of T is a model of ϕ, the *Completeness Theorem* if $T \vdash \phi$ holds exactly if every model of T is a model of ϕ, and the *Recursive Compactness Theorem* if every L-theory which is recursive in the set of axioms and rules, every finite subset of which which has a model, itself has a model.

Literature of logic has numerous examples of logic frames satisfying the Completeness Theorem (see Figure 5). In the case of extensions of first order logic the question of completeness depends in many cases on set theory. One of the oldest open problems concerning extensions of first order logic is the question whether the extension $L_{\omega\omega}(Q_2)$ of first order logic by the quantifier "*there exist at least* $\aleph_2$ *many*" is effectively axiomatizable or satisfies the Compactness Theorem (restricted to countable vocabularies). The answer is "yes" if the Generalized Continuum Hypothesis is assumed [8] but remains otherwise open (see however [33]). The concept of logic frames helps us here.

Theorem 5.2. [35] *The fully classical logic frame $L = (S, L_{\omega\omega}(Q_2), \models, A_K)$, where A_K is the set of axioms and rules of Keisler [20], satisfies the Completeness Theorem if and only if L satisfies the Compactness Theorem (for countable vocabularies).*

There is a more complicated axiomatization A (the details of A are omitted) of an arbitrary $L_{\omega\omega}(Q_{\alpha_1}, \ldots, Q_{\alpha_n})$ with the same property as A_K above.

Theorem 5.3. [35] *The extension $L(\vec{Q}) = L_{\omega\omega}(Q_{\alpha_1}, \ldots, Q_{\alpha_n})$ of first order logic has a canonically defined set A of axioms and rules such that the fully classical logic frame $L = (S, L(\vec{Q}), \models, A)$ satisfies: If L satisfies the Completeness Theorem, then it satisfies the Compactness Theorem (and vice versa).*

Why is this interesting? The point is that we cannot decide on the basis of ZFC whether L satisfies the Compactness Theorem or not, but we can decide on the basis of ZFC alone that all we need to care about is the Completeness Theorem. We can also prove a general result about logics of the form

$$L(\vec{Q}) = L_{\omega\omega}(Q_{\alpha_1}, Q_{\alpha_2}, \ldots)$$

as long as no α_n is a limit of some of the other ordinals α_i.

Theorem 5.4. [35] *The extension $L(\vec{Q}) = L_{\omega\omega}(Q_{\alpha_1}, Q_{\alpha_2}, \ldots)$ of first order logic has a canonically defined set A of axioms and rules such that the fully classical logic frame $L = (S, L(\vec{Q}), \models, A)$ satisfies: If L satisfies the Recursive Compactness Theorem, then it satisfies the Compactness Theorem.*

What is interesting in the above theorem is that we cannot decide on the basis of ZFC whether L satisfies the Compactness Theorem or not, but we can decide on the basis of ZFC alone that if there is a counter-example to compactness, it is recursive in the axioms.

The results about logic frames up to now have been about connections between completeness and compactness. But we can also ask, are there Lindström theorems for logic frames. In particular, no answer to the following question is known even in the case of first order logic:

Open Question. Is there a characterization of any of the known complete logic frames $L = (S, F, \models, A)$ in terms of natural conditions on S, F, and A?

References

[1] Hajnal Andréka, István Németi, and Johan van Benthem. Modal languages and bounded fragments of predicate logic. *J. Philos. Logic*, 27(3):217–274, 1998.

[2] K. Jon Barwise. Axioms for abstract model theory. *Ann. Math. Logic*, 7:221–265, 1974.

[3] J. Barwise and S. Feferman, editors. *Model-theoretic logics*. Springer-Verlag, New York, 1985.

[4] L. P. Belluce and C. C. Chang. A weak completeness theorem for infinite valued first-order logic. *J. Symbolic Logic*, 28:43–50, 1963.

[5] Garrett Birkhoff. *Lattice theory*. Third edition. American Mathematical Society Colloquium Publications, Vol. XXV. American Mathematical Society, Providence, R.I., 1967.

[6] Andreas Blass. A game semantics for linear logic. *Ann. Pure Appl. Logic*, 56(1-3):183–220, 1992.

[7] Xavier Caicedo. On extensions of $L_{\omega\omega}(Q_1)$. *Notre Dame J. Formal Logic*, 22(1):85–93, 1981.

[8] C. C. Chang. A note on the two cardinal problem. *Proc. Amer. Math. Soc.*, 16:1148–1155, 1965.

[9] C. C. Chang. Modal model theory. In *Cambridge Summer School in Mathematical Logic (Cambridge, 1971)*, pages 599–617. Lecture Notes in Math., Vol. 337. Springer, Berlin, 1973.

[10] J. Flum. Characterizing logics. In *Model-theoretic logics*, Perspect. Math. Logic, pages 77–120. Springer, New York, 1985.

[11] Marta García-Matos. Abstract model theory without negation. 2005. Doctoral thesis, University of Helsinki.

[12] Kurt Gödel. Die vollstandigkeit der axiome des logischen funktionenkalklils. *Monatshefte fur Mathematk und Physik,*, 37:349 – 360, 1930.

[13] Petr Hájek. *Metamathematics of fuzzy logic*, volume 4 of *Trends in Logic—Studia Logica Library*. Kluwer Academic Publishers, Dordrecht, 1998.

[14] Lars Hansen. Formalized token models and duality in semantics: an algebraic approach. *J. Symbolic Logic*, 69(2):443–477, 2004.

[15] Leon Henkin. Completeness in the theory of types. *J. Symbolic Logic*, 15:81–91, 1950.

[16] Jaakko Hintikka. *The principles of mathematics revisited*. Cambridge University Press, Cambridge, 1996. With an appendix by Gabriel Sandu.

[17] T. Hyttinen. Model theory for infinite quantifier languages. *Fundamenta Mathematicae*, 134:125–142, 1990.

[18] Jose Iovino. On the maximality of logics with approximations. *Journal of Symbolic Logic*, 66(4):1909–1918, 2001.

[19] Carol R. Karp. *Languages with expressions of infinite length*. North–Holland Publishing Co., Amsterdam, 1964.

[20] H. Jerome Keisler. Logic with the quantifier "there exist uncountably many". *Ann. Math. Logic*, 1:1–93, 1970.

[21] Saul A. Kripke. A completeness theorem in modal logic. *J. Symb. Logic*, 24:1–14, 1959.

[22] Per Lindström. On extensions of elementary logic. *Theoria*, 35:1–11, 1969.

[23] Steven Orey. On ω-consistency and related properties. *J. Symb. Logic*, 21:246–252, 1956.

[24] R. Parikh and J. Väänänen. Finite information logic. *Annals of Pure and Applied Logic*. to appear.

[25] E. L. Post. Introduction to a general theory of elementary propositions. *American Journal Mathematics*, 43:163 – 185, 1921.

[26] Maarten de Rijke. A Lindström's theorem for modal logic. In *Modal Logic and Process Algebra, Lecture Notes 53*, pages 217–230. CSLI Publications, Stanford, 1995.

[27] J. H. Schmerl. Transfer theorems and their applications to logics. In *Model-theoretic logics*, Perspect. Math. Logic, pages 177–209. Springer, New York, 1985.

[28] A. Schmidt. Über deduktive theorien mit mehren sorten von grunddingen. *Mathematische Annalen*, 115:485–506, 1928.

[29] D. S. Scott. A decision method for validity of sentences in two variables. *J. Symbolic Logic*, 27:477, 1962.

[30] Joseph Sgro. Maximal logics. *Proceedings of the American Mathematical Society*, 63(2):291–298, 1977.

[31] Saharon Shelah. Two cardinal compactness. *Israel Journal of Mathematics*, 9:193–198, 1971.

[32] Saharon Shelah. Generalized quantifiers and compact logic. *Trans. Amer. Math. Soc.*, 204:342–364, 1975.

[33] Saharon Shelah. The pair $(\aleph_n, \aleph_0)$ may fail $\aleph_0$–compactness. In *Proceedings of LC'2001*, volume submitted of *Lecture Notes in Logic*. ASL.

[34] Saharon Shelah and Jouko Väänänen. A new infinitary languages with interpolation. to appear.

[35] Saharon Shelah and Jouko Väänänen. Recursive logic frames. *Mathematical Logic Quarterly*, to appear.

[36] C. I. Steinhorn. Borel structures and measure and category logics. In *Model-theoretic logics*, Perspect. Math. Logic, pages 579–596. Springer, New York, 1985.

[37] Leslie H. Tharp. The characterization of monadic logic. *J. Symb. Log.*, 38(3):481–488, 1973.

[38] Jouko Väänänen. Pseudo-finite model theory. *Mat. Contemp.*, 24:169–183, 2003. 8th Workshop on Logic, Language, Informations and Computation—WoLLIC'2001 (Brasília).

[39] M. Ziegler. A language for topological structures which satisfies a Lindström-theorem. *Bull. Amer. Math. Soc.*, 82(4):568–570, 1976.

Marta García-Matos
Department of Mathematics and Statistics
University of Helsinki
Helsinki
Finland
e-mail: `marta.garcia-matos@helsinki.fi`

Jouko Väänänen
Department of Mathematics and Statistics
University of Helsinki
Helsinki
Finland
e-mail: `jouko.vaananen@helsinki.fi`

J.-Y. Beziau (Ed.), *Logica Universalis*, 2nd edition, 35–61
© 2007 Birkhäuser Verlag Basel/Switzerland

A Topological Approach to Universal Logic: Model-Theoretical Abstract Logics

Steffen Lewitzka

Abstract. In this paper we develop a topological approach to a theory of model-theoretical abstract logics. A model-theoretical abstract logic is given by a set of expressions (formulas), a class of interpretations (models) and a satisfaction relation between interpretations and expressions. Notions such as theory, consequence, etc. are derived in a natural way. We define topologies on the space of (prime) theories and on the space of models. Then structural properties of a logic are mirrored in the respective topological spaces and can be studied now by topological means. We introduce the notion of logic-homomorphism, a map between model-theoretical abstract logics that preserve structural (topological) properties. We study in detail conditions under which logic-homomorphisms determine continuous or/and open functions between the respective topological spaces. We define a logic-isomorphism as a bijective logic-homomorphism. Such a map forces a homeomorphism on the corresponding topological spaces.

Moreover, we show that certain maps between logics lead to a condition which has the same form as the satisfaction axiom of institutions. This promising result may serve in future research to establish a connection between our approach and the well-known category-theoretical concept of institution.

Mathematics Subject Classification (2000). Primary 03B22; Secondary 03G30, 03C95.

Keywords. Abstract logic, universal logic, general logic theory, topology, institutions.

1. Introduction

A precise definition of the concept "logic" depends on what one considers as the central components (consequence relation, calculus, model-theoretic semantics,

This research was supported by CNPq grants 150309/2003-1 and 350092/2006-0.

category-theoretic components etc.). So the nature of a general theory of logics depends on the choice between these different alternatives.

We would like to mention two important research lines which deal with the concept of a general logic. The first one was founded by A. Tarski and others and continued by the School of Warsaw (see [7]). A logic is here defined by a set L of expressions (over a given alphabet) together with a consequence relation $\Vdash \subseteq Pow(L) \times L$, which satisfies the following conditions for all $A, B \subseteq L$:

(i) For all $a \in A$ it holds that $A \Vdash a$.

(ii) If $A \Vdash b$ for all $b \in B$, and if $B \Vdash c$, then holds $A \Vdash c$.

(In particular, it follows that the consequence relation is monotonic: If $B \subseteq A$ and $B \Vdash c$, then follows $A \Vdash c$.) Many logics satisfy these two intuitive conditions (an exception are non-monotonic logics). Basic notions, such as validity or consistence, maximal consistence etc. now can be expressed by means of this consequence relation.

The other research line that we have in mind is based on the notion of institution (see [2],[6]), which has model-theoretical and category-theoretical components. An institution describes a larger structure than the above mentioned concept of logics (for example, classical first order logic can be seen as one institution, whereas the first approach yields a logic for each first order signature). Institutions have found fruitful applications in the investigation of integrations of different logical systems and in computer science (software specification) [2],[6].

Our approach is partially based on the first one sketched above. However, to a set of expressions we add models and a satisfaction relation. So we give priority to a model-theoretic approach which in the usual way determines a consequence relation. Moreover, the model-theoretic aspect is helpful in order to establish a connection to the concept of institution. In this way, one aim of our research is to find in some sense a coherence of the two research lines mentioned above.

A possible research program on abstract logics may be roughly speaking the following one: In a first step one tries to give a very general definition of a concept of abstract logic in order to capture the biggest possible class of concrete logics. Now it is essential to find invariant criteria and conditions in order to distinguish and classify logics. In our case, these conditions are given by the topological structure of a logic.

Another important aspect is the question for the relationships between logics. Here a theory of abstract logics has to clarify basic concepts. What is an extension of a given logic, what is a reduction or an expansion of a model of some logic to a model of another one? What are the possible maps between logics, which properties are preserved under these maps?

In order to deal with these questions we develop the fundamentals of a theory of model-theoretical abstract logics using methods and concepts from general topology. This enables us to study and to classify properties of logics inside a general and abstract framework. In this sense, our work can be seen as a contribution to Universal Logic — an area in which different proposals for a general theory of logics are studied and compared. We hope that the ideas developed in this paper give rise to a further and more extensive study of abstract logics using the powerful and sophisticated tools of general topology. In fact, there is ongoing work that generalizes and extends the here presented model-theoretic approach (see [4]).

We would like to mention that this paper is the slightly modified form of an earlier version ([3]) which appeared in the first edition of the book Logica Universalis. New insights and discoveries helped to simplify and to extend some results and to improve the exposition.

2. Model-theoretical abstract logics

We start by defining a model-theoretical abstract logic and some other basic concepts.

Definition 2.1. A model-theoretical abstract logic is a structure

$$\mathcal{L} = (Expr_{\mathcal{L}}, Int_{\mathcal{L}}, \vDash_{\mathcal{L}}).$$

(To simplify matters, we sometimes say "abstract logic" or only "logic".)

(i) $Expr_{\mathcal{L}}$ is a set of formulas over a given alphabet. We call the elements of $Expr_{\mathcal{L}}$ expressions or $\mathcal{L}$-expressions.

(ii) $Int_{\mathcal{L}}$ is a class of interpretations in which the $\mathcal{L}$-expressions receive their denotations. The element of $Int_{\mathcal{L}}$ are also called models.

(iii) $\vDash_{\mathcal{L}}$ is a relation, called satisfaction relation, on $Int_{\mathcal{L}} \times Expr_{\mathcal{L}}$. The satisfaction relation determines the truth values ("true" or "false") of the expressions in the interpretations.

- If $\mathcal{A} \vDash_{\mathcal{L}} a$ holds, then we say that the expression a is true in the interpretation $\mathcal{A}$. In this case, $\mathcal{A}$ is called a model of a. Otherwise, we say that a is false in $\mathcal{A}$. We can extend the satisfaction relation to sets of expressions and classes of interpretations. Let $\mathcal{K} \subseteq Int_{\mathcal{L}}$ and $A \subseteq Expr_{\mathcal{L}}$. Then we define:
  $\mathcal{A} \vDash_{\mathcal{L}} A$ if and only if $\mathcal{A} \vDash_{\mathcal{L}} a$ for all $a \in A$,
  $\mathcal{K} \vDash_{\mathcal{L}} a$ if and only if $\mathcal{A} \vDash_{\mathcal{L}} a$ for all $\mathcal{A} \in \mathcal{K}$.
  Finally, $\mathcal{K} \vDash_{\mathcal{L}} A$ is defined in the obvious way.
  The class of models of A is $Mod_{\mathcal{L}}(A) := \{\mathcal{A} \mid \mathcal{A} \vDash_{\mathcal{L}} A\}$.

- We say that an expression b follows from a set of expressions B, written $B \Vdash_{\mathcal{L}} b$, if $Mod_{\mathcal{L}}(B) \subseteq Mod_{\mathcal{L}}(\{b\})$ holds. The relation $\Vdash_{\mathcal{L}} \subseteq Pow(Expr_{\mathcal{L}}) \times Expr_{\mathcal{L}}$ is called consequence relation. A set B of expressions is called (deductively) closed, if from $B \Vdash_{\mathcal{L}} b$ follows $b \in B$.

- A set B of expressions is called valid, if $Mod_{\mathcal{L}}(B) = Int_{\mathcal{L}}$. That is, the expressions in B are true in all interpretations, which is equivalent to the condition $\varnothing \Vdash_{\mathcal{L}} B$, by short $\Vdash_{\mathcal{L}} B$. An expression is valid, if it is contained in a valid set. A set B of expressions is satisfiable, if $Mod_{\mathcal{L}}(B) \neq \varnothing$. A maximal satisfiable set of expressions is a satisfiable set B such that there is no satisfiable set B' with $B \subsetneq B'$. A contradiction is an expression that is not

38   S. Lewitzka

satisfiable. A contradictory set is a set of expressions that is not satisfiable. A theory (or an $\mathcal{L}$-theory) is a satisfiable and deductively closed set of $\mathcal{L}$-expressions. A maximal theory is a theory T, such that $T \subsetneq T'$ implies that T' is not a theory.

For $\mathcal{A} \in Int_\mathcal{L}$, we define $Th_\mathcal{L}(\mathcal{A}) := \{a \in Expr_\mathcal{L} \mid \mathcal{A} \vDash_\mathcal{L} a\}$. If $\mathcal{K}$ is a class of $\mathcal{L}$-interpretations, then we define $Th_\mathcal{L}(\mathcal{K}) := \bigcap_{\mathcal{A} \in \mathcal{K}} Th_\mathcal{L}(\mathcal{A})$ ($= \{a \in Expr_\mathcal{L} \mid \mathcal{K} \vDash_\mathcal{L} a\}$).[1]

We introduce two equivalence relations. The first one is defined on expressions, the second one on interpretations. To simplify matters, we call both by the same name "$\mathcal{L}$-equivalence" and write for both the same symbol "$=_\mathcal{L}$". The differentiation between them should be clear by the context. The expressions a and b are called $\mathcal{L}$-equivalent, written $a =_\mathcal{L} b$, if $Mod_\mathcal{L}(\{a\}) = Mod_\mathcal{L}(\{b\})$. The interpretations $\mathcal{A}$ and $\mathcal{B}$ are called $\mathcal{L}$-equivalent, written $\mathcal{A} =_\mathcal{L} \mathcal{B}$, if $Th_\mathcal{L}(\mathcal{A}) = Th_\mathcal{L}(\mathcal{B})$ holds. We denote the equivalence class of a modulo $=_\mathcal{L}$ by $[a]_\mathcal{L}$ and the equivalence class of $\mathcal{A}$ modulo $=_\mathcal{L}$ by $[\mathcal{A}]_\mathcal{L}$.

The sets $A, B \subseteq Expr_\mathcal{L}$ are $\mathcal{L}$-equivalent, written $A =_\mathcal{L} B$, if $Mod_\mathcal{L}(A) = Mod_\mathcal{L}(B)$.

- A logic $\mathcal{L}$ is called finitary, if for all $X \subseteq Expr_\mathcal{L}$ the following holds: If every finite subset of X is satisfiable, then X is satisfiable.
- We say that a logic $\mathcal{L}$ has (classical) negation, if for every expression $a \in Expr_\mathcal{L}$ there exists another expression $b \in Expr_\mathcal{L}$ with the following property (neg):
  For all $\mathcal{A} \in Int_\mathcal{L}$: $\mathcal{A} \vDash_\mathcal{L} b$ if and only if $\mathcal{A} \nvDash_\mathcal{L} a$.
  ($\mathcal{A} \nvDash_\mathcal{L} a$ means: not $\mathcal{A} \vDash_\mathcal{L} a$.) If $\mathcal{L}$ has negation and $a \in Expr_\mathcal{L}$, then we denote the set of expressions b satisfying the property (neg) by $neg(a)$.[2]
- We say that a logic $\mathcal{L}$ has conjunction, if for every pair of expressions a, b there exists some expression c with the following property (conjunct):
  For all $\mathcal{A} \in Int_\mathcal{L}$: $\mathcal{A} \vDash_\mathcal{L} c$ if and only if ($\mathcal{A} \vDash_\mathcal{L} a$ and $\mathcal{A} \vDash_\mathcal{L} b$).[3]
  If $\mathcal{L}$ has conjunction and a, b are expressions, then we denote the set of expressions c satisfying the condition (conjunct) by $conjunct(a, b)$. The set $conjunct(a_0, ..., a_n)$ is defined in analogous way.[4]
- An abstract logic is called compact or boolean, if it is a finitary logic with classical negation and conjunction.

In the following lemma we collect some simple observations concerning theories.

---

[1] It is easy to see (Lemma 2.2), that the sets $Th_\mathcal{L}(\mathcal{A})$ and $Th_\mathcal{L}(\mathcal{K})$ are in fact theories, as the notation suggests.
[2] Notice that expressions in $neg(a)$ are $\mathcal{L}$-equivalent.
[3] By induction on $n < \omega$ it is easy to see that if $\mathcal{L}$ has conjunction, then for every finite sequence $a_0, ..., a_n$ of expressions there is some expression c satisfying the condition (conjunct).
[4] The expressions in $conjunct(a_0, ...a_n)$ are $\mathcal{L}$-equivalent.

Lemma 2.2. *Let $\mathcal{L}$ be an abstract logic and T a satisfiable set of expressions. Then the following holds:*

- *If $T = Th_{\mathcal{L}}(\mathcal{A})$ for some interpretation $\mathcal{A}$, then T is a theory.*
- *The intersection of a nonempty set of theories is again a theory.*
- *The set of all valid $\mathcal{L}$-expressions is a theory. Moreover, it is the smallest theory (that is, it is contained in any theory).*

The conditions (i)–(iii) *are equivalent. If in addition $\mathcal{L}$ has negation, then all conditions* (i)–(v) *are equivalent.*

(i) *T is a maximally satisfiable set of expressions.*
(ii) *T is a maximal theory.*
(iii) *$T = Th_{\mathcal{L}}(\mathcal{A})$ for all $\mathcal{A} \vDash_{\mathcal{L}} T$.*
(iv) *$T = Th_{\mathcal{L}}(\mathcal{A})$ for some $\mathcal{A} \vDash_{\mathcal{L}} T$.*
(v) *For all $a \in Expr_{\mathcal{L}}$: $a \in T$ or $neg(a) \subseteq T$.*

Proof. If $T = Th_{\mathcal{L}}(\mathcal{A})$, then T is satisfiable (by $\mathcal{A}$). We show that T is deductively closed. So suppose that $T \Vdash_{\mathcal{L}} a$. Since $\mathcal{A} \vDash_{\mathcal{L}} T$, $\mathcal{A} \vDash_{\mathcal{L}} a$. Hence, $a \in Th_{\mathcal{L}}(\mathcal{A}) = T$.

In order to show the second assertion, suppose that $T = \bigcap_{i<\alpha} T_i$, where every T_i is a theory and $\alpha > 1$ is any ordinal. Then T is clearly satisfiable since it is a subset of satisfiable sets. Suppose that $T \Vdash_{\mathcal{L}} a$. The consequence relation is monotonic (we have $Mod(T_i) \subseteq Mod(T)$), so it follows that $T_i \Vdash_{\mathcal{L}} a$ for every $i < \alpha$. Since the T_i are deductively closed, $a \in \bigcap_{i<\alpha} T_i = T$. Whence, T is deductively closed too and therefore a theory.

Let T be the set of all valid expressions. Then T is satisfiable (by all interpretations!). Suppose that $T \Vdash_{\mathcal{L}} a$. Every $\mathcal{L}$-interpretation is a model of T, whence, a model of a. Then a is a valid expression and must be contained in T. Whence, T is deductively closed and a theory. If T' is any theory, then clearly $T' \Vdash_{\mathcal{L}} a$ for all valid expressions a. So $a \in T'$ for all valid expressions a, whence, $T \subseteq T'$ and T is the smallest theory.

(i)$\rightarrow$ (ii): If T is maximally satisfiable and $T \Vdash_{\mathcal{L}} a$, then $T \cup \{a\}$ is satisfiable, whence $a \in T$, by maximality of T. Hence, T is deductively closed and therefore a theory. If $T \subseteq T'$ and T' is a theory, then T' is satisfiable. Since T is maximally satisfiable, it follows that $T = T'$. Hence, T is a maximal theory.

(ii)$\rightarrow$ (iii): Let T be a maximal theory and suppose that $\mathcal{A}$ is any model of T. Then clearly $T \subseteq Th_{\mathcal{L}}(\mathcal{A})$. The fact that $Th_{\mathcal{L}}(\mathcal{A})$ is a theory that contains the *maximal* theory T forces $T = Th_{\mathcal{L}}(\mathcal{A})$.

(iii)$\rightarrow$(i): Suppose that (iii) holds, $T \subseteq T'$ and T' is satisfiable. Every model $\mathcal{A}$ of T' is a model of T. Whence, $T = Th_{\mathcal{L}}(\mathcal{A})$ for every model $\mathcal{A}$ of T', by (iii). Since $T \subseteq T' \subseteq Th_{\mathcal{L}}(\mathcal{A})$ it follows that $T = T'$, and T is maximally satisfiable.

Now we suppose that $\mathcal{L}$ has negation.

If (iii) holds, then follows (iv), since T is satisfiable and therefore there exists some model of T.

Suppose that (iv) holds and $a \notin T = Th_{\mathcal{L}}(\mathcal{A})$. Then $\mathcal{A} \nvDash_{\mathcal{L}} a$ and $\mathcal{A} \vDash_{\mathcal{L}} b$, for all $b \in neg(a)$. Hence, $neg(a) \subseteq T = Th_{\mathcal{L}}(\mathcal{A})$.

Now suppose that (v) holds. In order to show (i), we assume that there is some satisfiable set $T' \supseteq T$ and some $a \in T' \smallsetminus T$. From (v) follows that $neg(a) \subseteq T \subseteq T'$. By definition of $neg(a)$, T' can not be satisfiable — a contradiction. Hence, such a T' does not exist and T is maximally satisfiable. $\square$

Some theories are of particular importance:

Definition 2.3. Let $\mathcal{L}$ be a logic. A prime theory is a theory T of the form $T = Th_{\mathcal{L}}(\mathcal{A})$ for some interpretation $\mathcal{A}$. The set of all prime theories is $PTh(\mathcal{L}) := \{Th_{\mathcal{L}}(\mathcal{A}) \mid \mathcal{A} \in Int_{\mathcal{L}}\}$. The set of all maximal theories is $MTh(\mathcal{L}) := \{T \mid T$ is a maximal $\mathcal{L}$-theory$\}$.

If $\mathcal{L}$ has negation, then we call a maximal theory also a complete theory. This is justified by the equivalence of (ii) and (v) of the preceding Lemma.

Notice that from Lemma 2.2 it follows that $MTh(\mathcal{L}) \subseteq PTh(\mathcal{L})$, and if $\mathcal{L}$ has negation, then $MTh(\mathcal{L}) = PTh(\mathcal{L})$.

Before we continue we reach the following agreement: Let X be a family of sets. We say that X is closed under finite intersections, if for each $n \geq 1$ and any $Y_1, ..., Y_n \in X$ we have $Y_1 \cap ... \cap Y_n \in X$. We say that X is closed under complement, if for every $Y \in X, \bigcup X \smallsetminus Y \in X$.

Definition 2.4. Let $\mathcal{L}$ be an abstract logic. We define:

$$Mod(\mathcal{L}) := \{Mod_{\mathcal{L}}(a) \mid a \in Expr_{\mathcal{L}}\}, \text{ and}$$
$$(Mod(\mathcal{L})/_{=_{\mathcal{L}}}) := \{(Mod_{\mathcal{L}}(a)/_{=_{\mathcal{L}}}) \mid a \in Expr_{\mathcal{L}}\},$$

where $(Mod_{\mathcal{L}}(a)/_{=_{\mathcal{L}}})$ denotes the set of equivalence classes $[\mathcal{A}]_{\mathcal{L}}$ modulo $=_{\mathcal{L}}$ for every $\mathcal{A} \in Mod_{\mathcal{L}}(a)$. Let $U_{\mathcal{L}}$ be the set of all finite, non-empty intersections of elements of $Mod(\mathcal{L})$.[5] Let $V_{\mathcal{L}}$ be the set of all finite, non-empty intersections of elements of $(Mod(\mathcal{L})/_{=_{\mathcal{L}}})$ We define

$$\sigma_{\mathcal{L}}^1 := \{\bigcup \delta \mid \delta \subseteq U_{\mathcal{L}}\} \text{ and}$$
$$\sigma_{\mathcal{L}}^2 := \{\bigcup \delta \mid \delta \subseteq V_{\mathcal{L}}\}.$$

The tuple $(Int_{\mathcal{L}}, \sigma_{\mathcal{L}}^1)$ is called the space of models and the tuple $((Int_{\mathcal{L}}/_{=_{\mathcal{L}}}), \sigma_{\mathcal{L}}^2)$ is called the space of models modulo $\mathcal{L}$-equivalence.

For every $a \in Expr_{\mathcal{L}}$, we define $a^{*_{\mathcal{L}}} := \{T \in PTh(\mathcal{L}) \mid a \in T\}$. [6] If the context is clear, we write a^* instead of $a^{*_{\mathcal{L}}}$. Let $W_{\mathcal{L}}$ be the set of all finite, non-empty intersections of elements of $\mathcal{T}_{\mathcal{L}} := \{a^{*_{\mathcal{L}}} \mid a \in Expr_{\mathcal{L}}\}$. We define

$$\rho_{\mathcal{L}} := \{\bigcup \delta \mid \delta \subseteq W_{\mathcal{L}}\}\}.$$

The tuple $(PTh(\mathcal{L}), \rho_{\mathcal{L}})$ is called the space of prime theories of $\mathcal{L}$.

---

[5]That is, $u \in U_{\mathcal{L}} \iff u = Mod_{\mathcal{L}}(a_1) \cap ... \cap Mod_{\mathcal{L}}(a_n)$ for some $n \geq 1$ and some $a_1, ..., a_n \in Expr_{\mathcal{L}}$.

[6]Notice that $\mathcal{A} \in Mod_{\mathcal{L}}(a)$ if and only if $Th_{\mathcal{L}}(\mathcal{A}) \in a^{*_{\mathcal{L}}}$.

The previous Definition contains notions which recall topological concepts. A justification is given by the next Proposition. This Proposition also shows in which way one can classify abstract logics in terms of topological spaces. The Proposition yields a particular example: every boolean abstract logic corresponds to a boolean topological space.

Proposition 2.5. *Let $\mathcal{L}$ be an abstract logic. Then the following holds:*
The sets $Mod(\mathcal{L})$, $(Mod(\mathcal{L})/_{=_{\mathcal{L}}})$ are subbases of the topologies $\sigma_{\mathcal{L}}^1$ on $Int_{\mathcal{L}}$, $\sigma_{\mathcal{L}}^2$ on $(Int_{\mathcal{L}}/_{=_{\mathcal{L}}})$, respectively.
The following conditions are equivalent:

- *$\mathcal{L}$ has conjunction.*
- *$Mod(\mathcal{L})$ is closed under finite intersections and is a basis of the topology $\sigma_{\mathcal{L}}^1$ on $Int_{\mathcal{L}}$.*
- *$(Mod(\mathcal{L})/_{=_{\mathcal{L}}})$ is closed under finite intersections and is a basis of the topology $\sigma_{\mathcal{L}}^2$ on $(Int_{\mathcal{L}}/_{=_{\mathcal{L}}})$.*

The following conditions are equivalent:

(i) *$\mathcal{L}$ is a logic with conjunction and negation.*
(ii) *$(Int_{\mathcal{L}}, \sigma_{\mathcal{L}}^1)$ is a topological space with the basis $Mod(\mathcal{L})$ which is closed under finite intersections and under complement.[7]*
(iii) *$((Int_{\mathcal{L}}/_{=_{\mathcal{L}}}), \sigma_{\mathcal{L}}^2)$ is a topological space with the basis $(Mod(\mathcal{L})/_{=_{\mathcal{L}}})$ which is closed under finite intersections and under complement. The topological space is Hausdorff.*

Finally: $\mathcal{L}$ is a boolean logic if and only if $(Int_{\mathcal{L}}, \sigma_{\mathcal{L}}^1)$ is a compact, topological space with the basis $Mod(\mathcal{L})$ which is closed under finite intersection and under complement if and only if $((Int_{\mathcal{L}}/_{=_{\mathcal{L}}}), \sigma_{\mathcal{L}}^2)$ is a boolean space with the basis $(Mod(\mathcal{L})/_{=_{\mathcal{L}}})$ which is closed under finite intersection and under complement.[8]

Proof. Recall that $U_{\mathcal{L}}$ is the set of all finite, non-empty intersections of elements of $Mod(\mathcal{L})$. It is clear that $Int_{\mathcal{L}} = \bigcup U_{\mathcal{L}}$. Furthermore, $U_{\mathcal{L}}$ is (by definition) closed under finite intersections. These two conditions are sufficient for $U_{\mathcal{L}}$ being a basis of the topology $\sigma_{\mathcal{L}}^1$ on $Int_{\mathcal{L}}$. Hence, $Mod(\mathcal{L})$ is a subbasis of this topology. In a similar way follows that $V_{\mathcal{L}}$ is a basis of the topology $\sigma_{\mathcal{L}}^2$ on $(Int_{\mathcal{L}}/_{=_{\mathcal{L}}})$ and therefore $(Mod(\mathcal{L})/_{=_{\mathcal{L}}})$ is a subbasis of this topology. We have proved the first assertion of the Proposition.

Now suppose that $\mathcal{L}$ has conjunction. Then for every natural number n and every sequence of expressions $a_0, a_1, ..., a_n$ we have
$$Mod_{\mathcal{L}}(c) = \bigcap_{i \leq n} Mod_{\mathcal{L}}(a_i), \text{ for any } c \in conjunct(a_0, ..., a_n).$$
Hence, $Mod(\mathcal{L})$ is closed under finite intersections and therefore $Mod(\mathcal{L}) = U_{\mathcal{L}}$. Similarly, $(Mod(\mathcal{L})/_{=_{\mathcal{L}}})$ is closed under finite intersections and $(Mod(\mathcal{L})/_{=_{\mathcal{L}}}) = V_{\mathcal{L}}$. Thus, if $\mathcal{L}$ has conjunction, then $Mod(\mathcal{L})$ and $(Mod(\mathcal{L})/_{=_{\mathcal{L}}})$ are bases of the

[7]Note that if a basis of a topology is closed under complement, then, in particular, it is a basis of clopen sets.
[8]By a boolean space we mean a topological space which is compact, Hausdorff and has a basis of clopen sets.

respective topologies. Conversely, if $Mod(\mathcal{L})$ is closed under finite intersections, then for any two expressions a, b there is an expression c such that $Mod_{\mathcal{L}}(a) \cap Mod_{\mathcal{L}}(b) = Mod_{\mathcal{L}}(c)$. Hence, the logic has conjunction.

The same argumentations hold if we factorize $Mod(\mathcal{L})$ and $Int_{\mathcal{L}}$ by $\mathcal{L}$-equivalence. Thus, the three statements are equivalent.

Now suppose that (i) holds. We show (iii): By the first part of the Proposition it is sufficient to show that the basis $(Mod(\mathcal{L})/_{=_{\mathcal{L}}})$ is closed under complement and that the topology $\sigma_{\mathcal{L}}^2$ is Hausdorff. Since $\mathcal{L}$ has negation, we have $(Mod_{\mathcal{L}}(b)/_{=_{\mathcal{L}}}) = (Int_{\mathcal{L}}/_{=_{\mathcal{L}}}) \smallsetminus (Mod_{\mathcal{L}}(a)/_{=_{\mathcal{L}}})$, for every expression a and every $b \in neg(a)$. This shows that $(Mod(\mathcal{L})/_{=_{\mathcal{L}}})$ is closed under complement. The topological space $((Int_{\mathcal{L}}/_{=_{\mathcal{L}}}), \sigma_{\mathcal{L}}^2)$ is also Hausdorff: Suppose that $\mathcal{A}, \mathcal{A}' \in Int_{\mathcal{L}}$ are not $\mathcal{L}$-equivalent interpretations, that is, the equivalence classes $[\mathcal{A}]_{\mathcal{L}}$ and $[\mathcal{A}']_{\mathcal{L}}$ are different in $Int_{\mathcal{L}}$ modulo $=_{\mathcal{L}}$. Then we may assume that there exists some expression a such that $\mathcal{A} \vDash_{\mathcal{L}} a$ and $\mathcal{A}' \nvDash_{\mathcal{L}} a$. Since $\mathcal{L}$ has negation, for any $b \in neg(a)$ holds $\mathcal{A}' \vDash b$. But this implies that $[\mathcal{A}]_{\mathcal{L}} \in (Mod_{\mathcal{L}}(a)/_{=_{\mathcal{L}}})$, $[\mathcal{A}']_{\mathcal{L}} \in (Mod_{\mathcal{L}}(b)/_{=_{\mathcal{L}}})$, and $(Mod_{\mathcal{L}}(a)/_{=_{\mathcal{L}}}) \cap (Mod_{\mathcal{L}}(b)/_{=_{\mathcal{L}}}) = \varnothing$. Hence, the space is Hausdorff and (iii) holds.

On the other hand, if (iii) holds, then it is easy to see that $(Int_{\mathcal{L}}, \sigma_{\mathcal{L}}^1)$ is a topological space with the basis $Mod(\mathcal{L})$ which is closed under finite intersections and under complement. This yields (ii).

Now suppose that (ii) is true. From the first part of the Proposition we know that $\mathcal{L}$ has conjunction. Since the basis is closed under finite intersections, for every expression a there is another expression b such that $Mod_{\mathcal{L}}(b) = Int_{\mathcal{L}} \smallsetminus Mod_{\mathcal{L}}(a)$. It follows that $b \in neg(a)$ and $\mathcal{L}$ has negation. Hence (i) holds.

Finally, let us show the last part of the Proposition. Suppose that $\mathcal{L}$ is a boolean (compact) logic, that is, $\mathcal{L}$ is finitary and has negation and conjunction. By what we have already shown we know that $(Int_{\mathcal{L}}, \sigma_{\mathcal{L}}^1)$ and $((Int_{\mathcal{L}}/_{=_{\mathcal{L}}}), \sigma_{\mathcal{L}}^2)$ are topological spaces with bases that are closed under finite intersection and under complement (in particular, the bases consist of clopen sets) and that the latter space is Hausdorff. So it is sufficient to show compactness. In order to show that the topological space $(Int_{\mathcal{L}}, \sigma_{\mathcal{L}}^1)$ is compact, we assume that $X := Int_{\mathcal{L}} = \bigcup_{i \in I} U_i$ for some system of open sets $U_i \in \sigma_{\mathcal{L}}^1$, $i \in I$, where every $U_i = \bigcup_{a \in A_i} Mod_{\mathcal{L}}(a)$ for some $A_i \subseteq Expr_{\mathcal{L}}$. Put $A := \bigcup_{i \in I} A_i$. We get $\varnothing = X \smallsetminus \bigcup_{i \in I} U_i = X \smallsetminus \bigcup_{a \in A} Mod_{\mathcal{L}}(a) = \bigcap_{a \in A}(X \smallsetminus Mod_{\mathcal{L}}(a)) = \bigcap\{Mod_{\mathcal{L}}(b) \mid b \in neg(a), a \in A\} = Mod_{\mathcal{L}}(\{b \mid b \in neg(a), a \in A\})$. In other words, the set $B := \{b \mid b \in neg(a), a \in A\}$ is not satisfiable. Since $\mathcal{L}$ is finitary, there is some finite subset $B_f \subseteq B$ such that B_f is not satisfiable. It follows that $B_f = \{b \mid b \in neg(a), a \in A_f\}$ for some $finite$ $A_f \subseteq A$. (Notice that $Mod_{\mathcal{L}}(b) = Mod_{\mathcal{L}}(b')$ for any $b, b' \in neg(a)$, $a \in A$.) Hence, $\bigcap\{Mod_{\mathcal{L}}(b) \mid b \in neg(a), a \in A_f\} = \varnothing$. It follows that $X = X \smallsetminus \bigcap\{Mod_{\mathcal{L}}(b) \mid b \in neg(a), a \in A_f\} = \bigcup\{X \smallsetminus Mod_{\mathcal{L}}(b) \mid b \in neg(a), a \in A_f\} = \bigcup\{Mod_{\mathcal{L}}(a) \mid a \in A_f\}$. Whence, there is also a finite subset $I_f \subseteq I$ such that $X = \bigcup_{i \in I_f} U_i$. This shows that $(Int_{\mathcal{L}}, \sigma_{\mathcal{L}}^1)$ is compact. Finally, if we factorize $Int_{\mathcal{L}}$ and $Mod_{\mathcal{L}}(a)$ (for any $a \in Expr_{\mathcal{L}}$) by $=_{\mathcal{L}}$, then the compactness of $((Int_{\mathcal{L}}/_{=_{\mathcal{L}}}), \sigma_{\mathcal{L}}^2)$ follows from the compactness of $(Int_{\mathcal{L}}, \sigma_{\mathcal{L}}^1)$.

Now suppose that $((Int_{\mathcal{L}}/_{=_{\mathcal{L}}}), \sigma_{\mathcal{L}}^2)$ is a boolean space (a topological space with a basis of clopen sets, compact and Hausdorff) with the basis $(Mod(\mathcal{L})/_{=_{\mathcal{L}}})$ which is closed under finite intersection and under complement. We have already proved that $\mathcal{L}$ is a logic with negation and conjunction. We show that $\mathcal{L}$ is a finitary logic: Suppose that $A \subseteq Expr_{\mathcal{L}}$ is any set of expressions, which is not satisfiable. It is sufficient to show that there exists a finite subset of A that is not satisfiable. Since A is not satisfiable, $Mod_{\mathcal{L}}(A) = \bigcap_{a \in A} Mod_{\mathcal{L}}(a) = \varnothing$. Then we get $Int_{\mathcal{L}} = Int_{\mathcal{L}} \smallsetminus \bigcap_{a \in A} Mod_{\mathcal{L}}(a) = \bigcup_{a \in A}(Int_{\mathcal{L}} \smallsetminus Mod_{\mathcal{L}}(a)) = \bigcup\{Mod_{\mathcal{L}}(b) \mid b \in neg(a), a \in A\}$. The last term is a union of open sets that covers $Int_{\mathcal{L}}$. Since the topology $\sigma_{\mathcal{L}}^2$ is compact, $\sigma_{\mathcal{L}}^1$ is also compact and there exists a finite $A_f \subseteq A$ such that $Int_{\mathcal{L}} = \bigcup\{Mod_{\mathcal{L}}(b) \mid b \in neg(a), a \in A_f\}$. Hence, $\varnothing = Int_{\mathcal{L}} \smallsetminus \bigcup\{Mod_{\mathcal{L}}(b) \mid b \in neg(a), a \in A_f\} = \bigcap(Int_{\mathcal{L}} \smallsetminus \{Mod_{\mathcal{L}}(b) \mid b \in neg(a), a \in A_f\}) = \bigcap_{a \in A_f} Mod_{\mathcal{L}}(a) = Mod_{\mathcal{L}}(A_f)$. That is, the finite subset $A_f \subseteq A$ is not satisfiable. Hence, $\mathcal{L}$ is finitary and therefore also compact. $\qquad\square$

Proposition 2.6. *Let $\mathcal{L}$ be an abstract logic. Then the function $h : (Int_{\mathcal{L}}/_{=_{\mathcal{L}}}) \to PTh(\mathcal{L})$ defined by $h([\mathcal{A}]_{\mathcal{L}}) =: Th_{\mathcal{L}}(\mathcal{A})$ is a bijection from the class of interpretations factorized by $\mathcal{L}$-equivalence onto the set of prime theories. This function is also a bijection from the subbasis $(Mod(\mathcal{L})/_{=_{\mathcal{L}}})$ onto $\mathcal{T}_{\mathcal{L}} = \{a^{*_{\mathcal{L}}} \mid a \in Expr_{\mathcal{L}}\}$, and therefore it is also a bijection from the topology $\sigma_{\mathcal{L}}^2$ onto $\rho_{\mathcal{L}}$. In particular, $\mathcal{T}_{\mathcal{L}}$ forms a subbasis of the topology $\rho_{\mathcal{L}}$ on $PTh_{\mathcal{L}}$ and the topological spaces $((Int_{\mathcal{L}}/_{=_{\mathcal{L}}}), \sigma_{\mathcal{L}}^2)$ and $(PTh(\mathcal{L}), \rho_{\mathcal{L}})$ are homeomorphic via h.*

Proof. It is easy to see that h is well-defined, injective and surjective.

h is also a bijection from $(Mod(\mathcal{L})/_{=_{\mathcal{L}}})$ onto $\{a^* \mid a \in Expr_{\mathcal{L}}\}$:
For any $a \in Expr_{\mathcal{L}}$ we have

$$
\begin{aligned}
h(Mod_{\mathcal{L}}(a)/_{=_{\mathcal{L}}}) &= \{h([\mathcal{A}]_{\mathcal{L}}) \mid [\mathcal{A}]_{\mathcal{L}} \in (Mod(a)/_{=_{\mathcal{L}}})\} \\
&= \{Th_{\mathcal{L}}(\mathcal{A}) \mid [\mathcal{A}]_{\mathcal{L}} \in (Mod_{\mathcal{L}}(a)/_{=_{\mathcal{L}}})\} \\
&= \{Th_{\mathcal{L}}(\mathcal{A}) \mid \mathcal{A} \in Mod_{\mathcal{L}}(a)\} \\
&= \{T \in PTh(\mathcal{L}) \mid T \in a^{*_{\mathcal{L}}}\} \\
&= a^*.
\end{aligned}
$$

It is clear that h is surjective. Suppose that $(Mod_{\mathcal{L}}(a)/_{=_{\mathcal{L}}}) \neq (Mod_{\mathcal{L}}(b)/_{=_{\mathcal{L}}})$. Then we may assume that there is some $[\mathcal{A}]_{\mathcal{L}} \in (Mod_{\mathcal{L}}(a)/_{=_{\mathcal{L}}}) \smallsetminus (Mod_{\mathcal{L}}(b)/_{=_{\mathcal{L}}})$. Thus, $\mathcal{A}$ is a model of a, but not a model of b. This is equivalent to the condition that $a \in Th_{\mathcal{L}}(\mathcal{A})$ and $b \notin Th_{\mathcal{L}}(\mathcal{A})$. Hence, $Th_{\mathcal{L}}(\mathcal{A}) \in a^* \smallsetminus b^*$. It follows that h is also injective.

Now it follows that h is a bijection from $V_{\mathcal{L}}$ to $W_{\mathcal{L}}$ (see Definition 2.4). Finally, it is easy to see that $h(\bigcup \tau) = \bigcup(h(\tau))$, for any $\tau \subseteq V_{\mathcal{L}}$. Since every element of $\sigma_{\mathcal{L}}^2$ is of the form $\bigcup \tau$ and every element of $\rho_{\mathcal{L}}$ is of the form $\bigcup h(\tau)$, (h is a bijection from $V_{\mathcal{L}}$ onto $W_{\mathcal{L}}$), it follows that h is a bijection from $\sigma_{\mathcal{L}}^2$ onto $\rho_{\mathcal{L}}$. $\qquad\square$

Summarizing we can say that each abstract logic $\mathcal{L}$ gives rise to three topological spaces $(Int_{\mathcal{L}}, \sigma_{\mathcal{L}}^1)$, $((Int_{\mathcal{L}}/_{=_{\mathcal{L}}}), \sigma_{\mathcal{L}}^2)$ and $(PTh_{\mathcal{L}}, \rho_{\mathcal{L}})$ such that the two last-mentioned are homeomorphic.

Definition 2.7. Suppose that $\mathcal{L} = (Expr_{\mathcal{L}}, Int_{\mathcal{L}}, \vDash_{\mathcal{L}})$ and $\mathcal{L}' = (Expr_{\mathcal{L}'}, Int_{\mathcal{L}'}, \vDash_{\mathcal{L}'})$ are model-theoretical abstract logics.

(i) Let $f : Int_{\mathcal{L}} \to Int_{\mathcal{L}'}$ and $g : Expr_{\mathcal{L}} \to Expr_{\mathcal{L}'}$ be functions. We say that f is $\mathcal{L}$-injective, if from $\mathcal{A} \neq_{\mathcal{L}} \mathcal{B}$ follows that $f(\mathcal{A}) \neq_{\mathcal{L}'} f(\mathcal{B})$. g is called $\mathcal{L}$-injective, if from $a \neq_{\mathcal{L}} b$ follows that $g(a) \neq_{\mathcal{L}'} g(b)$. f is $\mathcal{L}$-surjective, if for any $[\mathcal{A}']_{\mathcal{L}'} \in (Int_{\mathcal{L}'}/_{=_{\mathcal{L}'}})$ there is some $\mathcal{A} \in Int_{\mathcal{L}}$ such that $f(\mathcal{A}) \in [\mathcal{A}']_{\mathcal{L}'}$. g is $\mathcal{L}$-surjective, if for any $[a']_{\mathcal{L}'} \in (Expr_{\mathcal{L}'}/_{=_{\mathcal{L}'}})$ there is some expression $a \in Expr_{\mathcal{L}}$ such that $g(a) \in [a']_{\mathcal{L}'}$.

The function f (the function g) is $\mathcal{L}$-bijective, if it is both $\mathcal{L}$-injective and $\mathcal{L}'$-surjective, respectively.

f is regular, if for all $\mathcal{A}, \mathcal{B} \in Int_{\mathcal{L}}$ holds:

$$\mathcal{A} =_{\mathcal{L}} \mathcal{B} \Longrightarrow f(\mathcal{A}) =_{\mathcal{L}'} f(\mathcal{B}).$$

g is regular, if for all $a, b \in Expr_{\mathcal{L}}$ holds:

$$a =_{\mathcal{L}} b \Longrightarrow g(a) =_{\mathcal{L}'} g(b).$$

(ii) A logic-homomorphism from $\mathcal{L}$ to $\mathcal{L}'$ is a pair (f, g) of functions $f : Int_{\mathcal{L}} \to Int_{\mathcal{L}'}$ and $g : Expr_{\mathcal{L}} \to Expr_{\mathcal{L}'}$ that satisfy the following condition: For every $a' \in Expr_{\mathcal{L}'}$ there exists an $A_{a'} \subseteq Expr_{\mathcal{L}}$ such that the following holds:
- $f^{-1}(Mod_{\mathcal{L}'}(a')) = \bigcup\{(Mod_{\mathcal{L}}(a)) \mid a \in A_{a'}\}$ and
- $g^{-1}(a'^{*\mathcal{L}'}) = \bigcup\{a^{*\mathcal{L}} \mid a \in A_{a'}\}$.

We write $(f, g) : \mathcal{L} \to \mathcal{L}'$. We say that a logic-homomorphism (f, g) is injective (surjective, bijective), if f and g are injective (surjective, bijective) functions, respectively.

(iii) A logic-homomorphism $(f, g) : \mathcal{L} \to \mathcal{L}'$ is called strong, if for all $a' \in Expr_{\mathcal{L}'}$, $A_{a'} = \{a \mid g(a)^{*\mathcal{L}'} \subseteq a'^{*\mathcal{L}'}\}$, i.e., if the following holds:
- $f^{-1}(Mod_{\mathcal{L}'}(a')) = \bigcup\{Mod_{\mathcal{L}}(a) \mid g(a)^{*\mathcal{L}'} \subseteq a'^{*\mathcal{L}'}\}$ and
- $g^{-1}(a'^{*\mathcal{L}'}) = \bigcup\{a^{*\mathcal{L}} \mid g(a)^{*\mathcal{L}'} \subseteq a'^{*\mathcal{L}'}\}$. [9]

If g satisfies the defining condition of a (strong) logic-homomorphism, then we say that g has the (strong) homomorphism property. In an analogous way we say that f has the (strong) homomorphism property, if the function f satisfies the defining equation of a (strong) logic-homomorphism. For short, we write SHP (HP) for (strong) homomorphism property, respectively.

(iv) A logic-isomorphism from $\mathcal{L}$ to $\mathcal{L}'$ is a bijective logic-homomorphism from $\mathcal{L}$ to $\mathcal{L}'$.

We say that $\mathcal{L}$ and $\mathcal{L}'$ are isomorphic, written $\mathcal{L} \simeq \mathcal{L}'$, if there exists some logic-isomorphism from $\mathcal{L}$ to $\mathcal{L}'$.

---

[9] Notice that $g(a)^{*\mathcal{L}'} \subseteq a'^{*\mathcal{L}'} \iff Mod_{\mathcal{L}'}(g(a)) \subseteq Mod_{\mathcal{L}'}(a') \iff g(a) \Vdash_{\mathcal{L}'} a'$, since for any interpretation $\mathcal{A}$ and any expression a: $\mathcal{A} \in Mod(a)$ if and only if $Th(\mathcal{A}) \in a^*$.

Remark 2.8.   (i) Our definition of a logic-homomorphism is motivated by the intuition that the functions g and f should preserve topological structure of the respective spaces and therefore should behave like continuous maps, although g is not a function from $PTh(\mathcal{L})$ to $PTh(\mathcal{L}')$ and therefore can not be a continuous map between these spaces. However, our results will reveal that a logic-homomorphism under certain conditions gives rise to a continuous function $G : PTh(\mathcal{L}) \to PTh(\mathcal{L}')$. In particular, a logic-isomorphism forces G to be an homeomorphism between the respective spaces (see Lemma 2.9(xi)).

The functions g and f of a logic-homomorphism (f, g) should also behave in a homogeneous way: If g gives rise to a continuous map G from the space $(PTh(\mathcal{L}), \rho_{\mathcal{L}})$ to the space $(PTh(\mathcal{L}'), \rho_{\mathcal{L}'})$, then there should also exist a function F (that depends on f) such that F is a continuous map from space $((Int_{\mathcal{L}}/_{=_{\mathcal{L}}}), \sigma_{\mathcal{L}}^2)$ to space $((Int_{\mathcal{L}'}/_{=_{\mathcal{L}'}}), \sigma_{\mathcal{L}'}^2)$ (see Corollary 2.27).

(ii) We have seen that the elements of $U_{\mathcal{L}}$, $V_{\mathcal{L}}$ and $W_{\mathcal{L}}$ are the basic open sets of the topologies $\sigma_{\mathcal{L}}^1$, $\sigma_{\mathcal{L}}^2$, $\rho_{\mathcal{L}}$, respectively, i.e., these sets are bases of the respective topologies, whereas $Mod(\mathcal{L})$, $(Mod(\mathcal{L})/_{=_{\mathcal{L}}})$ and $\{a^* \mid a \in Expr_{\mathcal{L}}\}$ are subbases of these topologies, respectively. Only if the logic has conjunction, then the subbases and bases coincide. Thus, following our motivation given in (i) above, we should define a logic-homomorphism on the set of basic open sets of the respective topologies, rather than on the elements of the subbases. However, in order to simplify matter we prefer the definition given above. In fact, one can verify that this is no restriction: all the results that appear in this paper do not depend on this detail.

Lemma 2.9. *Suppose that (f, g) is a logic-homomorphism from $\mathcal{L}$ to $\mathcal{L}'$. Then the following holds:*

(i) $\{g^{-1}(T') \mid T' \in PTh(\mathcal{L}')\} = PTh(\mathcal{L})$.

(ii) *f is regular.*

(iii) *For each $a \in Expr_{\mathcal{L}}$ it holds the following:*

$$g^{-1}(g(a)^{*\mathcal{L}'}) = a^{*\mathcal{L}} \text{ and}$$

$$f^{-1}(Mod_{\mathcal{L}'}(g(a))) = Mod_{\mathcal{L}}(a).$$

(iv) *g is $\mathcal{L}$-injective.*

(v) *The following equivalent conditions hold:*

(a) *$f(\mathcal{A}) \models_{\mathcal{L}'} g(a)$ if and only if $\mathcal{A} \models_{\mathcal{L}} a$, for all $\mathcal{L}$-interpretations $\mathcal{A}$ and all $\mathcal{L}$-expression a.*

(b) *$g^{-1}(Th_{\mathcal{L}'}(f(\mathcal{A}))) = Th_{\mathcal{L}}(\mathcal{A})$, for all $\mathcal{L}$-interpretations $\mathcal{A}$.*

(vi) *f is $\mathcal{L}$-injective.*

(vii) *For all $A \subseteq Expr_{\mathcal{L}}$ and all $a \in Expr_{\mathcal{L}}$ it holds the following:*

$$A \Vdash_{\mathcal{L}} a \iff g(A) \Vdash_{\mathcal{L}'} g(a).$$

(viii) *g is regular.*

(ix) *If g is surjective, then $g(Th_{\mathcal{L}}(\mathcal{A})) = Th_{\mathcal{L}'}(f(\mathcal{A}))$, for all $\mathcal{A} \in Int_{\mathcal{L}}$. If f is surjective, then $f(Mod_{\mathcal{L}}(a)) = Mod_{\mathcal{L}'}(g(a))$, for all $a \in Expr_{\mathcal{L}}$.*

46 S. Lewitzka

(x) *If f and g are surjective, then*
- $g(a^{*_{\mathcal{L}}}) = g(a)^{*_{\mathcal{L}'}}$,
- $f([\mathcal{A}]_{\mathcal{L}}) = [f(\mathcal{A})]_{\mathcal{L}'}$,
- $g([a]_{\mathcal{L}}) = [g(a)]_{\mathcal{L}'}$,

for all $a \in Expr_{\mathcal{L}}$ and all $\mathcal{A} \in Int_{\mathcal{L}}$.

(xi) *There exists an injective function $G : PTh(\mathcal{L}) \to PTh(\mathcal{L}')$ such that*

$$G^{-1}(a'^{*_{\mathcal{L}'}}) \subseteq g^{-1}(a'^{*_{\mathcal{L}'}}),$$

for all $a' \in Expr_{\mathcal{L}'}$. We may define G by $G(Th_{\mathcal{L}}(\mathcal{A})) := Th_{\mathcal{L}'}(f(\mathcal{A}))$.
If f is $\mathcal{L}'$-surjective, then

$$G^{-1}(a'^{*_{\mathcal{L}'}}) = g^{-1}(a'^{*_{\mathcal{L}'}}),$$

for all $a' \in Expr_{\mathcal{L}'}$. That is, if f is $\mathcal{L}'$-surjective, then G is a continuous map from the topological space $(PTh_{\mathcal{L}}, \rho_{\mathcal{L}})$ to the space $(PTh_{\mathcal{L}'}, \rho_{\mathcal{L}'})$.
If f and g are surjective, then the function $G : PTh(\mathcal{L}) \to PTh(\mathcal{L}')$, defined by $Th_{\mathcal{L}}(\mathcal{A}) \mapsto Th_{\mathcal{L}'}(f(\mathcal{A}))$, is an homeomorphism from space $(PTh_{\mathcal{L}}, \rho_{\mathcal{L}})$ to the space $(PTh_{\mathcal{L}'}, \rho_{\mathcal{L}'})$.
If $G' : (Int_{\mathcal{L}}/{=_{\mathcal{L}}}) \to (Int_{\mathcal{L}'}/{=_{\mathcal{L}'}})$ is the function defined by $[\mathcal{A}]_{\mathcal{L}} \mapsto [f(\mathcal{A})]_{\mathcal{L}'}$, then the analogous statements hold with respect to G' and the topological spaces $((Int_{\mathcal{L}}/{=_{\mathcal{L}}}), \sigma_{\mathcal{L}}^2)$ and $((Int_{\mathcal{L}'}/{=_{\mathcal{L}'}}), \sigma_{\mathcal{L}'}^2)$.

Proof. (i) Let $T' \in PTh(\mathcal{L}')$. Then for any $a' \in T'$, $T' \in a'^{*_{\mathcal{L}'}}$. Hence, $g^{-1}(T') \in g^{-1}(a'^{*_{\mathcal{L}'}}) = \bigcup\{a^{*_{\mathcal{L}}} \mid a \in A_{a'}\}$. Then there is some $a \in A_{a'}$, such that $g^{-1}(T') \in a^{*_{\mathcal{L}}}$. Hence, $g^{-1}(T')$ must be a theory in $PTh(\mathcal{L})$.

Now let $T \in PTh(\mathcal{L})$. There is some $\mathcal{A} \in Int_{\mathcal{L}}$ such that $T = Th_{\mathcal{L}}(\mathcal{A})$. We choose some $a' \in Th_{\mathcal{L}'}(f(\mathcal{A}))$. Then $f(\mathcal{A}) \in Mod_{\mathcal{L}'}(a')$ and $\mathcal{A} \in f^{-1}(Mod_{\mathcal{L}'}(a')) = \bigcup\{(Mod_{\mathcal{L}}(a)) \mid a \in A_{a'}\}$. Thus, there is some $a \in A_{a'}$ such that $\mathcal{A} \in Mod_{\mathcal{L}}(a)$. Then $Th_{\mathcal{L}}(\mathcal{A}) \in a^{*_{\mathcal{L}}} \subseteq \bigcup\{a^{*_{\mathcal{L}}} \mid a \in A_{a'}\} = g^{-1}(a'^{*_{\mathcal{L}'}})$. Hence, there must exist some prime theory $T' \in a'^{*_{\mathcal{L}'}}$ such that $g^{-1}(T') = T$.

(ii) Suppose that $f(\mathcal{A}) \neq_{\mathcal{L}'} f(\mathcal{B})$. We may assume that there is some a' such that $f(\mathcal{A}) \vDash_{\mathcal{L}'} a'$ and $f(\mathcal{B}) \nvDash_{\mathcal{L}'} a'$. Hence, $\mathcal{A} \in f^{-1}(Mod_{\mathcal{L}'}(a'))$ and $\mathcal{B} \notin f^{-1}(Mod_{\mathcal{L}'}(a'))$. Since $f^{-1}(Mod_{\mathcal{L}'}(a')) = \bigcup\{Mod_{\mathcal{L}}(a) \mid a \in A_{a'}\}$, there is some $a \in A_{a'}$ such that $\mathcal{A} \vDash_{\mathcal{L}} a$ and $\mathcal{B} \nvDash_{\mathcal{L}} a$, thus $\mathcal{A} \neq_{\mathcal{L}} \mathcal{B}$.

(iii) Suppose $T \in g^{-1}(g(a)^{*_{\mathcal{L}'}})$. Then clearly $a \in T$, thus $T \in a^{*_{\mathcal{L}}}$. Conversely, suppose $T \in a^{*_{\mathcal{L}}}$. By (i), there is some $T' \in Th(\mathcal{L}')$ such that $g^{-1}(T') = T$. Then $g(a) \in T'$ and $T' \in g(a)^{*_{\mathcal{L}}}$. Thus, $T \in g^{-1}(g(a)^{*_{\mathcal{L}'}})$.
Since (f, g) is a logic-homomorphism and $g^{-1}(g(a)^{*_{\mathcal{L}'}}) = a^{*_{\mathcal{L}}}$, it follows that $f^{-1}(Mod_{\mathcal{L}'}(g(a))) = Mod_{\mathcal{L}}(a)$, for all $a \in Expr_{\mathcal{L}}$.

(iv) Suppose that $g(a) =_{\mathcal{L}'} g(b)$. This is equivalent to the condition $g(a)^{*_{\mathcal{L}'}} = g(b)^{*_{\mathcal{L}'}}$. By the preceding item, $a^{*_{\mathcal{L}}} = g^{-1}(g(a)^{*_{\mathcal{L}'}}) = g^{-1}(g(b)^{*_{\mathcal{L}'}} = b^{*_{\mathcal{L}}}$, that is, $a =_{\mathcal{L}} b$.

(v) $f(\mathcal{A}) \in Mod_{\mathcal{L}'}(g(a)) \iff \mathcal{A} \in f^{-1}(Mod_{\mathcal{L}'}(g(a))) = Mod_{\mathcal{L}}(a)$, by (iii). Thus, (a) holds. Now it is easy to see that (a) and (b) are equivalent statements.

(vi) Suppose that $f(\mathcal{A}) =_{\mathcal{L}'} f(\mathcal{B})$. Then it follows that $Th_{\mathcal{L}'}(f(\mathcal{A})) = Th_{\mathcal{L}'}(f(\mathcal{B}))$ and, by (v),

$$Th_{\mathcal{L}}(\mathcal{A}) = g^{-1}(Th_{\mathcal{L}'}(f(\mathcal{A}))) = g^{-1}(Th_{\mathcal{L}'}(f(\mathcal{B}))) = Th_{\mathcal{L}}(\mathcal{B}).$$

Thus, $\mathcal{A} =_{\mathcal{L}} \mathcal{B}$, and f is $\mathcal{L}$-injective.

(vii) First, we observe that for any logic $\mathcal{L}$ and any $A \subseteq Expr_{\mathcal{L}}$, $a \in Expr_{\mathcal{L}}$ the following holds: $A \Vdash_{\mathcal{L}} a$ if and only if $Mod_{\mathcal{L}}(A) \subseteq Mod_{\mathcal{L}}(a)$ if and only if $a \in \bigcap\{T \in PTh(\mathcal{L}) \mid A \subseteq T\}$.

Suppose that $A \Vdash_{\mathcal{L}} a$. Let $T' \in PTh(\mathcal{L}')$ such that $g(A) \subseteq T'$. Then $A \subseteq g^{-1}(g(A)) \subseteq g^{-1}(T') = T$ for some $T \in PTh(\mathcal{L})$, by (i). From $A \Vdash_{\mathcal{L}} a$ and the preceding observation follows that $a \in T = g^{-1}(T')$, hence $g(a) \in T'$. Since T' was an arbitrary prime theory, it follows that $g(a) \in \bigcap\{T' \in PTh(\mathcal{L}') \mid g(A) \subseteq T'\}$.

The converse implication follows from (v)(a).

(viii) $a =_{\mathcal{L}} b$ is equivalent to the condition $a \Vdash_{\mathcal{L}} b$ and $b \Vdash_{\mathcal{L}} a$. Now the assertion follows immediately from (vii).

(ix) Let $\mathcal{A} \in Int_{\mathcal{L}}$. By (v)(b), $g(Th_{\mathcal{L}}(\mathcal{A})) = g(g^{-1}(Th_{\mathcal{L}}(f(\mathcal{A}))) \subseteq Th_{\mathcal{L}}(f(\mathcal{A}))$. Now let $a' \in Th_{\mathcal{L}}(f(\mathcal{A}))$. Since g is surjective, there is some $a \in Expr_{\mathcal{L}}$ such that $g(a) = a'$, thus, $f(\mathcal{A}) \vDash_{\mathcal{L}'} g(a)$. By (v)(a), $a \in Th_{\mathcal{L}}(\mathcal{A})$. Hence, $g(a) = a' \in g(Th_{\mathcal{L}}(\mathcal{A}))$ and $Th_{\mathcal{L}}(f(\mathcal{A})) \subseteq g(Th_{\mathcal{L}}(\mathcal{A}))$. The assertion concerning f follows in a similar way.

(x) Suppose that f and g are surjective. Then

$$
\begin{aligned}
g(a^{*_{\mathcal{L}}}) &= \{g(T) \mid a \in T \in PTh(\mathcal{L})\} \\
&= \{g(T) \mid T = Th_{\mathcal{L}}(\mathcal{A}), \mathcal{A} \in Mod_{\mathcal{L}}(a)\} \\
&= \{Th_{\mathcal{L}'}(f(\mathcal{A})) \mid \mathcal{A} \in Mod_{\mathcal{L}}(a)\}, \text{ by (ix)} \\
&= \{Th_{\mathcal{L}'}(\mathcal{A}') \mid \mathcal{A}' \in f(Mod_{\mathcal{L}}(a))\} \\
&= \{Th_{\mathcal{L}'}(\mathcal{A}') \mid \mathcal{A}' \in Mod_{\mathcal{L}'}(g(a)))\}, \text{ by (ix)} \\
&= \{T \in PTh(\mathcal{L}') \mid T' \in g(a)^{*_{\mathcal{L}'}}\} \\
&= g(a)^{*_{\mathcal{L}'}}.
\end{aligned}
$$

We have proved the first item of (x). The remaining assertions follow easily from regularity, surjectivity and $\mathcal{L}$-injectivity of f and g.

(xi) For each $T \in PTh(\mathcal{L})$ there is some $T' \in PTh(\mathcal{L}')$ such that $g^{-1}(T') = T$. By the axiom of choice, there exists a function G that assigns to each $T \in PTh(\mathcal{L})$ such a $T' \in PTh(\mathcal{L}')$. Moreover, by (v)(b), we may define G by $Th_{\mathcal{L}}(\mathcal{A}) \mapsto Th_{\mathcal{L}'}(f(\mathcal{A}))$, for every $\mathcal{A} \in Int_{\mathcal{L}}$. This map is well-defined, since f is regular. It is easy to see that G is injective: If $T_1 \neq T_2$, then we may assume that there is some $a \in T_1 \smallsetminus T_2$. Then $g(a) \in G(T_1) \smallsetminus G(T_2)$, thus $G(T_1) \neq G(T_2)$.

We show that $G^{-1}(a'^{*_{\mathcal{L}'}}) \subseteq g^{-1}(a'^{*_{\mathcal{L}'}})$ holds, for all $a' \in Expr_{\mathcal{L}'}$: Let $a' \in Expr_{\mathcal{L}'}$ and suppose $T \in G^{-1}(a'^{*_{\mathcal{L}'}})$. Then $T = G^{-1}(T') = g^{-1}(T')$, for some $T' \in a'^{*_{\mathcal{L}'}}$. Therefore, $T \in g^{-1}(a'^{*_{\mathcal{L}'}})$, and the assertion follows.

Now let us suppose that f is $\mathcal{L}'$-surjective and that G is defined by $Th_{\mathcal{L}}(\mathcal{A}) \mapsto Th_{\mathcal{L}'}(f(\mathcal{A}))$. Let $T \in g^{-1}(a'^{*_{\mathcal{L}'}})$. There is some $T' \in a'^{*_{\mathcal{L}'}}$ such that $g^{-1}(T') = T$. Furthermore, $T' = Th_{\mathcal{L}'}(\mathcal{A}')$, for some $\mathcal{A}' \in Int_{\mathcal{L}'}$. By $\mathcal{L}'$-surjectivity of f, there is some $\mathcal{A} \in Int_{\mathcal{L}}$ such that $f(\mathcal{A}) =_{\mathcal{L}'} \mathcal{A}'$. Thus, $T' = Th_{\mathcal{L}'}(f(\mathcal{A}))$ and $T = g^{-1}(T') = g^{-1}(Th_{\mathcal{L}'}(f(\mathcal{A}))) = Th_{\mathcal{L}}(\mathcal{A}) = G^{-1}((Th_{\mathcal{L}'}(f(\mathcal{A})))$. Hence, $T \in G^{-1}(a'^{*_{\mathcal{L}'}})$. This shows that $g^{-1}(a'^{*_{\mathcal{L}'}}) \subseteq G^{-1}(a'^{*_{\mathcal{L}'}})$.

It is easy to see that for any two $\mathcal{L}'$-expressions a' and b', $G^{-1}(a'^{*_{\mathcal{L}'}} \cap b'^{*_{\mathcal{L}'}}) = G^{-1}(a'^{*_{\mathcal{L}'}}) \cap G^{-1}(b'^{*_{\mathcal{L}'}})$. From this it follows that the inverse of G maps basic open sets of $\rho_{\mathcal{L}'}$ to open sets of $\rho_{\mathcal{L}'}$. (Recall that the basic open sets of $\rho_{\mathcal{L}'}$ are finite intersections of elements of the subbasis $\{a'^{*_{\mathcal{L}'}} \mid a' \in Expr_{\mathcal{L}'}\}$.) Thus, G is a continuous map from the topology $\rho_{\mathcal{L}}$ to the topology $\rho_{\mathcal{L}'}$.

Finally, suppose that f and g are surjective. By (ix), we may define G by

$$G(Th_{\mathcal{L}}(\mathcal{A})) := g(Th_{\mathcal{L}}(\mathcal{A})) = Th_{\mathcal{L}'}(f(\mathcal{A})).$$

Then from our previous results it follows that G is a bijective function. Furthermore, from (iii) and from the first item of (x) it follows that G is a bijection between the respective subbases of the topologies $\rho_{\mathcal{L}}$ and $\rho_{\mathcal{L}'}$. Now it is easy to see that G maps basic open sets of $\rho_{\mathcal{L}}$ to basic open sets of $\rho_{\mathcal{L}'}$ and that the inverse of G maps basic open sets of $\rho_{\mathcal{L}'}$ to basic open sets of $\rho_{\mathcal{L}}$. Hence, G is also a bijective open and continuous map, thus, an homeomorphism between the respective topological spaces.

The last observation follows immediately from Proposition 2.6. $\square$

The most important properties of a strong logic-homomorphism are collected in the following Lemma:

Lemma 2.10. *Suppose that (f, g) is a strong logic-homomorphism from $\mathcal{L}$ to $\mathcal{L}'$. Then holds the following:*

(i) *If $T_1', T_2' \in PTh(\mathcal{L}')$ and $g^{-1}(T_1') = g^{-1}(T_2')$, then $T_1' = T_2'$.*

(ii) *f is $\mathcal{L}'$-surjective.*

(iii) *There exists exactly one bijective function $G : PTh(\mathcal{L}) \to PTh(\mathcal{L}')$ such that*

$$G^{-1}(T') = g^{-1}(T')$$

for all $T' \in PTh(\mathcal{L}')$. Furthermore, G satisfies the equations

$$G(a^{*_{\mathcal{L}}}) = g(a)^{*_{\mathcal{L}'}},$$

$$G^{-1}(a'^{*_{\mathcal{L}'}}) = g^{-1}(a'^{*_{\mathcal{L}'}})$$

for all $a \in Expr_{\mathcal{L}}$ and for all $a' \in Expr_{\mathcal{L}'}$. In particular, G is an homeomorphism from the topological space $(PTh(\mathcal{L}), \rho_{\mathcal{L}})$ onto the topological space $(PTh(\mathcal{L}'), \rho_{\mathcal{L}'})$.

Proof. (i) Suppose $g^{-1}(T_1') = g^{-1}(T_2')$ and let $a' \in T_1'$, that is, $T_1' \in a'^{*_{\mathcal{L}'}}$. Then $T := g^{-1}(T_1') \in g^{-1}(a'^{*_{\mathcal{L}'}}) = \bigcup\{a^{*_{\mathcal{L}}} \mid g(a) \Vdash_{\mathcal{L}'} a'\}$, thus $T \in a^{*_{\mathcal{L}}}$ for some a with $g(a) \Vdash_{\mathcal{L}'} a'$. Since $a \in T$, $g(a) \in g(T) \subseteq T_2'$. T_2' is deductively closed and it follows that $a' \in T_2'$. Hence, $T_1' \subseteq T_2'$. The converse inclusion follows in the same way.

(ii) Let $\mathcal{A}' \in Int_{\mathcal{L}'}$. Then $g^{-1}(Th_{\mathcal{L}'}(\mathcal{A}')) = Th_{\mathcal{L}}(\mathcal{A})$ for some $\mathcal{A} \in Int_{\mathcal{L}}$, by Lemma 2.9(i). On the other hand, by Lemma 2.9(v)(b) we get $g^{-1}(Th_{\mathcal{L}'}(f(\mathcal{A}))) = Th_{\mathcal{L}}(\mathcal{A})$. From (i) of this Lemma it follows that $Th_{\mathcal{L}'}(\mathcal{A}') = Th_{\mathcal{L}'}(f(\mathcal{A}))$, thus $\mathcal{A}' =_{\mathcal{L}'} f(\mathcal{A})$. This proves the assertion.

(iii) We define G by $G(Th_{\mathcal{L}}(\mathcal{A})) := Th_{\mathcal{L}'}(f(\mathcal{A}))$. Since f is regular, G is well-defined. G is injective, since f is $\mathcal{L}$-injective, and G is surjective, since f is $\mathcal{L}'$-surjective.

Now let $T' \in PTh(\mathcal{L}')$. Then $T' = Th_{\mathcal{L}'}(\mathcal{A}')$ for some $\mathcal{A}' \in Int_{\mathcal{L}'}$. Since f is $\mathcal{L}'$-surjective, there is some $\mathcal{A} \in Int_{\mathcal{L}}$ such that $f(\mathcal{A}) =_{\mathcal{L}'} \mathcal{A}'$. Then

$$G^{-1}(Th_{\mathcal{L}'}(\mathcal{A}')) = G^{-1}(Th_{\mathcal{L}'}(f(\mathcal{A}))$$
$$= Th_{\mathcal{L}}(\mathcal{A})$$
$$= g^{-1}(Th_{\mathcal{L}'}(f(\mathcal{A}))), \text{ by Lemma 2.9(v)(b)}$$
$$= g^{-1}(Th_{\mathcal{L}'}(\mathcal{A}')).$$

Hence, $G^{-1}(T') = g^{-1}(T')$. This shows the existence of G.

Now let us assume that $G' : PTh(\mathcal{L}) \to PTh(\mathcal{L}')$ is any bijective function such that $G'^{-1}(T') = g^{-1}(T')$ $(= G^{-1}(T'))$ for all $T' \in PTh(\mathcal{L}')$. Then clearly G' equals G on all $T \in PTh(\mathcal{L})$. Hence, there is exactly one such a function G.

In order to show the "furthermore-clause", let $T \in PTh(\mathcal{L})$. Then $T = Th_{\mathcal{L}}(\mathcal{A})$ for some $\mathcal{A} \in Int_{\mathcal{L}}$. By the definition of G: $a \in T$ if and only if $g(a) \in G(T)$. Thus, for any $T \in PTh(\mathcal{L})$ we get: $T \in a^{*_{\mathcal{L}}}$ if and only if $G(T) \in g(a)^{*_{\mathcal{L}'}}$. Since G is surjective, $G(a^{*_{\mathcal{L}}}) = g(a)^{*_{\mathcal{L}'}}$ for all $a \in Expr_{\mathcal{L}}$. Since $G^{-1}(T') = g^{-1}(T')$ for each $T' \in PTh(\mathcal{L}')$, it is also clear that $G^{-1}(a'^{*_{\mathcal{L}'}}) = g^{-1}(a'^{*_{\mathcal{L}'}})$ for all $a' \in Expr_{\mathcal{L}'}$.

Now one easily shows that G maps basic open sets to basic open sets and that the inverse of G maps basic open sets to open sets. Thus, G is an homeomorphism from the space $(PTh(\mathcal{L}), \rho_{\mathcal{L}})$ to the space $(PTh(\mathcal{L}'), \rho_{\mathcal{L}'})$. $\square$

Proposition 2.11. *Suppose that $\mathcal{L}$ and $\mathcal{L}'$ are abstract logics and let $f : Int_{\mathcal{L}} \to Int_{\mathcal{L}'}$ and $g : Expr_{\mathcal{L}} \to Expr_{\mathcal{L}'}$ be functions such that g is $\mathcal{L}$-surjective. If for any $a \in Expr_{\mathcal{L}}$, $g^{-1}(g(a)^{*_{\mathcal{L}'}}) = a^{*_{\mathcal{L}}}$ and $f^{-1}(Mod_{\mathcal{L}'}(g(a))) = Mod_{\mathcal{L}}(a)$, then (f, g) is a strong logic-homomorphism from $\mathcal{L}$ to $\mathcal{L}'$.*

Proof. Since g is $\mathcal{L}$-surjective, for every $b' \in Expr_{\mathcal{L}'}$ there is some $a \in Expr_{\mathcal{L}}$ such that $g(a) =_{\mathcal{L}'} b'$. Hence, $g^{-1}(b'^{*_{\mathcal{L}'}}) = g^{-1}(g(a)^{*_{\mathcal{L}'}}) = a^{*_{\mathcal{L}}}$ and $f^{-1}(Mod_{\mathcal{L}'}(b')) = f^{-1}(Mod_{\mathcal{L}'}(g(a))) = Mod_{\mathcal{L}}(a)$, by hypothesis. Now it is sufficient to show that for each $a \in Expr_{\mathcal{L}}$, $a^{*_{\mathcal{L}}} = \bigcup\{b^{*_{\mathcal{L}}} \mid g(b)^{*_{\mathcal{L}'}} \subseteq g(a)^{*_{\mathcal{L}'}}\}$ and $Mod_{\mathcal{L}}(a) = \bigcup\{Mod_{\mathcal{L}}(b) \mid g(b)^{*_{\mathcal{L}'}} \subseteq g(a)^{*_{\mathcal{L}'}}\}$:

Suppose that $a \in Expr_{\mathcal{L}}$. Clearly, $a^{*_{\mathcal{L}}} \subseteq \bigcup\{b^{*_{\mathcal{L}}} \mid g(b)^{*_{\mathcal{L}'}} \subseteq g(a)^{*_{\mathcal{L}'}}\}$, since $g(a)^{*_{\mathcal{L}'}} \subseteq g(a)^{*_{\mathcal{L}'}}$. For the converse suppose that $g(b)^{*_{\mathcal{L}'}} \subseteq g(a)^{*_{\mathcal{L}'}}$. By hypothesis, we have $b^{*_{\mathcal{L}}} = g^{-1}(g(b)^{*_{\mathcal{L}'}}) \subseteq g^{-1}(g(a)^{*_{\mathcal{L}'}}) = a^{*_{\mathcal{L}}}$. This yields $\bigcup\{b^{*_{\mathcal{L}}} \mid g(b)^{*_{\mathcal{L}'}} \subseteq g(a)^{*_{\mathcal{L}'}}\} \subseteq a^{*_{\mathcal{L}}}$. Thus, $a^{*_{\mathcal{L}}} = \bigcup\{b^{*_{\mathcal{L}}} \mid g(b)^{*_{\mathcal{L}'}} \subseteq g(a)^{*_{\mathcal{L}'}}\}$. This is equivalent to $Mod_{\mathcal{L}}(a) = \bigcup\{Mod_{\mathcal{L}}(b) \mid g(b)^{*_{\mathcal{L}'}} \subseteq g(a)^{*_{\mathcal{L}'}}\}$. Hence, (f, g) is a strong logic-homomorphism. $\square$

50 S. Lewitzka

Corollary 2.12. *Let $\mathcal{L}$ and $\mathcal{L}'$ be abstract logics. The logic-isomorphisms from $\mathcal{L}$ to $\mathcal{L}'$ are exactly the bijective strong logic-homomorphisms from $\mathcal{L}$ to $\mathcal{L}'$.*

Proof. If (f,g) is a strong logic-homomorphism such that f and g are bijective functions, then (f,g) is a logic-isomorphism, by definition. Conversely, suppose that (f,g) is a logic-isomorphism. Then f and g are bijective functions. Furthermore, by Lemma 2.9(iii), $a^{*_\mathcal{L}} = g^{-1}(g(a)^{*_{\mathcal{L}'}})$ and $Mod_\mathcal{L}(a) = f^{-1}(Mod_{\mathcal{L}'}(g(a))$, for all $a \in Expr_\mathcal{L}$. Now, Proposition 2.11 says that (f,g) is a strong logic-homomorphism. $\square$

Corollary 2.13. *Suppose that (f,g) is a logic-homomorphism from logic $\mathcal{L}$ to logic $\mathcal{L}'$ such that g is $\mathcal{L}'$-surjective. Then (f,g) is a strong logic-homomorphism.*

Proof. By Lemma 2.9(iii), we may apply Proposition 2.11. $\square$

Definition 2.14. Suppose that $\mathcal{L}$ is an abstract logic. Let $\sim_I$ be an equivalence relation on $Int_\mathcal{L} \times Int_\mathcal{L}$ that refines $=_\mathcal{L}$ and let $\sim_E$ be an equivalence relation on $Expr_\mathcal{L} \times Expr_\mathcal{L}$ that refines $=_\mathcal{L}$. Then we define the abstract factor logic of $\mathcal{L}$ modulo $\sim := (\sim_I, \sim_E)$ in the following way:

- $(Int_\mathcal{L}/_{\sim_I})$ is the set of equivalence classes of $Int_\mathcal{L}$ modulo $\sim_I$, $(Expr_\mathcal{L}/_{\sim_E})$ is the set of equivalence classes of $Expr_\mathcal{L}$ modulo $\sim_E$. The equivalence classes of any interpretation $\mathcal{A}$ and any expression a are denoted by $[\mathcal{A}]_{\sim_I}$ and $[a]_{\sim_E}$, respectively.
- The satisfaction relation $\vDash_\sim$ of the abstract factor logic is defined as follows:
  For any $\mathcal{L}$-interpretation $\mathcal{A}$ and any $\mathcal{L}$-expression a holds $[\mathcal{A}]_{\sim_I} \vDash_\sim [a]_{\sim_E}$ if and only if $\mathcal{A} \vDash_\mathcal{L} a$.
- Finally, we denote the abstract factor logic by $(\mathcal{L}/_\sim)$ and put
  $$(\mathcal{L}/_\sim) := ((Expr_\mathcal{L}/_{\sim_E}), (Int_\mathcal{L}/_{\sim_I}), \vDash_\sim).$$

Remark 2.15.   (i) The satisfaction relation of the abstract factor logic is well-defined: Suppose that $\mathcal{A} \vDash_\mathcal{L} a$ and $\mathcal{A} \sim_I \mathcal{B}$ and $a \sim_E b$. By definition, these equivalence relations refine $\mathcal{L}$-equivalence. Hence, $\mathcal{A} =_\mathcal{L} \mathcal{B}$ and $a =_\mathcal{L} b$. Then follows that $\mathcal{B} \vDash_\mathcal{L} b$. Whence, $[\mathcal{A}]_{\sim_I} \vDash_\sim [a]_{\sim_E}$ if and only if $[\mathcal{B}]_{\sim_I} \vDash_\sim [b]_{\sim_E}$.

(ii) Since the equivalence relation $\sim_E$ refines $\mathcal{L}$-equivalence, for any theory T in $\mathcal{L}$ and any $a \sim_E b$ holds: $a \in T$ if and only if $b \in T$. Moreover, $T = Th_\mathcal{L}(\mathcal{A})$ if and only if $T/_{\sim_E} = Th_{(\mathcal{L}/_\sim)}([\mathcal{A}]_{\sim_I})$ and $a \in T$ if and only if $[a]_{\sim_E} \in T/_{\sim_E}$. Hence, the mapping $a \mapsto [a]_{\sim_E}$ induces a mapping $T \mapsto T/_{\sim_E}$, which is a bijection between the prime theories of $\mathcal{L}$ and the prime theories of the factor logic.

For similar reasons: If $\mathcal{A} \sim_I \mathcal{B}$ and $\mathcal{A} \in Mod_\mathcal{L}(a)$, then $\mathcal{B} \in Mod_\mathcal{L}(a)$; $\mathcal{A} \in Mod_\mathcal{L}(a)$ if and only if $[\mathcal{A}]_{\sim_I} \in Mod_{(\mathcal{L}/_\sim)}([a]_{\sim_E})$.

The following result is an adaption of a well-known fact from universal algebra (the Homomorphism Lemma):

Proposition 2.16. *Let (f,g) be a surjective logic-homomorphism from $\mathcal{L}$ onto $\mathcal{L}'$. The equivalence relations $Ker(f):=\{(\mathcal{A},\mathcal{B}) \mid f(\mathcal{A}) = f(\mathcal{B})\}$ and $Ker(g):=\{(a,b) \mid g(a) = g(b)\}$ refine $\mathcal{L}$-equivalence.*
Put $\sim:=(Ker(f),Ker(g))$. Then the tuple (f_1,g_1), where $f_1 : Int_\mathcal{L} \to (Int_\mathcal{L}/_{Ker(f)})$ is defined by $\mathcal{A} \mapsto [\mathcal{A}]_{Ker(f)}$ and $g_1 : Expr_\mathcal{L} \to (Expr_\mathcal{L}/_{Ker(g)})$ is defined by $a \mapsto [a]_{Ker(g)}$, is a surjective strong logic-homomorphism from $\mathcal{L}$ onto $(\mathcal{L}/_\sim)$ (called the canonical logic-homomorphism).

Furthermore, the abstract factor logic $(\mathcal{L}/_\sim)$ and the abstract logic $\mathcal{L}'$ are isomorphic.

Proof. Since f and g are is $\mathcal{L}$-injective functions, it follows that $Ker(f)$ and $Ker(g)$ refine $\mathcal{L}$-equivalence.

By Remark 2.15, g_1 is a bijection from $PTh(\mathcal{L})$ onto $PTh(\mathcal{L}/_\sim)$ and it follows that

$$g_1^{-1}([a]^*_{Ker(g)}) = \{g_1^{-1}(T/_{Ker(g)}) \mid [a]_{Ker(g)} \in (T/_{Ker(g)})\}$$
$$= \{T \mid a \in T \in PTh(\mathcal{L})\}$$
$$= a^*$$

Furthermore, by the definitions it is clear that

$$f_1^{-1}(Mod_{(\mathcal{L}/_\sim)}([a]_{Ker(g)}) = Mod_\mathcal{L}(a).$$

It follows that (f_1,g_1) is a surjective strong logic-homomorphism (Proposition 2.11).
We define (f_2,g_2) with $f_2 : (Int_\mathcal{L}/_{Ker(f)}) \to Int_{\mathcal{L}'}$ and $g_2 : (Expr_\mathcal{L}/_{Ker(g)}) \to Expr_{\mathcal{L}'}$ by: $f_2([\mathcal{A}]_{Ker(f)}) := f(\mathcal{A})$ and $g_2([a]_{Ker(g)}) := g(a)$. Then f_2, g_2 are clearly surjective, since f and g are surjective. By definition, they are also injective. Furthermore, by definition, $g = g_2 \circ g_1$ and $f = f_2 \circ f_1$.

Let $a' \in Expr_{\mathcal{L}'}$. Since g is surjective, there is some $a \in Expr_\mathcal{L}$ with $g(a) = a'$. By Lemma 2.9(iii), it holds the following:
$g_2^{-1}(g(a)^*) = g_1 \circ g^{-1}(g(a)^*) = g_1(a^*) = [a]^*_{Ker(g)}$, since, by Remark 2.15., g_1 is a bijective map between $PTh(\mathcal{L})$ and $PTh(\mathcal{L}/_\sim)$.

Furthermore, $f_2^{-1}(Mod_{\mathcal{L}'}(g(a))) = f_1 \circ f^{-1}(Mod_{\mathcal{L}'}(g(a))) = f_1(Mod_\mathcal{L}(a)) = Mod_{(\mathcal{L}/_\sim)}([a]_{Ker(g)})$.

Now, by Proposition 2.11, (f_2,g_2) is a bijective strong logic-homomorphism. By Corollary 2.12, (f_2,g_2) is a logic-isomorphism from $(\mathcal{L}/_\sim)$ to $\mathcal{L}'$. $\square$

We define further relationships between logics:

Definition 2.17. Suppose that $\mathcal{L},\mathcal{L}'$ are abstract logics with $Expr_\mathcal{L} \subseteq Expr_{\mathcal{L}'}$.

- $\mathcal{L}'$ is called an extension of $\mathcal{L}$, written $\mathcal{L} \le \mathcal{L}'$, if $PTh(\mathcal{L}) = \{T' \cap Expr_\mathcal{L} \mid T' \in PTh(\mathcal{L}')\}$.
- $\mathcal{L}'$ is a weak extension of $\mathcal{L}$, written $\mathcal{L} \le_w \mathcal{L}'$, if $PTh(\mathcal{L}) \subseteq \{T' \cap Expr_\mathcal{L} \mid T' \in PTh(\mathcal{L}')\}$.

- $\mathcal{L}'$ is a definable extension of $\mathcal{L}$, if $\mathcal{L} \leq \mathcal{L}'$ and for every $a' \in Expr_{\mathcal{L}'}$ there is some $a \in Expr_{\mathcal{L}}$ such that $a' =_{\mathcal{L}'} a$. We write $\mathcal{L} \leq_{def} \mathcal{L}'$.
- $\mathcal{L}'$ is called a ∞-definable (or an almost-definable) extension of $\mathcal{L}$, if $\mathcal{L} \leq \mathcal{L}'$ and for any $a' \in Expr_{\mathcal{L}'}$, there is some set $A \subseteq Expr_{\mathcal{L}}$ such that $a' =_{\mathcal{L}'} A$. We write $\mathcal{L} \leq_{\infty} \mathcal{L}'$.

A connection between logic-homomorphisms and extensions is given by the following observation:

Proposition 2.18. *Let $\mathcal{L}, \mathcal{L}'$ be logics with $Expr_{\mathcal{L}} \subseteq Expr_{\mathcal{L}'}$ and let $i : Expr_{\mathcal{L}} \to Expr_{\mathcal{L}'}$ be the identity map $a \mapsto a$ on $Expr_{\mathcal{L}}$. If $(f, i) : \mathcal{L} \to \mathcal{L}'$ is a logic-homomorphism, then $\mathcal{L} \leq \mathcal{L}'$.*

Proof. If the hypothesis hold, then for any $\mathcal{A} \in Int_{\mathcal{L}}$, $Th_{\mathcal{L}'}(f(\mathcal{A})) \cap Expr_{\mathcal{L}} = i^{-1}(Th_{\mathcal{L}'}(f(\mathcal{A}))) = Th_{\mathcal{L}}(\mathcal{A})$, by Lemma 2.9(v). Hence, $PTh(\mathcal{L}) \subseteq \{T' \cap Expr_{\mathcal{L}} \mid T' \in PTh(\mathcal{L}')\}$. The converse inclusion also follows from Lemma 2.9(i), since $i^{-1}(T') = T' \cap Expr_{\mathcal{L}}$ for $T' \in PTh(\mathcal{L}')$. $\qquad\square$

Definition 2.19. Suppose that $\mathcal{L} \leq_w \mathcal{L}'$. Let $\mathcal{A}$ be an $\mathcal{L}$-interpretation and let $\mathcal{A}'$ be an $\mathcal{L}'$-interpretation. We say that $\mathcal{A}$ is an $\mathcal{L}$-reduction of $\mathcal{A}'$ and $\mathcal{A}'$ is an $\mathcal{L}'$-expansion of $\mathcal{A}$ and write $(\mathcal{A}' \restriction_{\mathcal{L}}) =_{\mathcal{L}} \mathcal{A}$, if $Th_{\mathcal{L}}(\mathcal{A}) = Th_{\mathcal{L}'}(\mathcal{A}') \cap Expr_{\mathcal{L}}$.

Remark 2.20. Let $\mathcal{L} \leq_w \mathcal{L}'$. Every $\mathcal{L}$-interpretation $\mathcal{A}$ has an $\mathcal{L}'$-expansion. If $\mathcal{L} \leq \mathcal{L}'$, then every $\mathcal{L}'$-interpretation $\mathcal{A}'$ has an $\mathcal{L}$-reduction. In this case, there is — up to $\mathcal{L}$-equivalence — exactly one $\mathcal{L}$-reduction of every $\mathcal{A}' \in Int_{\mathcal{L}'}$.

Proof. This follows easily from the definitions. $\qquad\square$

Definition 2.21. The abstract logic $\mathcal{L}$ is a sublogic of the abstract logic $\mathcal{L}'$, if

- $Int_{\mathcal{L}} \subseteq Int_{\mathcal{L}'}$,
- $Expr_{\mathcal{L}} = Expr_{\mathcal{L}'}$ and
- $\models_{\mathcal{L}} = \models_{\mathcal{L}'} \restriction (Int_{\mathcal{L}} \times Expr_{\mathcal{L}})$.

We write $\mathcal{L} \subseteq \mathcal{L}'$.

If $\mathcal{L} \subseteq \mathcal{L}'$, then we call $\mathcal{L}'$ a superlogic of $\mathcal{L}$.

Lemma 2.22.   (i) *If $\mathcal{L} \subseteq \mathcal{L}'$, then $\mathcal{L} \leq_w \mathcal{L}'$.*

(ii) $\mathcal{L} \subseteq \mathcal{L}' \iff Expr_{\mathcal{L}} = Expr_{\mathcal{L}'}$, $Int_{\mathcal{L}} \subseteq Int_{\mathcal{L}'}$ *and for all $a \in Expr_{\mathcal{L}}$ holds: $a^{*\mathcal{L}} = a^{*\mathcal{L}'} \cap PTh(\mathcal{L})$.*

Proof. Both statements are clear by the definitions. $\qquad\square$

Definition 2.23. Let $\mathcal{L}, \mathcal{L}'$ be abstract logics.

- Suppose that (f_1, g_1) and (f_2, g_2) are logic-homomorphisms from $\mathcal{L}$ to $\mathcal{L}'$. We say that (f_1, g_1) and (f_2, g_2) are equivalent and write $(f_1, g_1) \approx (f_2, g_2)$, if for all $\mathcal{L}$-interpretations $\mathcal{A}$ and all $\mathcal{L}$-expressions a the following holds:
  (i) $f_1(\mathcal{A}) =_{\mathcal{L}'} f_2(\mathcal{A})$ and
  (ii) $g_1(a) =_{\mathcal{L}'} g_2(a)$.

- We fix some function $g : Expr_{\mathcal{L}} \to Expr_{\mathcal{L}'}$ and define a refinement $\approx_g \subseteq \approx$ by $(f_1, g) \approx_g (f_2, g)$, if for all $\mathcal{L}$-interpretations $\mathcal{A}$ holds:

$$f_1(\mathcal{A}) =_{\mathcal{L}'} f_2(\mathcal{A}).$$

For short, we write $f_1 \approx_g f_2$ instead of $(f_1, g) \approx_g (f_2, g)$.

- Let $F : (Int_{\mathcal{L}}/_{=_{\mathcal{L}}}) \to (Int_{\mathcal{L}'}/_{=_{\mathcal{L}'}})$ be a function. We say that a function $f : Int_{\mathcal{L}} \to Int_{\mathcal{L}'}$ is of type F and write $f : F$, if $f(\mathcal{A}) \in F([\mathcal{A}]_{\mathcal{L}})$, for all $\mathcal{A} \in Int_{\mathcal{L}}$.[10]

Lemma 2.24. *Let $\mathcal{L}, \mathcal{L}'$ be abstract logics.*

- *Equivalence of logic-homomorphisms $\approx$ is an equivalence relation on the set of all logic-homomorphisms from $\mathcal{L}$ to $\mathcal{L}'$. For any $g : Expr_{\mathcal{L}} \to Expr_{\mathcal{L}'}$, $\approx_g$ is an equivalence relation on the set of all logic-homomorphisms $(f, g) : \mathcal{L} \to \mathcal{L}'$, where g is fixed.*

- *For any pair of regular functions $f_1, f_2 : Int_{\mathcal{L}} \to Int_{\mathcal{L}'}$ the following conditions are equivalent:*

  (i) *There is an injective function $F : (Int_{\mathcal{L}}/_{=_{\mathcal{L}}}) \to (Int_{\mathcal{L}'}/_{=_{\mathcal{L}'}})$ such that $f_1 : F$ and $f_2 : F$ hold.*

  (ii) *For all $\mathcal{L}$-interpretations $\mathcal{A}$ it holds that $f_1(\mathcal{A}) =_{\mathcal{L}'} f_2(\mathcal{A})$ and f_1, f_2 are $\mathcal{L}$-injective.*

  (iii) *For all $\mathcal{L}$-interpretations $\mathcal{A}, \mathcal{B}$ it holds the following:*

$$\mathcal{A} =_{\mathcal{L}} \mathcal{B} \iff f_1(\mathcal{A}) =_{\mathcal{L}'} f_2(\mathcal{B}),$$

  and f_1, f_2 are $\mathcal{L}$-injective.

Proof. The assertions of the first point are clear.

Now suppose that $\mathcal{L}, \mathcal{L}'$ are logics and $f_1, f_2 : Int_{\mathcal{L}} \to Int_{\mathcal{L}'}$ are regular functions. (i)→(ii) is clear.

(ii)→(iii): Suppose that (ii) holds. Suppose $\mathcal{A} =_{\mathcal{L}} \mathcal{B}$. By hypothesis, $f_1(\mathcal{A}) =_{\mathcal{L}'} f_2(\mathcal{A})$, and $f_2(\mathcal{A}) =_{\mathcal{L}'} f_2(\mathcal{B})$, since f_2 is regular. Hence, $f_1(\mathcal{A}) =_{\mathcal{L}'} f_2(\mathcal{B})$.

Now suppose $f_1(\mathcal{A}) =_{\mathcal{L}'} f_2(\mathcal{B})$. By hypothesis, $f_2(\mathcal{B}) =_{\mathcal{L}'} f_1(\mathcal{B})$. Hence, $f_1(\mathcal{A}) =_{\mathcal{L}'} f_1(\mathcal{B})$ and $\mathcal{A} =_{\mathcal{L}} \mathcal{B}$, since f_1 is $\mathcal{L}$-injective.

(iii)→(i): Suppose that (iii) holds. Then for all $\mathcal{A}, \mathcal{B}$: $\mathcal{A} =_{\mathcal{L}} \mathcal{B} \iff f_1(\mathcal{A}) =_{\mathcal{L}'} f_2(\mathcal{B})$. Hence, $f_2(\mathcal{B}) =_{\mathcal{L}'} f_1(\mathcal{B})$, since $\mathcal{B} =_{\mathcal{L}} \mathcal{B}$ always holds. In particular, it follows for all $\mathcal{A}, \mathcal{B}$: $\mathcal{A} =_{\mathcal{L}} \mathcal{B}$ iff $f_1(\mathcal{A}) =_{\mathcal{L}'} f_1(\mathcal{B})$, since f_1 is $\mathcal{L}$-injective.

Now we define $F : (Int_{\mathcal{L}}/_{=_{\mathcal{L}}}) \to (Int_{\mathcal{L}'}/_{=_{\mathcal{L}'}})$ by $F([\mathcal{A}]_{\mathcal{L}}) = [f_1(\mathcal{A})]_{\mathcal{L}'}$. By the previous consideration, F is well-defined. Now it is very easy to see that F is injective and satisfies the condition (i). $\square$

Definition 2.25. Suppose that $\mathcal{L}, \mathcal{L}'$ are abstract logics and $g : Expr_{\mathcal{L}} \to Expr_{\mathcal{L}'}$ is a function.

---

[10]Notice that for any given function $F : (Int_{\mathcal{L}}/_{=_{\mathcal{L}}}) \to (Int_{\mathcal{L}'}/_{=_{\mathcal{L}'}})$ the axiom of choice guarantees the existence of a function $f : F$.

- Suppose that a function $F : (Int_{\mathcal{L}}/=_{\mathcal{L}}) \rightarrow (Int_{\mathcal{L}'}/=_{\mathcal{L}'})$ satisfies the condition $g^{-1}(Th_{\mathcal{L}'}(F([\mathcal{A}]_{\mathcal{L}}))) = Th_{\mathcal{L}}(\mathcal{A})$, for all $\mathcal{L}$-interpretations $\mathcal{A}$. Then we call F a big complement of g.
- A function $f : Int_{\mathcal{L}} \rightarrow Int_{\mathcal{L}'}$ satisfying the condition $g^{-1}(Th_{\mathcal{L}'}(f(\mathcal{A}))) = Th_{\mathcal{L}}(\mathcal{A})$, for all $\mathcal{L}$-interpretations $\mathcal{A}$, is called a small complement of g.

Theorem 2.26. *Suppose that $\mathcal{L}, \mathcal{L}'$ are abstract logics and $g : Expr_{\mathcal{L}} \rightarrow Expr_{\mathcal{L}'}$ is a function such that $\{g^{-1}(T') \mid T' \in PTh(\mathcal{L}')\} = PTh(\mathcal{L})$. Then holds the following:*

(i) *There exists a function $F : (Int_{\mathcal{L}}/=_{\mathcal{L}}) \rightarrow (Int_{\mathcal{L}'}/=_{\mathcal{L}'})$ such that F is a big complement of g. F is injective.*

(ii) *Let F be a big complement of g. Then every function $f : Int_{\mathcal{L}} \rightarrow Int_{\mathcal{L}'}$ with the property $f : F$ is a regular function and a small complement of g.*

   Now let $f : Int_{\mathcal{L}} \rightarrow Int_{\mathcal{L}'}$ be a regular function and a small complement of g. Then there exists a function $F : (Int_{\mathcal{L}}/=_{\mathcal{L}}) \rightarrow (Int_{\mathcal{L}'}/=_{\mathcal{L}'})$, such that $f : F$ and F is a big complement of g.

Proof. (i) By hypothesis, $\{g^{-1}(T') \mid T' \in PTh(\mathcal{L}')\} = PTh(\mathcal{L})$. Hence, for any $[\mathcal{A}]_{\mathcal{L}} \in (Int_{\mathcal{L}}/=_{\mathcal{L}})$ there is some $[\mathcal{A}']_{\mathcal{L}'} \in (Int_{\mathcal{L}'}/=_{\mathcal{L}'})$ such that $g^{-1}(Th_{\mathcal{L}'}(\mathcal{A}')) = Th_{\mathcal{L}}(\mathcal{A})$. Notice that such an $[\mathcal{A}']_{\mathcal{L}'}$ is in general not unique. However, the axiom of choice guarantees the existence of a function $F : (Int_{\mathcal{L}}/=_{\mathcal{L}}) \rightarrow (Int_{\mathcal{L}'}/=_{\mathcal{L}'})$ assigning to any $[\mathcal{A}]_{\mathcal{L}}$ an $[\mathcal{A}']_{\mathcal{L}'}$ such that $g^{-1}(Th_{\mathcal{L}'}(\mathcal{A}')) = Th_{\mathcal{L}}(\mathcal{A})$. This function F is a big complement of g.

F is injective: If $[\mathcal{A}]_{\mathcal{L}} \neq [\mathcal{B}]_{\mathcal{L}}$, then $Th_{\mathcal{L}}(\mathcal{A}) \neq Th_{\mathcal{L}}(\mathcal{B})$. If F maps $[\mathcal{A}]_{\mathcal{L}}$ to $[\mathcal{A}']_{\mathcal{L}'}$ and $[\mathcal{B}]_{\mathcal{L}}$ to $[\mathcal{B}']_{\mathcal{L}}$, then

$$g^{-1}(Th_{\mathcal{L}'}(\mathcal{A}')) = Th_{\mathcal{L}}(\mathcal{A}) \neq Th_{\mathcal{L}}(\mathcal{B}) = g^{-1}(Th_{\mathcal{L}'}(\mathcal{B}')).$$

Now it follows that $Th_{\mathcal{L}'}(\mathcal{A}') \neq Th_{\mathcal{L}'}(\mathcal{B}')$, whence $[\mathcal{A}']_{\mathcal{L}'} \neq [\mathcal{B}']_{\mathcal{L}'}$ and F is injective.

(ii) If F is a big complement of g and $f : F$, then f is clearly regular. f is also a small complement of g, since $Th_{\mathcal{L}'}(f(\mathcal{A})) = Th_{\mathcal{L}'}(F([\mathcal{A}]_{\mathcal{L}}))$.

Now suppose that f is regular and a small complement of g. We define $F : (Int_{\mathcal{L}}/=_{\mathcal{L}}) \rightarrow (Int_{\mathcal{L}'}/=_{\mathcal{L}'})$ by $F([\mathcal{A}]_{\mathcal{L}}) := [f(\mathcal{A})]_{\mathcal{L}'}$. This definition is well-defined, since f is regular. Then $f : F$ and it is easy to see that F is a big complement of g, since $Th_{\mathcal{L}'}(f(\mathcal{A})) = Th_{\mathcal{L}'}(F([\mathcal{A}]_{\mathcal{L}}))$. $\qquad\square$

Corollary 2.27. *Let (f, g) be a logic-homomorphism from $\mathcal{L}$ to $\mathcal{L}'$.*

(i) *There exists a big complement F of g such that $f : F$ and the set*

$$[f]^F_{\approx_g} := \{(f', g) \mid f' : F\}$$

is exactly an equivalence class on the class of all logic-homomorphisms $(., g) : \mathcal{L} \rightarrow \mathcal{L}'$ modulo $\approx_g$. Hence, there is a bijection $[f]^F_{\approx_g} \mapsto F$ between these equivalence classes and the set of all big complements F of g.

(ii) *If f is $\mathcal{L}'$-surjective, then there exists a big complement F of g such that $f : F$ and*

$$F^{-1}(Mod_{\mathcal{L}'}(a')/=_{\mathcal{L}'}) = \bigcup\{(Mod_{\mathcal{L}}(a)/=_{\mathcal{L}}) \mid a \in A_{a'}\},$$

for all $a' \in Expr_{\mathcal{L}'}$.

Furthermore, this F is a continuous map from the space $((Int_{\mathcal{L}}/_{=_{\mathcal{L}}}), \sigma_{\mathcal{L}}^2)$ to the space $((Int_{\mathcal{L}}/_{=_{\mathcal{L}}}), \sigma_{\mathcal{L}}^2)$.

Proof. (i) If (f,g) is a logic-homomorphism, then, by Lemma 2.9, f is regular, a small complement of g and $\{g^{-1}(T') \mid T' \in PTh(\mathcal{L}')\} = PTh(\mathcal{L})$. By item (ii) of the preceding Theorem, there is a big complement F of g with $f : F$.

Suppose that $f' \in [f]_{\approx_g}^F$. Then $f'(\mathcal{A}) =_{\mathcal{L}'} f(\mathcal{A})$ for any $\mathcal{A} \in Int_{\mathcal{L}}$. It follows that $f'^{-1}(Mod_{\mathcal{L}'}(a')) = f^{-1}(Mod_{\mathcal{L}'}(a'))$, for all $a' \in Expr_{\mathcal{L}'}$. Whence, (f',g) is a logic-homomorphism and $(f',g) \approx_g (f,g)$. Hence, $[f]_{\approx_g}^F = \{(f',g) \mid f' : F\}$ is a set of $\approx_g$-equivalent logic-homomorphisms.

Now suppose that (f',g) is a logic-homomorphism and $f' \approx_g f$. Since $f : F$, we get $f' : F$ and $(f',g) \in [f]_{\approx_g}^F$. This proves the assertion.

(ii) If (f,g) is a logic-homomorphism, then f is regular and a small complement of g, by Lemma 2.9. Now, as in the proof of the previous Theorem, we may define $F : (Int_{\mathcal{L}}/_{=_{\mathcal{L}}}) \to (Int_{\mathcal{L}'}/_{=_{\mathcal{L}'}})$ by $F([\mathcal{A}]_{\mathcal{L}}) := [f(\mathcal{A})]_{\mathcal{L}'}$. Then $f : F$ and F is a big complement of g. Suppose that f is $\mathcal{L}'$-surjective. Then for each $\mathcal{A}' \in Int_{\mathcal{L}'}$ there is $\mathcal{A} \in Int_{\mathcal{L}}$ such that $F^{-1}([\mathcal{A}']_{\mathcal{L}'}) = F^{-1}([f(\mathcal{A})]_{\mathcal{L}'}) = [\mathcal{A}]_{\mathcal{L}}$. Now the equation follows from the fact that $f : F$ and $f^{-1}(Mod_{\mathcal{L}'}(a')) = \bigcup\{(Mod_{\mathcal{L}}(a)) \mid a \in A_{a'}\}$ holds.

By Lemma 2.9(xi), the function $G : PTh(\mathcal{L}) \to PTh(\mathcal{L}')$, defined by $Th_{\mathcal{L}}(\mathcal{A}) \mapsto Th_{\mathcal{L}'}(f(\mathcal{A}))$, is a continuous map from space $(PTh_{\mathcal{L}}, \rho_{\mathcal{L}})$ to the space $(PTh_{\mathcal{L}'}, \rho_{\mathcal{L}'})$. From the definitions of F and G and Proposition 2.6 it follows that F is a continuous map from the space $((Int_{\mathcal{L}}/_{=_{\mathcal{L}}}), \sigma_{\mathcal{L}}^2)$ to the space $((Int_{\mathcal{L}'}/_{=_{\mathcal{L}'}}), \sigma_{\mathcal{L}'}^2)$. $\square$

Theorem 2.28. *Suppose that $\mathcal{L}, \mathcal{L}'$ are abstract logics and $g : Expr_{\mathcal{L}} \to Expr_{\mathcal{L}'}$ is a function with SHP. Then the following holds:*

(i) *There exists exactly one big complement $F : (Int_{\mathcal{L}}/_{=_{\mathcal{L}}}) \to (Int_{\mathcal{L}'}/_{=_{\mathcal{L}'}})$ of g.*

(ii) *For any function $F : (Int_{\mathcal{L}}/_{=_{\mathcal{L}}}) \to (Int_{\mathcal{L}'}/_{=_{\mathcal{L}'}})$ the following two conditions are equivalent:*
   - *F is the big complement of g.*
   - *$F^{-1}(Mod_{\mathcal{L}'}(a')/_{=_{\mathcal{L}'}}) = \bigcup\{(Mod_{\mathcal{L}}(a)/_{=_{\mathcal{L}}}) \mid g(a)^* \subseteq a'^*\}$ holds, for all $\mathcal{L}'$-expressions a'.*

(iii) *Let F be a big complement of g. Then for every function $f : Int_{\mathcal{L}} \to Int_{\mathcal{L}'}$ with $f : F$, $(f,g) : \mathcal{L} \to \mathcal{L}'$ is a strong logic-homomorphism. On the other hand, if $(f,g) : \mathcal{L} \to \mathcal{L}'$ is a strong logic-homomorphism, then there exists a function F, such that $f : F$ and F is a big complement of g.*

(iv) *Any two strong logic-homomorphisms (f_1,g) and (f_2,g) from $\mathcal{L}$ to $\mathcal{L}'$ are $\approx_g$-equivalent.*

(v) *The big complement $F : (Int_{\mathcal{L}}/_{=_{\mathcal{L}}}) \to (Int_{\mathcal{L}'}/_{=_{\mathcal{L}'}})$ of g is a bijection.*

Proof. (i) The existence of a big complement of g follows from the preceding Theorem. Now suppose that F' is another big complement of g. Then

$$g^{-1}(Th_{\mathcal{L}'}(F([\mathcal{A}]_{\mathcal{L}}))) = Th_{\mathcal{L}}(\mathcal{A}) = g^{-1}(Th_{\mathcal{L}'}(F'([\mathcal{A}]_{\mathcal{L}}))),$$

for any $\mathcal{A} \in Int_{\mathcal{L}}$. Then from Lemma 2.10(i) follows that F and F' coincide on $[\mathcal{A}]_{\mathcal{L}}$, hence $F = F'$ and there exists exactly one big complement of g.

(ii) First, suppose that F is the big complement of g, that is, we have $g^{-1}(Th_{\mathcal{L}'}(F([\mathcal{A}]_{\mathcal{L}}))) = Th_{\mathcal{L}}(\mathcal{A}))$ for all interpretations $\mathcal{A}$.

Let $[\mathcal{A}]_{\mathcal{L}} \in F^{-1}(Mod_{\mathcal{L}'}(a')/_{=_{\mathcal{L}'}})$. Then there is some $\mathcal{A}' \in Int_{\mathcal{L}'}$ with $F([\mathcal{A}]_{\mathcal{L}}) = [\mathcal{A}']_{\mathcal{L}'} \in (Mod_{\mathcal{L}'}(a')/_{=_{\mathcal{L}'}})$. Since $Th_{\mathcal{L}'}(\mathcal{A}') \in a'^{*\mathcal{L}'}$, we conclude $g^{-1}(Th_{\mathcal{L}'}(\mathcal{A}')) = Th_{\mathcal{L}}(\mathcal{A}) \in g^{-1}(a'^{*\mathcal{L}'}) = \bigcup\{a^{*\mathcal{L}'} \mid g(a)^* \subseteq a'^{*\mathcal{L}'}\}$. It follows that $[\mathcal{A}]_{\mathcal{L}} \in \bigcup\{(Mod_{\mathcal{L}}(a)/_{=_{\mathcal{L}}}) \mid g(a)^{*\mathcal{L}'} \subseteq a'^{*\mathcal{L}'}\}$ and we have shown the inclusion "$\subseteq$".

Now let

$$[\mathcal{A}]_{\mathcal{L}} \in \bigcup\{(Mod_{\mathcal{L}}(a)/_{=_{\mathcal{L}})}) \mid g(a)^{*\mathcal{L}'} \subseteq a'^{*\mathcal{L}'}\}$$
$$= \bigcup\{(Mod_{\mathcal{L}}(a)/_{=_{\mathcal{L}}}) \mid Mod_{\mathcal{L}'}(g(a)) \subseteq Mod_{\mathcal{L}'}(a')\}.$$

Then $\mathcal{A} \models_{\mathcal{L}} a$, for some a with the property that $Mod_{\mathcal{L}'}(g(a)) \subseteq Mod_{\mathcal{L}'}(a')$. Since $a \in Th_{\mathcal{L}}(\mathcal{A}) = g^{-1}(Th_{\mathcal{L}'}(F([\mathcal{A}]_{\mathcal{L}}))$, it follows that $g(a) \in Th_{\mathcal{L}'}(F([\mathcal{A}]_{\mathcal{L}}))$, that is, $F([\mathcal{A}]_{\mathcal{L}}) \in (Mod_{\mathcal{L}'}(g(a))/_{=_{\mathcal{L}'}})$. Since this set is contained in $(Mod_{\mathcal{L}'}(a')/_{=_{\mathcal{L}'}})$, it follows that $[\mathcal{A}]_{\mathcal{L}} \in F^{-1}(Mod_{\mathcal{L}'}(a')/_{=_{\mathcal{L}'}})$. So we have also proved the inclusion "$\supseteq$".

Now suppose that F satisfies the equation

$$F^{-1}(Mod_{\mathcal{L}'}(a')/_{=_{\mathcal{L}'}}) = \bigcup\{(Mod_{\mathcal{L}}(a)/_{=_{\mathcal{L}}}) \mid g(a)^{*\mathcal{L}'} \subseteq a'^{*\mathcal{L}'}\},$$

for all $\mathcal{L}'$-expressions a'. Similarly as in the proof of Lemma 2.9(v) (now we have to factorize by $\mathcal{L}$-equivalence) one can show that $Th_{\mathcal{L}}(\mathcal{A}) = g^{-1}(Th_{\mathcal{L}'}(F([\mathcal{A}]_{\mathcal{L}'})))$, for all $\mathcal{A} \in Int_{\mathcal{L}}$. Thus, F is a big complement of g.

(iii) We suppose that F is a big complement of g and f is a function such that $f : F$. By (ii), F satisfies $F^{-1}(Mod_{\mathcal{L}'}(a')/_{=_{\mathcal{L}'}}) = \bigcup\{(Mod_{\mathcal{L}}(a)/_{=_{\mathcal{L}}}) \mid g(a)^{*\mathcal{L}'} \subseteq a'^{*\mathcal{L}'}\}$. We show that this equation is still true, if we replace F by f. It is sufficient to show $F^{-1}(Mod_{\mathcal{L}'}(a')/_{=_{\mathcal{L}'}}) = f^{-1}(Mod_{\mathcal{L}'}(a')/_{=_{\mathcal{L}'}})$ for any $a' \in Expr_{\mathcal{L}'}$. That is, we show that for every $a' \in Expr_{\mathcal{L}'}$ and every $[\mathcal{A}']_{\mathcal{L}'} \in (Mod_{\mathcal{L}'}(a')/_{=_{\mathcal{L}'}})$ holds $X := f^{-1}([\mathcal{A}']_{\mathcal{L}'}) = F^{-1}([\mathcal{A}']_{\mathcal{L}'}) =: [\mathcal{A}]_{\mathcal{L}}$, for some set X of $\mathcal{L}$-interpretations and some $\mathcal{L}$-interpretation $\mathcal{A}$. So suppose that $[\mathcal{A}']_{\mathcal{L}'} \in (Mod_{\mathcal{L}'}(a')/_{=_{\mathcal{L}'}})$. For every $\mathcal{B} \in X$, $f(\mathcal{B}) \in [\mathcal{A}']_{\mathcal{L}'}$. It follows that $[\mathcal{A}']_{\mathcal{L}'} = F([\mathcal{B}]_{\mathcal{L}})$, for every $\mathcal{B} \in X$, since $f : F$. Then $\mathcal{B} \in F^{-1}([\mathcal{A}']_{\mathcal{L}})$, for every $\mathcal{B} \in X$. Hence, $X \subseteq F^{-1}([\mathcal{A}']_{\mathcal{L}'}) = [\mathcal{A}]_{\mathcal{L}}$. On the other hand, if $\mathcal{B} \in [\mathcal{A}]_{\mathcal{L}}$, then $f(\mathcal{B}) \in F([\mathcal{B}]_{\mathcal{L}}) = F([\mathcal{A}]_{\mathcal{L}}) = [\mathcal{A}']_{\mathcal{L}'}$. It follows that $\mathcal{B} \in X$, whence $[\mathcal{A}]_{\mathcal{L}} \subseteq X$, and finally $X = [\mathcal{A}]_{\mathcal{L}}$. We have shown, that (f, g), for $f : F$, satisfies the definition of a strong logic-homomorphism.

The second part of the assertion follows from the preceding Theorem.

(iv) If (f_1, g) and (f_2, g) are strong logic-homomorphisms, then by (iii) and (i) of this Theorem there is exactly one function F such that $f_1 : F$ and $f_2 : F$ and

F is a big complement of g. Now the assertion follows from the previous Corollary or from Lemma 2.24.

(v) F is injective, by Theorem 2.26(i). Let f be a function such that $f : F$. By (iii), (f, g) is a strong logic-homomorphism. By Lemma 2.10, f is $\mathcal{L}$-surjective. It follows that F is surjective. $\square$

The following result is a direct consequence of the previous Theorem.

Corollary 2.29. *Suppose that $\mathcal{L}, \mathcal{L}'$ are abstract logics and $g : Expr_{\mathcal{L}} \to Expr_{\mathcal{L}'}$ is a function with SHP. Then there exists — up to $\approx_g$-equivalence — exactly one function $f : Int_{\mathcal{L}} \to Int_{\mathcal{L}'}$ such that (f, g) is a strong logic-homomorphism.*

The preceding results give us an alternative defining condition for a strong logic-homomorphism:

Corollary 2.30. *Let $\mathcal{L}, \mathcal{L}'$ be abstract logics and suppose that $g : Expr_{\mathcal{L}} \to Expr_{\mathcal{L}'}$ is a function with SHP. Let $f : Int_{\mathcal{L}} \to Int_{\mathcal{L}'}$ be a regular function. Then the following conditions are equivalent:*

(i) *(f, g) is a strong logic-homomorphism.*
(ii) *$\mathcal{A} \vDash_{\mathcal{L}} a \iff f(\mathcal{A}) \vDash_{\mathcal{L}'} g(a)$, for all $\mathcal{A} \in Int_{\mathcal{L}}$ and for all $a \in Expr_{\mathcal{L}}$.*
(iii) *$f^{-1}(Mod_{\mathcal{L}'}(g(a))) = Mod_{\mathcal{L}}(a)$, for all $a \in Expr_{\mathcal{L}}$*
(iv) *$g^{-1}(Th_{\mathcal{L}'}(f(\mathcal{A}))) = Th_{\mathcal{L}}(\mathcal{A})$, for all $\mathcal{L}$-interpretations $\mathcal{A}$.*
(v) *f is a small complement of g.*

Proof. (i)→(ii) follows from Lemma 2.9. The proof of (ii)→(iii)→(iv) is straightforeward. (iv)→(v) is the definition of a small complement. Finally, suppose that (v) holds. By the second statement of Theorem 2.26(ii), there is F such that $f : F$ and F is a big complement of f. By Theorem 2.28(iii), (f, g) is a strong logic-homomorphism. $\square$

Definition 2.31. Let $\mathcal{L}, \mathcal{L}'$ be abstract logics and suppose that $g : Expr_{\mathcal{L}} \to Expr_{\mathcal{L}'}$ is a function such that $\{g^{-1}(T) \mid T \in PTh(\mathcal{L}')\} \subseteq PTh(\mathcal{L})$. We define a function $\mathcal{G} : (Int_{\mathcal{L}'}/_{=_{\mathcal{L}'}}) \to (Int_{\mathcal{L}}/_{=_{\mathcal{L}}})$ by

$$\mathcal{G}([\mathcal{A}']_{\mathcal{L}'}) := [\mathcal{A}]_{\mathcal{L}} :\iff g^{-1}(Th_{\mathcal{L}'}(\mathcal{A}')) = Th_{\mathcal{L}}(\mathcal{A}).$$

$\mathcal{G}$ is called the inverse complement of g.

It is clear by the definition that $\mathcal{G}$ is well-defined.

Theorem 2.32. *Let (f, g) be a logic-homomrphism from $\mathcal{L}$ to $\mathcal{L}'$. Then the inverse complement $\mathcal{G}$ of g is a surjective continuous map from the space $((Int_{\mathcal{L}'}/_{=_{\mathcal{L}'}}), \sigma_{\mathcal{L}'}^2)$ to the space $((Int_{\mathcal{L}}/_{=_{\mathcal{L}}}), \sigma_{\mathcal{L}}^2)$. If (f, g) is a strong logic-homomorphism, then $\mathcal{G}$ is an homeomorphism between these two spaces and $\mathcal{G} = F^{-1}$, where F is the (unique) complement of g.*

Proof. Let $\mathcal{A} \in Int_{\mathcal{L}}$. Since (f, g) is a logic-homomorphism, by Lemma 2.9, there is some $T' \in PTh(\mathcal{L}')$ such that $T' = Th_{\mathcal{L}'}(\mathcal{A}')$, for some $\mathcal{A}' \in Int_{\mathcal{L}'}$, and $g^{-1}(T') = Th_{\mathcal{L}}(\mathcal{A})$. By definition of $\mathcal{G}$, $\mathcal{G}([\mathcal{A}']_{\mathcal{L}'}) = [\mathcal{A}]_{\mathcal{L}}$, thus, $\mathcal{G}$ is surjective. $\mathcal{G}$ is continuous:

58 S. Lewitzka

It is enough to show that $\mathcal{G}^{-1}(Mod_{\mathcal{L}}(a)/_{=_{\mathcal{L}}}) = (Mod_{\mathcal{L}'}(g(a))/_{=_{\mathcal{L}}})$, for each $a \in Expr_{\mathcal{L}}$. We have:

$[\mathcal{A}']_{\mathcal{L}'} \in \mathcal{G}^{-1}(Mod_{\mathcal{L}}(a)/_{=_{\mathcal{L}}})$

$\Longleftrightarrow \mathcal{G}([\mathcal{A}']_{\mathcal{L}'}) \in (Mod_{\mathcal{L}}(a)/_{=_{\mathcal{L}}})$

$\Longleftrightarrow Th_{\mathcal{L}}(\mathcal{G}([\mathcal{A}']_{\mathcal{L}'})) \in a^{*_{\mathcal{L}}}$

$\Longleftrightarrow Th_{\mathcal{L}}(\mathcal{A}) \in a^{*_{\mathcal{L}}}$, where $\mathcal{G}([\mathcal{A}']_{\mathcal{L}'}) = [\mathcal{A}]_{\mathcal{L}}$

$\Longleftrightarrow g^{-1}(Th_{\mathcal{L}'}(\mathcal{A}')) = Th_{\mathcal{L}}(\mathcal{A}) \in a^{*_{\mathcal{L}}}$, by def. of the inverse complement

$\Longleftrightarrow a \in g^{-1}(Th_{\mathcal{L}'}(\mathcal{A}'))$

$\Longleftrightarrow g(a) \in Th_{\mathcal{L}'}(\mathcal{A}')$

$\Longleftrightarrow Th_{\mathcal{L}'}(\mathcal{A}') \in g(a)^{*_{\mathcal{L}'}}$

$\Longleftrightarrow [\mathcal{A}']_{\mathcal{L}'} \in (Mod_{\mathcal{L}'}(g(a))/_{=_{\mathcal{L}'}})$

Now let us suppose that (f,g) is strong. Then the injectivity of $\mathcal{G}$ follows from Lemma 2.10(i). Thus, $\mathcal{G}$ is bijective and continuous, by the first part of the Theorem. It remains to show that $\mathcal{G}$ is also an open map: Since g has (SHP), by Theorem 2.28, there is exactly one big complement F of g. F is a bijective function. Now, by the definition of the inverse complement $\mathcal{G}$, one can see that $F = \mathcal{G}^{-1}$, i.e., $\mathcal{G} = F^{-1}$. In particular, $\mathcal{G}$ is a bijection. Since, by Theorem 2.28, F satisfies

$$F^{-1}(Mod_{\mathcal{L}'}(a')/_{=_{\mathcal{L}'}}) = \bigcup\{(Mod_{\mathcal{L}}(a)/_{=_{\mathcal{L}}}) \mid g(a)^* \subseteq a'^*\},$$

we get

$$\mathcal{G}(Mod_{\mathcal{L}'}(a')/_{=_{\mathcal{L}'}}) = \bigcup\{Mod_{\mathcal{L}}(a)/_{=_{\mathcal{L}'}} \mid g(a)^{*_{\mathcal{L}'}} \subseteq a'^{*_{\mathcal{L}'}}\},$$

for all $a' \in Expr_{\mathcal{L}'}$. From this it follows that $\mathcal{G}$ is an open map from the space $(Int_{\mathcal{L}'}/_{=_{\mathcal{L}'}})$ to the space $(Int_{\mathcal{L}}/_{=_{\mathcal{L}}})$. Hence, $\mathcal{G}$ is an homeomorphism between these spaces. $\qquad\square$

Corollary 2.33. *Let* $\mathcal{L}, \mathcal{L}'$ *be abstract logics and suppose that* $g : Expr_{\mathcal{L}} \to Expr_{\mathcal{L}'}$ *is a function such that* $\{g^{-1}(T) \mid T \in PTh(\mathcal{L}')\} \subseteq PTh(\mathcal{L})$. *Let* $\mathcal{G}$ *be the inverse complement of* g. *Then for all* $\mathcal{A}' \in Int_{\mathcal{L}'}$ *and for all* $a \in Expr_{\mathcal{L}}$ *the following holds:*

$$\mathcal{A}' \vDash_{\mathcal{L}'} g(a) \Longleftrightarrow \mathcal{G}([\mathcal{A}']_{\mathcal{L}'}) \vDash_{\mathcal{L}} a.$$

In particular, the above equivalence holds for any function g *such that* (f,g) *is a logic-homomorphism from* $\mathcal{L}$ *to* $\mathcal{L}'$.

Proof. Simply by the definitions we get:
$g(a) \in Th_{\mathcal{L}'}(\mathcal{A}') \Longleftrightarrow a \in g^{-1}(Th_{\mathcal{L}'}([\mathcal{A}']_{\mathcal{L}'})) \Longleftrightarrow a \in Th_{\mathcal{L}}(\mathcal{G}([\mathcal{A}']_{\mathcal{L}'}))$, for all $a \in Expr_{\mathcal{L}}$ and all $\mathcal{A}' \in Int_{\mathcal{L}'}$. The last assertion follows from Lemma 2.9(i). $\quad\square$

Notice that the equivalence stated in the above Corollary has the same form as the satisfaction axiom of institutions ([2],[6]), see the short discussion in section 4 below.

3. Two examples

Let us outline two simple examples of abstract logics and logic-homomorphisms. The study of further and more complicated examples we must postpone here to later works.

Example.

Let $\Sigma \subseteq \Sigma'$ be first order signatures and let $\mathcal{L}, \mathcal{L}'$ be the following logics:

- $Expr_{\mathcal{L}}$ is the set of first order formulas over Σ and $Expr_{\mathcal{L}'}$ is the set of first order formulas over Σ'.
- $Int_{\mathcal{L}}, Int_{\mathcal{L}'}$ are the classes of all Σ-structures (Σ'-structures) together with variable assignments, respectively.
- $\vDash_{\mathcal{L}}, \vDash_{\mathcal{L}'}$ are the respective satisfaction relations, defined in the usual way.

Then $\mathcal{L} \leq \mathcal{L}'$. If $\mathcal{L}'$ is a definable extension of $\mathcal{L}$, that is, for every $a' \in Expr_{\mathcal{L}'}$ there is some $a \in Expr_{\mathcal{L}}$ such that $a =_{\mathcal{L}'} a'$, then it is easy to see that the identity $i : Expr_{\mathcal{L}} \to Expr_{\mathcal{L}'}$ has SHP. By Theorem 2.28, there is a function $f : Int_{\mathcal{L}} \to Int_{\mathcal{L}'}$ such that $(f,i) : \mathcal{L} \to \mathcal{L}'$ is a strong logic-homomorphism. Moreover, any other strong logic-homomorphism is equivalent to (f,i).

Example. Let $Expr_{\mathcal{L}}$ be the set of first order formulas in the language of graphs, and let $Int_{\mathcal{L}} = Int_{\mathcal{L}'}$ be the class of finite graphs together with variable assignments. Let $Expr_{\mathcal{L}'}$ be the set of first order formulas in the language of graphs extended by infinite disjunctions (and infinite conjunctions). Suppose that $\vDash_{\mathcal{L}}$ and $\vDash_{\mathcal{L}'}$ are defined in the usual way.

Then clearly $\mathcal{L} \leq \mathcal{L}'$. Though, this extension is of another nature than the extension of example (i). (In (i) we extend the non-logical part of the language, whereas here we extend the logical part.) For example, connectivity of (finite) graphs is definable in $\mathcal{L}'$ but not in $\mathcal{L}$.

There is a strong logic homomorphism $(i_1, i_2) : \mathcal{L} \to \mathcal{L}'$, where i_1 and i_2 are the identity functions on $Int_{\mathcal{L}}$ and $Expr_{\mathcal{L}}$, respectively. Let us outline a proof: Every finite model (finite graph) is characterizable — up to ismorphism — by a single first order formula. This formula a isolates the theory T of the finite model, i.e., $\Vdash_{\mathcal{L}} a \to b$ for all $b \in T$, that is, $a^{*_{\mathcal{L}}} = \{T\}$. Since $Int_{\mathcal{L}} = Int_{\mathcal{L}'}$, for any theory T of logic $\mathcal{L}$ and any theory T' of logic $\mathcal{L}'$ such that $T = T' \cap Expr_{\mathcal{L}}$ it is easy to see that $a \in Expr_{\mathcal{L}}$ isolates T in $\mathcal{L}$ if and only if a isolates T' in $\mathcal{L}'$. Furthermore, for every $T' \in PTh(\mathcal{L}')$ there is some $a \in Expr_{\mathcal{L}}$ such that a isolates T'. Notice also that $i_2^{-1}(T') = T' \cap Expr_{\mathcal{L}} \in PTh(\mathcal{L})$ and that for each $T \in PTh(\mathcal{L})$ there is exactly one $T' \in PTh(\mathcal{L}')$ such that $T' \cap Expr_{\mathcal{L}} = T$. Then for each $a' \in Expr_{\mathcal{L}'}$ we get the following:

$$i_2^{-1}(a'^{*_{\mathcal{L}'}}) = \bigcup\{a^{*_{\mathcal{L}}} \mid a \text{ isolates some } T' \in a'^{*_{\mathcal{L}'}}\} = \bigcup\{a^{*_{\mathcal{L}}} \mid i_2(a) \Vdash_{\mathcal{L}'} a'\} = \bigcup\{a^{*_{\mathcal{L}}} \mid i_2(a)^{*_{\mathcal{L}'}} \subseteq a'^{*_{\mathcal{L}'}}\}.$$

Thus, i_2 has SHP. The rest follows from Theorem 2.28.

4. Future work

One aim of future research on model-theoretical abstract logics is the study of specific examples of logics and their logic-homomorphisms, extensions, etc., using the concepts and tools developed in this paper. We hope that our work will be helpful in order to study in a systematic way concrete logics and their structural properties and expressive power. Our topological approach may offer new possibilities for such a systematic study. A promising idea seems to be the study of the category of (model-theoretical abstract) logics together with logic-homomorphisms (or a generalized form of logic maps) as morphisms. Since every logic gives rise to a topological space, we obtain a functor from this category to the category of topological spaces. Now, results and methods from topology may be useful to get new insights into properties of logics.

Another promising connection to the category-theoretical level is given by Corollary 2.33. The equivalence exposed in the Corollary has the same form as the satisfaction axiom of the well-known category-theoretical concept of institutions ([2], [6]). An institution can be considered in some sense as a global description of a logic. For example, the whole first order logic (with models and languages over all signatures) forms an institution. On the other hand, an abstract logic, in our sense, gives a local perspective: Each first order signature gives rise to a single model-theoretical abstract logic. Now, the result of Corollary 2.33 is interesting in this context. It suggests that a category of model-theoretical abstract logics together with logic-homomorphisms (or more general, logic maps g as given in the hypothesis of the Corollary) forms a structure that obeys the satisfaction axiom of institutions. In other words, from the local point of view of model-theoretical abstract logics it is possible to arrive at the global level of institutions. It would be interesting to elaborate and to study these coherences in detail.

There is an obvious generalization of our concept of logic: If we omit the model-theoretical component and consider only a set of (abstract) theories which is closed under (non-empty) intersections, then we obtain a notion of abstract logic which is more general (note that a logic may be given in a purely proof-theoretical way, that is, as a deduction system, independent of any model-theoretic semantics). In particular, each model-theoretical abstract logic gives rise to such a general abstract logic. In this broader setting, which is studied in [4], it is also possible to define a more general form of logic-homomorphisms. The model-theoretical approach turns out to be a special case.

Another task may be the attempt to define further logical connectivities (in particular, non-classical ones) inside an abstract logic by means of theories. We gave here only a few examples (Definition 2.1) concerning classical negation and conjunction. The possibility to give such definitions then can be understood as a confirmation that we are on the right way with our abstract approach.

References

[1] R. Engelking, *General Topology*. Second edition, Heldermann Verlag, 1989.

[2] J. A. Goguen and R. M. Burstall, *Institutions: Abstract Model Theory for Specification and Programming*. Journal of the ACM, 39(1):95-146, January 1992.

[3] Steffen Lewitzka, *A Topological Approach to Universal Logic: Model-Theoretical Abstract Logics*. In: [5]

[4] Steffen Lewitzka, *Abstract Logics and Logic Maps*. To appear in Proceedings of the First World Congress of Universal Logic 2005.

[5] Jean-Yves Beziau (Ed.), *Logica Universalis: Towards a General Theory of Logic*. First edition, Birkhäuser Verlag, Basel, Switzerland, 2005.

[6] T. Mossakowski, J. A. Goguen, R. Diaconescu, A. Tarlecki, *What is a Logic?*. In: [5]

[7] R. Wójcicki, *Theory of Logical Calculi*. Kluwer Academic Publishers, 1988.

Steffen Lewitzka
Federal University of Bahia – UFBA
Instituto de Matemática
Departamento de Ciência da Computação
Laboratório de Sistemas Distribuídos – LaSiD
Av. Ademar de Barros, CEP 40170-110
Salvador da Bahia, BA
Brazil
e-mail: steffenlewitzka@web.de

J.-Y. Beziau (Ed.), *Logica Universalis*, 2nd edition, 63–86
© 2007 Birkhäuser Verlag Basel/Switzerland

Selfextensional Logics with Implication

Ramon Jansana

Abstract. The aim of this paper is to develop the theory of the selfextensional logics with an implication for which it holds the deduction-detachment theorem, as presented in [8], but avoiding the use of Gentzen-systems to prove the main results as much as possible.

Mathematics Subject Classification (2000). 03B22, 03C05, 03G25.

Keywords. Abstract algebraic logic, logics with a deduction theorem, Hilbert algebras.

1. Introduction

Abstract Algebraic Logic (AAL) is the area of algebraic logic which studies the process of algebraization of the different logical systems. For information on AAL the reader is addressed to [10]. The concept of logic that is taken as primary in the AAL field is that of a consequence relation between sets of formulas and formulas which has the substitution-invariance property; informally speaking this means that if Γ is a set of formulas and φ is a formula that follows according to the logic from Γ, then for every pair (Δ, ψ) of the same form as (Γ, φ), ψ follows from Δ. A logic in this sense may have different replacement properties. The strongest one is shared by classical, intuitionistic and all the intermediate propositional logics. It says that if $\vdash_\mathcal{S}$ is the consequence relation of $\mathcal{S}$, for any set of formulas Γ, any formulas φ, ψ, δ and any variable p

if $\Gamma, \varphi \vdash_\mathcal{S} \psi$ and $\Gamma, \psi \vdash_\mathcal{S} \varphi$, then $\Gamma, \delta(p/\varphi) \vdash_\mathcal{S} \delta(p/\psi)$ and $\Gamma, \delta(p/\varphi) \vdash_\mathcal{S} \delta(p/\psi)$,

where $\delta(p/\varphi)$ and $\delta(p/\psi)$ are the formulas obtained by substituting φ for p and ψ for p in δ respectively. This strong replacement property can be seen as a formal counterpart of Frege's compositionality principle for truth. Logics satisfying this

The research in this paper was partially supported by Spanish DGESIC grant BFM2001-3329 and Catalan grant 2001SGR-00017. The study was begun during the author's stay at the ILLC of the University of Amsterdam in the academic year 1999–2000 supported by the Spanish DGESIC grant PR199-0179.

replacement property are called Fregean in [6]; the origin of the name comes from the studies by R. Suszko on his non-Fregean logic. Several important logics are not Fregean, for instance almost all the logics of the modal family. Many, like the so-called local consequence relation of the modal logic K, satisfy a weaker replacement property: for all formulas φ, ψ, δ,

$$\text{if } \varphi \vdash_{\mathcal{S}} \psi \text{ and } \psi \vdash_{\mathcal{S}} \varphi, \text{ then } \delta(p/\varphi) \vdash_{\mathcal{S}} \delta(p/\psi) \text{ and } \delta(p/\psi) \vdash_{\mathcal{S}} \delta(p/\varphi).$$

A logic is said to be *selfextensional* if it satisfies this weaker replacement property. In algebraic terms this means that the interderivability relation between formulas is a congruence relation of the formula algebra. R. Wójcicki coined the name in [17].

The class of protoalgebraic logics is the class of logics for which the theory of the algebraic-like semantics of its elements is the best understood in AAL. A logic is protoalgebraic if it has a generalizad implication, i.e. a set of formulas in two variables, which we denote as $\Rightarrow(p, q)$, with the generalized modus ponens rule (from p and $\Rightarrow(p, q)$ infer q) and such that for every $\varphi \in \Rightarrow(p, q)$, $\varphi(q/p)$ is a theorem. Roughly speaking protoalgebraic logics are the logics for which the semantics of logical matrices is well behaved from the point of view of universal algebra, in the sense that many of the results of universal algebra have counterparts of specific logical interest in the theory of logical matrices for protoalgebraic logics. A logical matrix is a pair $\langle \mathbf{A}, F \rangle$ where $\mathbf{A}$ is an algebra and F is a subset of the domain of $\mathbf{A}$; it is said to be a model of a logic $\mathcal{S}$ if $\mathbf{A}$ is of the type of $\mathcal{S}$ and F is closed under the interpretations in $\mathbf{A}$ of the inferences of $\mathcal{S}$, namely if for every set of formulas Γ, every formula φ and every interpretation v of the formulas in $\mathbf{A}$, if $\Gamma \vdash_{\mathcal{S}} \varphi$ and the interpretations by v of the elements in Γ belong to F, then the interpretation by v of φ belongs to F. If this is the case it is said that F is an $\mathcal{S}$-filter of $\mathbf{A}$.

Several interesting logics are not protoalgebraic. For non-protoalgebraic logics logical matrix semantics is not so well behaved. For instance, the class of algebras that the theory of logical matrices canonically associates with a non-protoalgebraic logic does not necessarily coincide with the class one would intuitively expect to be associated with it. An illustration of this phenomenon is found in the conjunction-disjunction fragment of classical logic. Here the expected class of algebras is the class of distributive lattices, but, as is shown in [11], this class is not the class of algebras the theory of matrices provides.

In [8] a general theory of the algebraization of logic is developed using generalized matrices (where they are called abstract logics) as possible models for logical systems. A generalized matrix is a pair $\langle \mathbf{A}, \mathcal{C} \rangle$ where $\mathbf{A}$ is an algebra and $\mathcal{C}$ the family of closed sets of some finitary closure operator on the domain A of $\mathbf{A}$. It is said to be a model of a logic $\mathcal{S}$ if $\mathbf{A}$ is of the type of $\mathcal{S}$ and $\mathcal{C}$ is a family of $\mathcal{S}$-filters of $\mathbf{A}$. Using generalized matrices, in [8] a canonical way is proposed to associate a class of algebras $\mathbf{Alg}\mathcal{S}$ with each logical system $\mathcal{S}$ that in the known non-protoalgebraic logics supplies the expected results and for protoalgebraic logics gives exactly the class of algebras the theory of logical matrices associates with them. In [8] several general results are proved that sustain the claim that the class of algebras $\mathbf{Alg}\mathcal{S}$ is the natural class of algebras that corresponds to a given deductive system $\mathcal{S}$, and a way to obtain $\mathbf{Alg}\mathcal{S}$ as the result of performing the Lindenbaum-Tarski method suitably generalized is given in [5] and [10].

Among the class of generalized matrices that are models of a given deductive system $\mathcal{S}$ we find the class of the full g-models of $\mathcal{S}$. A full g-model of $\mathcal{S}$ is a generalized matrix $\langle \mathbf{A}, \mathcal{C} \rangle$ which from the logical point of view is equivalent to a generalized matrix of the form $\langle \mathbf{A}, \mathrm{Fi}_{\mathcal{S}}\mathbf{A} \rangle$, where $\mathrm{Fi}_{\mathcal{S}}\mathbf{A}$ is the set of all $\mathcal{S}$-filters of $\mathbf{A}$. The generalized matrices of this last form are called basic full g-models of $\mathcal{S}$ and the study of the class of full g-models can thus be reduced to their study. The class of full g-models of a deductive system was singled out in [8] as an important class and their systematic study was started.

Given a generalized matrix $\mathcal{A} = \langle \mathbf{A}, \mathcal{C} \rangle$, its "interderivability" relation is defined as follows: two elements are related if they belong to the same elements of $\mathcal{C}$. This relation is called the Frege relation of $\mathcal{A}$. Hence a logic $\mathcal{S}$ is selfextensional iff the Frege relation of the generalizad matrix $\langle \mathbf{Fm}, \mathrm{Th}\mathcal{S} \rangle$, where $\mathbf{Fm}$ is the algebra of formulas and $\mathrm{Th}\mathcal{S}$ the family of the theories of $\mathcal{S}$, is a congruence of $\mathbf{Fm}$. Among the selfextensional logics there is an important class introduced in [8], the class of fully selfextensional logics (note that there they are called strongly selfextensional). A logic $\mathcal{S}$ is fully selfextensional if for every full model $\langle \mathbf{A}, \mathcal{C} \rangle$ of $\mathcal{S}$ the Frege relation of $\langle \mathbf{A}, \mathcal{C} \rangle$ is a congruence of $\mathbf{A}$, which is equivalent to saying that for every algebra $\mathbf{A}$ the relation of belonging to the same $\mathcal{S}$-filters in $\mathcal{C}$ is a congruence. The class of fully selfextensional logics is included properly in the class of selfextensional logics as shown in [1].

The present paper studies a class of protoalgebraic selfextensional deductive systems using the tools of the semantics of generalized matrices. It is the class of selfextensional deductive systems $\mathcal{S}$ with a binary formula, or term, $p \Rightarrow q$ which has the deduction-detachment property, that is, such that for every set of formulas Γ and all formulas φ, ψ,

$$\Gamma, \varphi \vdash_{\mathcal{S}} \psi \quad \text{iff} \quad \Gamma \vdash_{\mathcal{S}} \varphi \Rightarrow \psi.$$

Many deductive systems belong to this class, for instance the modal local consequence relations given by classes of Kripke frames in the standard language for many-modal logic. Hardly any of these are Fregean.

In [8] selfextensional deductive systems are studied using Gentzen systems as one of the main tools. The present paper develops part of the theory developed in [8] of the selfextensional deductive systems $\mathcal{S}$ with an implication $\Rightarrow$ with the deduction-detachment property without recourse to Gentzen systems. In this way we provide new and much simpler proofs of the following two results in [8].

1. For every deductive system $\mathcal{S}$ with an implication with the deduction-detachment property the class of algebras $\mathbf{Alg}\mathcal{S}$ is a variety (Theorems 4.27 of [8]).
2. Every selfextensional deductive system $\mathcal{S}$ with an implication with the deduction-detachment property is fully selfextensional (Theorems 4.31 and 4.46 of [8]).

In [13] it is proved that for every algebraic similarity type with a binary term $\wedge$ there is a dual isomorphism between the set of selfextensional deductive systems where $\wedge$ is a conjunction, ordered by the extension relation, and the set, ordered by inclusion, of all the subvarieties of the variety axiomatized by the semilattice equations $x \wedge x \approx x$, $x \wedge (y \wedge z) \approx (x \wedge y) \wedge z$ and $x \wedge y \approx y \wedge x$. We prove the parallel result for selfextensional logics with an implication that has the deduction-detachment property, namely:

3. for every algebraic similarity type and any of its binary terms $\Rightarrow$ there is a dual isomorphism between the set of selfextensional deductive systems where $\Rightarrow$ has the deduction-detachment property, ordered by the extension relation, and the set, ordered by inclusion, of all the subvarieties of the variety axiomatized by the Hilbert algebra equations H1-H4 below.

In our way to prove these results without recourse to Gentzen systems we characterize the selfextensional logics with a binary term $x \Rightarrow y$ that has the deduction-detachment property, as the logics $\mathcal{S}$ for which there is a class of algebras K such that the equations that define the Hilbert algebras

H1. $x \Rightarrow x \approx y \Rightarrow y$

H2. $(x \Rightarrow x) \Rightarrow x \approx x$

H3. $x \Rightarrow (y \Rightarrow z) \approx (x \Rightarrow y) \Rightarrow (x \Rightarrow z)$

H4. $(x \Rightarrow y) \Rightarrow ((y \Rightarrow x) \Rightarrow y) \approx (y \Rightarrow x) \Rightarrow ((x \Rightarrow y) \Rightarrow x)$.

hold for the term $\Rightarrow$ in K and the following two conditions are satisfied:

1. $\varphi_0, \dots, \varphi_{n-1} \vdash_{\mathcal{S}} \varphi$ iff $\forall \mathbf{A} \in \mathsf{K}\ \forall v \in \mathrm{Hom}(\mathbf{Fm}, \mathbf{A})$

$$v(\varphi_0 \Rightarrow (\dots \Rightarrow (\varphi_{n-1} \Rightarrow \varphi_n)\dots)) = 1.$$

2. $\emptyset \vdash_{\mathcal{S}_\mathsf{K}} \varphi$ iff $\forall \mathbf{A} \in \mathsf{K}\ \forall v \in \mathrm{Hom}(\mathbf{Fm}, \mathbf{A}) v(\varphi) = 1.$

The deductive systems with these properties are called Hilbert-based in this paper.

At the end of Section 3 we give a characterization of the Hilbert-based deductive systems which are regularly algebraizable: they are the Fregean ones which are Hilbert-based. In Section 4 we obtain some results on these systems and a different proof of a result of Czelakowski and Pigozzi in [6]. Finally, in Section 5 we deal with Gentzen systems and we give a different, simpler proof of Proposition 4.47 (iii) and Proposition 4.44 in [8] using the results obtained in Section 3.

2. Preliminaries

In this section we survey the elements of AAL that will be used in the paper and we fix notation. For detailed expositions we address the reader to [2], [4], [8], [10] and [18].

Let $\mathcal{L}$ be an algebraic similarity type (or set of connectives) that we fix throughout this section. All algebras considered, etc., will be of this type. The set of all homomorphisms from an algebra $\mathbf{A}$ to an algebra $\mathbf{B}$ is denoted by $\mathrm{Hom}(\mathbf{A}, \mathbf{B})$.

Let $\mathbf{Fm}$ be the absolutely free algebra of type $\mathcal{L}$ with a denumerable set Var of generators. The elements of Var will be called, as usual, propositional variables.

The algebra **Fm** is called the *formula algebra* of type $\mathcal{L}$ and the elements of its domain Fm are the *formulas* of type $\mathcal{L}$. A *deductive system* of type $\mathcal{L}$ is a pair $\mathcal{S} = \langle \mathbf{Fm}, \vdash_{\mathcal{S}} \rangle$ where $\vdash_{\mathcal{S}}$ is a relation between sets of formulas and formulas such that

1. If $\varphi \in \Gamma$, then $\Gamma \vdash_{\mathcal{S}} \varphi$.
2. If $\Gamma \vdash_{\mathcal{S}} \varphi$ and for every $\psi \in \Gamma$, $\Delta \vdash_{\mathcal{S}} \psi$, then $\Delta \vdash_{\mathcal{S}} \varphi$.
3. If $\Gamma \vdash_{\mathcal{S}} \varphi$, then for any substitution σ, $\sigma[\Gamma] \vdash_{\mathcal{S}} \sigma(\varphi)$, where a *substitution* is an homomorphism from the formula algebra **Fm** into itself.

From 1. and 2. it follows that:

4. If $\Gamma \vdash_{\mathcal{S}} \varphi$ then for any ψ, $\Gamma \cup \{\psi\} \vdash_{\mathcal{S}} \varphi$.

The relation $\vdash_{\mathcal{S}}$ is called the *consequence relation* of $\mathcal{S}$.

A deductive system $\mathcal{S}$ is said to be *finitary* if for every set of formulas $\Gamma \cup \{\varphi\}$, $\Gamma \vdash_{\mathcal{S}} \varphi$ implies that $\Gamma' \vdash_{\mathcal{S}} \varphi$ for some finite $\Gamma' \subseteq \Gamma$. All the deductive systems we deal with in the paper are finitary, so from now on *when we say 'deductive system' we understand finitary deductive system*. A *theory* of a deductive system $\mathcal{S}$, or $\mathcal{S}$-*theory* for short, is a set of formulas Γ that is closed under the consequence relation of $\mathcal{S}$, that is, for every formula φ, if $\Gamma \vdash_{\mathcal{S}} \varphi$, then $\varphi \in \Gamma$. The set of $\mathcal{S}$-theories will be denoted by $\mathbf{Th}\mathcal{S}$.

A deductive system $\mathcal{S}$ is said to be *selfextensional* if its interderivability relation, denoted by $\dashv_{\mathcal{S}}\vdash$, is a congruence of the formula algebra, and it is said to be *Fregean* if for every set of formulas Γ, the interderivability relation modulo Γ, namely the relation defined by $\Gamma, \varphi \vdash_{\mathcal{S}} \psi$ and $\Gamma, \psi \vdash_{\mathcal{S}} \varphi$, is a congruence of the formula algebra.

Given a deductive system $\mathcal{S}$ and an algebra **A** with universe A, a set $F \subseteq A$ is an $\mathcal{S}$-*filter* if for any homomorphism h from **Fm** into **A**, any set of formulas Γ and any formula φ, if $\Gamma \vdash_{\mathcal{S}} \varphi$ and $h[\Gamma] \subseteq F$, then $h(\varphi) \in F$. If the deductive system is finitary the condition can be replaced by the corresponding condition that requires in addition that Γ is finite. We denote the set of all $\mathcal{S}$-filters of an algebra **A** by $\mathrm{Fi}_{\mathcal{S}}\mathbf{A}$. The set of all $\mathcal{S}$-filters of the formula algebra **Fm** is exactly the set $\mathbf{Th}\mathcal{S}$ of all the theories of $\mathcal{S}$. A *logical matrix*, abbreviatedly a matrix, is a pair $\langle \mathbf{A}, F \rangle$ where **A** is an algebra and F is a subset of the universe of **A**. A matrix $\mathcal{M} = \langle \mathbf{A}, F \rangle$ is a (*matrix*) *model* of a deductive system $\mathcal{S}$ if F is an $\mathcal{S}$-filter of **A**. Therefore the matrix models of $\mathcal{S}$ on the formula algebra are the matrices of the form $\langle \mathbf{Fm}, T \rangle$ where T is an $\mathcal{S}$-theory.

A *finitary closed-set system* on a set A is a family $\mathcal{C}$ of subsets of A that contains A and is closed under arbitrary intersections and under unions of upward directed subfamilies with respect to the inclusion relation. If $\mathcal{C}$ is a finitary closed-set system on a set A the closure operator $\mathrm{Clo}_{\mathcal{C}}$ on A associated with $\mathcal{C}$ is the closure operator defined by

$$\mathrm{Clo}_{\mathcal{C}}(X) = \bigcap \{F \in \mathcal{C} : X \subseteq F\},$$

for each $X \subseteq A$. The closure operator $\mathrm{Clo}_{\mathcal{C}}$ is finitary in the following sense: if $a \in \mathrm{Clo}_{\mathcal{C}}(X)$, then there is a finite $Y \subseteq X$ such that $a \in \mathrm{Clo}_{\mathcal{C}}(Y)$. Moreover, given a finitary closure operator C on a set A, the family $\mathcal{C}_C$ of all C-closed subsets X of A, i.e. such that $C(X) = X$, is a finitary closed-set system. It is well known that $\mathrm{Clo}_{\mathcal{C}_C} = C$, and that if $\mathcal{C}$ is a finitary closed-set system, then $\mathcal{C}_{\mathrm{Clo}_{\mathcal{C}}} = \mathcal{C}$.

A *generalized matrix*, g-matrix for short, is a pair $\boldsymbol{\mathcal{A}} = \langle \mathbf{A}, \mathcal{C} \rangle$ where $\mathbf{A}$ is an algebra and $\mathcal{C}$ is a finitary closed-set system on the universe A. Usually we will denote the closure operator determined by $\mathcal{C}$ on A by $\mathrm{Clo}_{\boldsymbol{\mathcal{A}}}$. We will also refer to the closed-set system of a matrix $\boldsymbol{\mathcal{A}}$ by $\mathcal{C}_{\boldsymbol{\mathcal{A}}}$. Notice that for every finitary deductive system $\mathcal{S}$ the structure $\langle \mathbf{Fm}, \mathbf{Th}\mathcal{S} \rangle$ is a generalized matrix. Its associated closure operator can be identified with the consequence relation $\vdash_{\mathcal{S}}$. The finitarity of $\mathcal{S}$ is essential for obtaining that $\mathbf{Th}\mathcal{S}$ is closed under unions of upwards directed subfamilies (by the inclusion order). Generalized matrices are exactly the finitary abstract logics of the monograph [8].

A generalized matrix $\boldsymbol{\mathcal{A}} = \langle \mathbf{A}, \mathcal{C} \rangle$ is a *generalized model*, g-model for short, of a deductive system $\mathcal{S}$ if every element of $\mathcal{C}$ is an $\mathcal{S}$-filter, that is, if $\mathcal{C} \subseteq \mathrm{Fi}_{\mathcal{S}}\mathbf{A}$. The g-matrix $\langle \mathbf{Fm}, \mathbf{Th}\mathcal{S} \rangle$ is obviously a g-model of the deductive system $\mathcal{S}$.

Given a generalized matrix $\boldsymbol{\mathcal{A}} = \langle \mathbf{A}, \mathcal{C} \rangle$, its *Tarski congruence*, denoted by $\widetilde{\boldsymbol{\Omega}}_{\mathbf{A}}(\mathcal{C})$, is the greatest congruence of $\mathbf{A}$ compatible with every element of $\mathcal{C}$, that is, such that for every $F \in \mathcal{C}$ and every $a, b \in A$, if $\langle a, b \rangle \in \widetilde{\boldsymbol{\Omega}}_{\mathbf{A}}(\mathcal{C})$ and $a \in F$, then $b \in F$. Sometimes we will denote $\widetilde{\boldsymbol{\Omega}}_{\mathbf{A}}(\mathcal{C})$ by $\widetilde{\boldsymbol{\Omega}}(\boldsymbol{\mathcal{A}})$. A generalized matrix $\boldsymbol{\mathcal{A}} = \langle \mathbf{A}, \mathcal{C} \rangle$ is *reduced* if its Tarski congruence is the identity. The class of the algebraic reducts of the reduced g-matrix models of $\mathcal{S}$, denoted by $\mathbf{Alg}\mathcal{S}$, is the class of algebras that according to the general algebraic semantics for deductive systems developed in [8] deserves to be considered the canonical class of algeras of $\mathcal{S}$. This class turns out to have the following simpler description that is the best for working purposes in the present paper:

$$\mathbf{Alg}\mathcal{S} = \{ \mathbf{A} : \langle \mathbf{A}, \mathrm{Fi}_{\mathcal{S}}\mathbf{A} \rangle \text{ is reduced} \}.$$

A *strict homomorphism* from a g-matrix $\boldsymbol{\mathcal{A}} = \langle \mathbf{A}, \mathcal{C} \rangle$ to a g-matrix $\boldsymbol{\mathcal{B}} = \langle \mathbf{B}, \mathcal{D} \rangle$ is a homomorphism from $\mathbf{A}$ to $\mathbf{B}$ such that $\mathcal{C} = \{ h^{-1}[F] : F \in \mathcal{D} \}$. Bijective strict homomorphisms are called *isomorphisms*, and surjective strict homomorphisms are called *bilogical morphisms* in [8]. If there is a strict surjective homomorphism from $\boldsymbol{\mathcal{A}} = \langle \mathbf{A}, \mathcal{C} \rangle$ onto $\boldsymbol{\mathcal{B}} = \langle \mathbf{B}, \mathcal{D} \rangle$ we write $\boldsymbol{\mathcal{A}} \succeq \boldsymbol{\mathcal{B}}$. In other words this means that $\boldsymbol{\mathcal{B}}$ is a strict homomorphic image of $\boldsymbol{\mathcal{A}}$. The most typical surjective strict homomorphisms appear in the process of reducing a g-matrix. Given a g-matrix $\boldsymbol{\mathcal{A}} = \langle \mathbf{A}, \mathcal{C} \rangle$, its *reduction* is the g-matrix $\boldsymbol{\mathcal{A}}^* = \langle \mathbf{A}/\widetilde{\boldsymbol{\Omega}}(\boldsymbol{\mathcal{A}}), \mathcal{C}/\widetilde{\boldsymbol{\Omega}}(\boldsymbol{\mathcal{A}}) \rangle$, where $\mathbf{A}/\widetilde{\boldsymbol{\Omega}}(\boldsymbol{\mathcal{A}})$ is the quotient algebra and $\mathcal{C}/\widetilde{\boldsymbol{\Omega}}(\boldsymbol{\mathcal{A}}) = \{ F/\widetilde{\boldsymbol{\Omega}}(\boldsymbol{\mathcal{A}}) : F \in \mathcal{C} \}$. The projection homomorphism $\pi : \mathbf{A} \to \mathbf{A}/\widetilde{\boldsymbol{\Omega}}(\boldsymbol{\mathcal{A}})$ is a surjective strict homomorphism from $\boldsymbol{\mathcal{A}}$ onto $\boldsymbol{\mathcal{A}}^*$. It is known ([8] Proposition 1.14) that if $\boldsymbol{\mathcal{A}} \succeq \boldsymbol{\mathcal{B}}$, then $\boldsymbol{\mathcal{A}}^*$ is isomorphic to $\boldsymbol{\mathcal{B}}^*$.

The notion of full g-model of a deductive system is one of the main notions introduced in [8]. A generalized matrix $\boldsymbol{\mathcal{A}} = \langle \mathbf{A}, \mathcal{C} \rangle$ is said to be a *basic full g-model* of a deductive system $\mathcal{S}$ if $\mathcal{C} = \mathrm{Fi}_{\mathcal{S}}\mathbf{A}$ and it is said to be a *full g-model* of $\mathcal{S}$ if there is a basic full g-model $\boldsymbol{\mathcal{B}}$ of $\mathcal{S}$ such that $\boldsymbol{\mathcal{A}} \succeq \boldsymbol{\mathcal{B}}$, that is, if one of its strict homomorphic images is a basic full g-model of $\mathcal{S}$. In [8] it is proved that if $\mathcal{A} \succeq \mathcal{B}$, then $\mathcal{A}$ is a full g-model of $\mathcal{S}$ iff $\mathcal{B}$ is so. Since the logical properties of the g-matrices are the properties which are preserved under strict homomorphisms, the class of full g-models is then the natural class of models one has to deal with. Moreover, it has many interesting properties that make it a very useful tool in the study of deductive systems, in particular for relating the algebraic treatment of a deductive system with the algebraic treatment of the several Gentzen calculi that define it. We will see this in the last section of the paper. Another important feature is that $\mathbf{Alg}\mathcal{S}$ is the class of the algebraic reducts of the reduced full g-models of $\mathcal{S}$.

Given a g-matrix $\mathcal{A} = \langle \mathbf{A}, \mathcal{C} \rangle$, its *Frege relation* $\mathbf{\Lambda}(\mathcal{A})$ is defined by

$$\langle a, b \rangle \in \mathbf{\Lambda}(\mathcal{A}) \quad \text{iff} \quad \mathrm{Clo}_{\mathcal{A}}(\{a\}) = \mathrm{Clo}_{\mathcal{A}}(\{b\}).$$

It is easy to see that $\widetilde{\mathbf{\Omega}}(\mathcal{A})$ is the largest congruence of $\mathbf{A}$ included in $\mathbf{\Lambda}(\mathcal{A})$. We will also denote the Frege relation of $\langle \mathbf{A}, \mathcal{C} \rangle$ by $\mathbf{\Lambda}_{\mathbf{A}}(\mathcal{C})$. For any deductive system $\mathcal{S}$, the interderivability relation ($\varphi \dashv_{\mathcal{S}} \vdash \psi$) is the Frege relation of the g-matrix $\langle \mathbf{Fm}, \mathbf{Th}\mathcal{S} \rangle$. Thus, a deductive system $\mathcal{S}$ is selfextensional iff the Frege relation of $\langle \mathbf{Fm}, \mathbf{Th}\mathcal{S} \rangle$ is a congruence. We denote the Frege relation of this g-matrix by $\mathbf{\Lambda}(\mathcal{S})$.

A deductive system $\mathcal{S}$ is said to be *fully selfextensional* when the Frege relation of every of its full g-models is a congruence. Thus, every fully selfextensional deductive system is selfextensional. The converse is not true as is shown in [1]. A deductive system $\mathcal{S}$ is said to be *fully Fregean*, if for every algebra $\mathbf{A}$ and every $\mathcal{S}$-filter F of $\mathbf{A}$, the Frege relation of the g-matrix $\langle \mathbf{A}, \mathrm{Fi}_{\mathcal{S}}\mathbf{A}^{F} \rangle$, where $\mathrm{Fi}_{\mathcal{S}}\mathbf{A}^{F} = \{G \in \mathrm{Fi}_{\mathcal{S}}\mathbf{A} : F \subseteq G\}$, is a congruence of $\mathbf{A}$. Clearly every fully Fregean deductive system is Fregean. In [1] it is shown that not every Fregean deductive system is fully Fregean.

Each deductive system has an associated variety, the variety $\mathsf{K}_{\mathcal{S}}$ generated by the algebra $\mathbf{Fm}/\widetilde{\mathbf{\Omega}}(\mathcal{S})$, which is the free algebra over a denumerable set of generators of $\mathsf{K}_{\mathcal{S}}$ (see [8], [13]). The class $\mathsf{K}_{\mathcal{S}}$ is called the *intrinsic variety* of $\mathcal{S}$ in [13]. This variety plays an important role in the proof of the main theorems of the paper. The variety $\mathsf{K}_{\mathcal{S}}$ can be described as the variety whose valid equations are the equations $\varphi \approx \psi$ such that $\langle \varphi, \psi \rangle \in \widetilde{\mathbf{\Omega}}(\mathcal{S})$. Thus

$$\mathsf{K}_{\mathcal{S}} \models \varphi \approx \psi \quad \text{iff} \quad \forall \delta \in Fm \ \forall p \in Var \ \delta(p/\varphi) \dashv_{\mathcal{S}} \vdash \delta(p/\psi).$$

In particular, if $\mathcal{S}$ is selfextensional,

$$\mathsf{K}_{\mathcal{S}} \models \varphi \approx \psi \quad \text{iff} \quad \varphi \dashv_{\mathcal{S}} \vdash \psi \quad \text{iff} \quad \langle \varphi, \psi \rangle \in \mathbf{\Lambda}(\mathcal{S}).$$

The relation between the classes of algebras $\mathbf{Alg}\mathcal{S}$ and $\mathsf{K}_{\mathcal{S}}$ associated with a deductive system $\mathcal{S}$ is that of inclusion: $\mathbf{Alg}\mathcal{S} \subseteq \mathsf{K}_{\mathcal{S}}$. Moreover, $\mathsf{K}_{\mathcal{S}}$ is the variety generated by $\mathbf{Alg}\mathcal{S}$. Thus, when $\mathbf{Alg}\mathcal{S}$ is a variety, the two classes are equal. This is for instance the case for classical logic and for intuitionistic logic. But there are deductive systems $\mathcal{S}$ for which the inclusion is proper: for example the algebraizable logic BCK is such that $\mathbf{Alg}\mathrm{BCK} \subsetneq \mathsf{K}_{\mathrm{BCK}}$.

To conclude this section on preliminaries we recall the definitions of algebraizable deductive system and regularly algebraizable deductive system. A set of formulas $\Delta(p,q)$ in at most two variables is a *set of equivalence formulas* for a deductive system $\mathcal{S}$ if for every algebra $\mathbf{A}$ and every $\mathcal{S}$-filter F of $\mathbf{A}$, $\mathbf{\Omega_A}(F) = \{\langle a,b \rangle \in A \times A : \Delta^{\mathbf{A}}(a,b) \subseteq F\}$. A set of equations $\tau(p)$ in at most one variable is a *set of defining equations* for a deductive system $\mathcal{S}$ if for every algebra $\mathbf{A} \in \mathbf{Alg}\mathcal{S}$ the least $\mathcal{S}$-filter of $\mathbf{A}$ is the set of solutions in $\mathbf{A}$ of the equations in $\tau(p)$, that is the set $\{a \in A : \mathbf{A} \models \tau(p)[a]\}$. A deductive system $\mathcal{S}$ is *algebraizable* if it has a set of equivalence formulas and a set of defining equations. An algebraizable deductive system $\mathcal{S}$ is *regularly algebraizable* if for any set of equivalence formulas $\Delta(p,q)$ the corresponding G-rule holds, that is, $p, q \vdash_{\mathcal{S}} \Delta(p,q)$.

3. Hilbert-based deductive systems

Let $\mathcal{S}$ be a deductive system; we say that a binary term $\Rightarrow$ has *the deduction-detachment property*, or is a *deduction-detachment term*, if for every set of formulas Γ and every formulas φ, ψ,

$$\Gamma, \varphi \vdash_{\mathcal{S}} \psi \quad \text{iff} \quad \Gamma \vdash_{\mathcal{S}} \varphi \Rightarrow \psi.$$

A deductive system $\mathcal{S}$ is said to have the *uniterm deduction-detachement property* (u-DDP) relative to a binary term $\Rightarrow$ if the term $\Rightarrow$ has the deduction-detachment property, and it is said to have the *uniterm deduction-detachement property* if it has the uniterm deduction-detachement property relative to some binary term.

Notice that if $\mathcal{S}$ has the u-DDP relative to $\Rightarrow$ and relative to $\Rightarrow'$ then

$$p \Rightarrow q \dashv_{\mathcal{S}} \vdash p \Rightarrow' q.$$

Thus, if $\mathcal{S}$ is selfextensional, for all formulas φ, ψ, $\langle \varphi \Rightarrow \psi, \varphi \Rightarrow' \psi \rangle \in \widetilde{\mathbf{\Omega}}(\mathcal{S})$.

Definition 1. A class K of algebras is *Hilbert-based relative to a binary term* $\Rightarrow$ if the following equations are valid in K:

H1. $x \Rightarrow x \approx y \Rightarrow y$
H2. $(x \Rightarrow x) \Rightarrow x \approx x$
H3. $x \Rightarrow (y \Rightarrow z) \approx (x \Rightarrow y) \Rightarrow (x \Rightarrow z)$
H4. $(x \Rightarrow y) \Rightarrow ((y \Rightarrow x) \Rightarrow y) \approx (y \Rightarrow x) \Rightarrow ((x \Rightarrow y) \Rightarrow x)$.

Thus K is Hilbert-based relative to a binary term $\Rightarrow$ if for every $\mathbf{A} \in \mathsf{K}$ the algebra $\langle A, \Rightarrow^{\mathbf{A}} \rangle$ is a Hilbert algebra. We will refer to the equations (H1)-(H4) as the *Hilbert equations*.

Definition 2. We say that a class of algebras is *Hilbert-based* if it is Hilbert-based relative to some binary term.

A class of algebras Q is said to be *pointed* if there is a term $\varphi(x_0, \ldots, x_n)$ with the property that $\varphi(x_0, \ldots, x_n) \approx \varphi(y_0, \ldots, y_n)$ is valid in Q for all variables $y_0, \ldots, y_n$. Thus for every $\mathbf{A} \in \mathsf{Q}$ and any two valuations v, v' on $\mathbf{A}$, $v(\varphi) = v'(\varphi)$. Such a term is called a *constant term* since it behaves like a constant. Once fixed we will usually refer to it by $\top$. Any Hilbert-based class of algebras K is pointed, because the term $x \Rightarrow x$ is a constant term, that is, for every algebra $\mathbf{A} \in \mathsf{K}$ and all $a, b \in A$, $a \Rightarrow a = b \Rightarrow b$. Let us denote the constant interpretation of $x \Rightarrow x$ in $\mathbf{A}$ by $1^{\mathbf{A}}$ or simply by 1. Given a Hilbert-based class of algebras K and an algebra $\mathbf{A} \in \mathsf{K}$ we define the relation $\leq^{\mathbf{A}}$ on A by

$$a \leq^{\mathbf{A}} b \quad \text{iff} \quad a \Rightarrow b = 1. \tag{1}$$

We will omit the superscript in $\leq^{\mathbf{A}}$ when no confusion is likely.

Definition 3. Given a Hilbert-based class of algebras K relative to $\Rightarrow$ and an algebra $\mathbf{A} \in \mathsf{K}$, a set $F \subseteq A$ is an $\Rightarrow$-*implicative filter* of $\mathbf{A}$ if

1. $1 \in F$
2. for all $a, b \in A$, if $a \Rightarrow b \in F$ and $a \in F$, then $b \in F$.

Definition 4. A deductive system $\mathcal{S}$ is *Hilbert-based* relative to a binary term $\Rightarrow$ and a class of algebras K which is Hilbert-based relative to $\Rightarrow$ if for all formulas $\varphi_0, \ldots, \varphi_n, \varphi$,

$$\varphi_0, \ldots, \varphi_n \vdash_{\mathcal{S}} \varphi \quad \text{iff} \quad \forall \mathbf{A} \in \mathsf{K} \; \forall v \in \mathrm{Hom}(\mathbf{Fm}, \mathbf{A}), \tag{2}$$

$$v(\varphi_0 \Rightarrow (\ldots \Rightarrow (\varphi_{n-1} \Rightarrow (\varphi_n \Rightarrow \varphi)\ldots)) = 1$$

and

$$\vdash_{\mathcal{S}} \varphi \quad \text{iff} \quad \forall \mathbf{A} \in \mathsf{K} \; \forall v \in \mathrm{Hom}(\mathbf{Fm}, \mathbf{A}), v(\varphi) = 1. \tag{3}$$

Property (2) is independent of the order in which the formulas $\varphi_0, \ldots, \varphi_n$ are taken because for any permutation π of $\{0, \ldots, n\}$, $v(\varphi_0 \Rightarrow (\ldots \Rightarrow (\varphi_{n-1} \Rightarrow (\varphi_n \Rightarrow \varphi)\ldots)) = 1$ iff $v(\varphi_{\pi(0)} \Rightarrow (\ldots \Rightarrow (\varphi_{\pi(n-1)} \Rightarrow (\varphi_{\pi(n)} \Rightarrow \varphi)\ldots)) = 1$. In the sequel when we say $\mathcal{S}$ is Hilbert-based relative to $\Rightarrow$ and K we assume that K is Hilbert-based relative to $\Rightarrow$.

We say that $\mathcal{S}$ is *Hilbert-based* if there is a binary term $\Rightarrow$ and a Hilbert-based class of algebras relative to $\Rightarrow$ such that $\mathcal{S}$ is Hilbert-based relative to them.

If $\mathcal{S}$ is Hilbert-based relative to $\Rightarrow$ and K then it is also Hilbert-based relative to $\Rightarrow$ and the variety generated by K. The remark in the next proposition together with Corollary 8 show that if $\mathcal{S}$ is Hilbert-based there is only one variety relative to which it is Hilbert-based. We can denote it by $\mathsf{V}(\mathcal{S})$.

Proposition 5. *If $\mathcal{S}$ is a Hilbert-based deductive system relative to K and $\Rightarrow$ and relative to K' and $\Rightarrow'$ then the varieties generated by K and by K' are the same.*

Proof. Assume that $\varphi \approx \psi$ is an equation valid in K. Then for every $\mathbf{A} \in \mathsf{K}$ and every $v \in \mathrm{Hom}(\mathbf{Fm}, \mathbf{A})$, $v(\varphi) = v(\psi)$. Therefore $v(\varphi \Rightarrow \psi) = v(\psi \Rightarrow \varphi) = 1$. Hence, $\varphi \dashv_{\mathcal{S}} \vdash \psi$. Then for every $\mathbf{A} \in \mathsf{K}'$ and every $v \in \mathrm{Hom}(\mathbf{Fm}, \mathbf{A})$, $v(\varphi \Rightarrow' \psi) = v(\psi \Rightarrow' \varphi) = 1$. Therefore, $v(\varphi) = v(\psi)$. Hence, $\varphi \approx \psi$ is valid in K'. Analogously we obtain that the equations valid in K' are valid in K. $\square$

Remark 6. Condition (2) in the definition of Hilbert-based deductive system implies that if $\mathcal{S}$ is Hilbert-based relative to K, then $\varphi \dashv_{\mathcal{S}} \vdash \psi$ iff $\mathsf{K} \models \varphi \approx \psi$. Therefore, $\varphi \dashv_{\mathcal{S}} \vdash \psi$ iff $\mathsf{V}(\mathcal{S}) \models \varphi \approx \psi$.

Notice that the definition of Hilbert-based deductive system implies that there cannot be two different deductive systems which are Hilbert-based relative to the same variety.

Proposition 7. *If $\mathcal{S}$ is Hilbert-based relative to $\Rightarrow$, then*

1. *$\mathcal{S}$ is selfextensional,*
2. *$\Rightarrow$ has the deduction-detachment property in $\mathcal{S}$,*
3. *the variety $V(\mathcal{S})$ is the intrinsic variety $\mathsf{K}_{\mathcal{S}}$, thus $\mathcal{S}$ is Hilbert-based relative to its intrinsic variety.*

Proof. Assume that $\mathcal{S}$ is Hilbert-based relative to $\Rightarrow$ and the variety K. (1) Let us see that $\mathbf{\Lambda}(\mathcal{S})$ is a congruence. If $\varphi \dashv_{\mathcal{S}}\vdash \psi$ then $V(\mathcal{S}) \models \varphi \approx \psi$, therefore for every formula δ and every variable p, $V(\mathcal{S}) \models \delta(p/\varphi) = \delta(p/\psi)$, which, by the above remark, implies that $\delta(p/\psi) \dashv_{\mathcal{S}}\vdash \delta(p/\psi)$. (2) Let us show that $\mathcal{S}$ has u-DDP relative to $\Rightarrow$. Assume that $\Gamma, \varphi \vdash_{\mathcal{S}} \psi$. Let $\varphi_0, \ldots, \varphi_{n-1} \in \Gamma$ such that $\varphi_0, \ldots, \varphi_{n-1}, \varphi \vdash_{\mathcal{S}} \psi$ or $\varphi \vdash_{\mathcal{S}} \psi$. Therefore for every $\mathbf{A} \in \mathsf{K}$ and every $v \in \mathrm{Hom}(\mathbf{Fm}, \mathbf{A})$, $v(\varphi_0 \Rightarrow (\ldots \Rightarrow (\varphi \Rightarrow \psi)\ldots)) = 1$ or for every $\mathbf{A} \in \mathsf{K}$ and every $v \in \mathrm{Hom}(\mathbf{Fm}, \mathbf{A})$, $v(\varphi \Rightarrow \psi) = 1$. Hence, $\Gamma \vdash_{\mathcal{S}} \varphi \Rightarrow \psi$. On the other hand, if $\Gamma \vdash_{\mathcal{S}} \varphi \Rightarrow \psi$, it is also easy to seen that $\Gamma, \varphi \vdash_{\mathcal{S}} \psi$. (3) From the definition of the intrinsic variety of $\mathcal{S}$ and the selfextensionality of $\mathcal{S}$ we have

$$\mathsf{K}_{\mathcal{S}} \models \varphi \approx \psi \quad \text{iff} \quad \varphi \dashv_{\mathcal{S}}\vdash \psi.$$

Therefore, by the above remark, $V(\mathcal{S}) = \mathsf{K}_{\mathcal{S}}$. $\square$

Corollary 8. *If $\mathcal{S}$ is a Hilbert-based deductive system relative to $\Rightarrow$ and also relative to $\Rightarrow'$, then for every φ, ψ, $\varphi \Rightarrow \psi \dashv_{\mathcal{S}}\vdash \varphi \Rightarrow' \psi$, and $\langle \varphi \Rightarrow \psi, \psi \Rightarrow \varphi \rangle \in \widetilde{\mathbf{\Omega}}(\mathcal{S})$.*

Proof. The first part follows immediately from the fact that, under the assumptions, by the above theorem both $\Rightarrow$ and $\Rightarrow'$ are deduction-detachment terms for $\mathcal{S}$. The second part follows from the selfextensionality of $\mathcal{S}$. $\square$

Proposition 5 and Corollary 8 allow us to speak simply of Hilbert-based deductive systems when convenient.

Theorem 9. *A deductive system $\mathcal{S}$ is selfextensional and has the uniterm deduction-detachment property iff it is Hilbert-based.*

Proof. By the proposition above we have the implication from right to left. To prove the other implication assume that $\mathcal{S}$ is selfextensional and has u-DDP relative to $\Rightarrow$. Let us consider the algebra $\mathbf{Fm}/\mathbf{\Lambda}(\mathcal{S})$. It is not difficult to check that $\{\mathbf{Fm}/\mathbf{\Lambda}(\mathcal{S})\}$ is Hilbert-based relative to $\Rightarrow$. Moreover,

$$
\begin{aligned}
\varphi_0, \ldots, \varphi_{n-1} \vdash_{\mathcal{S}} \varphi \quad &\text{iff} \quad \vdash_{\mathcal{S}} \varphi_0 \Rightarrow (\ldots \Rightarrow (\varphi_{n-1} \Rightarrow \varphi)\ldots) \\
&\text{iff} \quad \varphi \Rightarrow \varphi \dashv_{\mathcal{S}}\vdash \varphi_0 \Rightarrow (\ldots \Rightarrow (\varphi_{n-1} \Rightarrow \varphi)\ldots) \\
&\text{iff} \quad \mathbf{Fm}/\mathbf{\Lambda}(\mathcal{S}) \models \varphi_0 \Rightarrow (\ldots \Rightarrow (\varphi_{n-1} \Rightarrow \varphi)\ldots) \approx 1 \\
&\text{iff} \quad \forall v \in \mathrm{Hom}(\mathbf{Fm}, \mathbf{Fm}/\mathbf{\Lambda}(\mathcal{S})), \\
&\qquad v(\varphi_0 \Rightarrow (\ldots \Rightarrow (\varphi_{n-1} \Rightarrow \varphi)\ldots)) = 1,
\end{aligned}
$$

and

$$\vdash_{\mathcal{S}} \varphi \quad \text{iff} \quad \varphi \Rightarrow \varphi \dashv_{\mathcal{S}} \vdash \varphi$$
$$\text{iff} \quad \mathbf{Fm}/\mathbf{\Lambda}(\mathcal{S}) \models \varphi \approx 1$$
$$\text{iff} \quad \forall v \in \mathrm{Hom}(\mathbf{Fm}, \mathbf{Fm}/\mathbf{\Lambda}(\mathcal{S})), v(\varphi) = 1.$$

Thus $\mathcal{S}$ is Hilbert-based relative to the variety $\mathsf{K}_{\mathcal{S}} = \mathsf{V}(\mathbf{Fm}/\mathbf{\Lambda}(\mathcal{S}))$ and $\Rightarrow$. $\qquad\square$

Let K be a Hilbert-based variety relative to $\Rightarrow$. We define the deductive system $\mathcal{S}_{\mathsf{K}}^{\Rightarrow}$ as follows:

$$\varphi_0, \ldots, \varphi_n \vdash_{\mathcal{S}_{\mathsf{K}}^{\Rightarrow}} \varphi \quad \text{iff} \quad \forall \mathbf{A} \in \mathsf{K} \; \forall v \in \mathrm{Hom}(\mathbf{Fm}, \mathbf{A})$$

$$v(\varphi_0 \Rightarrow (\ldots \Rightarrow (\varphi_{n-1} \Rightarrow (\varphi_n \Rightarrow \varphi))\ldots)) = 1$$

and

$$\vdash_{\mathcal{S}_{\mathsf{K}}^{\Rightarrow}} \varphi \quad \text{iff} \quad \forall \mathbf{A} \in \mathsf{K} \; \forall v \in \mathrm{Hom}(\mathbf{Fm}, \mathbf{A}) \; v(\varphi) = 1.$$

From the definitions it follows straightforwardly that:

Proposition 10. *For every Hilbert-based variety K relative to $\Rightarrow$ the deductive system $\mathcal{S}_{\mathsf{K}}^{\Rightarrow}$ is Hilbert-based relative to K and $\Rightarrow$ and $\mathsf{V}(\mathcal{S}_{\mathsf{K}}^{\Rightarrow}) = \mathsf{K}$.*

Let us fix a binary term $\Rightarrow$. From the results above it follows that there is a bijection between the Hilbert-based deductive systems relative to $\Rightarrow$ and the Hilbert-based varieties relative to $\Rightarrow$. This bijection is in fact a dual isomorphism when we order the deductive systems by extension and the varieties by the relation of being a subvariety.

A Hilbert-based deductive system $\mathcal{S}$ is determined exactly by its Frege relation, that is by the pairs of formulas $\langle \varphi, \psi \rangle$ which are interderivable in $\mathcal{S}$, and the extension relation between Hilbert-based deductive systems corresponds to the inclusion relation between their Frege relations.

Proposition 11. *Let $\mathcal{S}$ and $\mathcal{S}'$ be two Hilbert-based deductive systems. Then*

$$\Lambda(\mathcal{S}) \subseteq \Lambda(\mathcal{S}') \quad \text{iff} \quad \mathcal{S}' \text{ is an extension of } \mathcal{S}.$$

Therefore, if $\Lambda(\mathcal{S}) = \Lambda(\mathcal{S}')$, then $\mathcal{S} = \mathcal{S}'$.

Proof. It is clear that if $\mathcal{S}'$ is an extension of $\mathcal{S}$ then $\Lambda(\mathcal{S}) \subseteq \Lambda(\mathcal{S}')$. Assume that $\Lambda(\mathcal{S}) \subseteq \Lambda(\mathcal{S}')$. Then

$$
\begin{aligned}
\varphi_0, \ldots, \varphi_{n-1} \vdash_{\mathcal{S}} \varphi \quad &\text{iff} \quad \vdash_{\mathcal{S}} \varphi_0 \Rightarrow (\ldots \Rightarrow (\varphi_{n-1} \Rightarrow \varphi)\ldots) \\
&\text{iff} \quad \varphi_0 \Rightarrow (\ldots \Rightarrow (\varphi_{n-1} \Rightarrow \varphi)\ldots) \dashv_{\mathcal{S}} \vdash \varphi \Rightarrow \varphi \\
&\text{then} \quad \varphi_0 \Rightarrow (\ldots \Rightarrow (\varphi_{n-1} \Rightarrow \varphi)\ldots) \dashv_{\mathcal{S}'} \vdash \varphi \Rightarrow \varphi \\
&\text{iff} \quad \vdash_{\mathcal{S}'} \varphi_0 \Rightarrow (\ldots \Rightarrow (\varphi_{n-1} \Rightarrow \varphi)\ldots). \\
&\text{iff} \quad \varphi_0, \ldots, \varphi_{n-1} \vdash_{\mathcal{S}'} \varphi
\end{aligned}
$$

and

$$\vdash_{\mathcal{S}} \varphi \quad \text{iff} \quad \varphi \dashv_{\mathcal{S}} \vdash \varphi \Rightarrow \varphi$$
$$\text{then} \quad \varphi \dashv_{\mathcal{S}'} \vdash \varphi \Rightarrow \varphi$$
$$\text{iff} \quad \vdash_{\mathcal{S}'} \varphi.$$
$$\text{iff} \quad \vdash_{\mathcal{S}'} \varphi.$$

Thus $\mathcal{S}'$ is an extension of $\mathcal{S}$. $\qquad\qquad\square$

To state the theorem, given an algebraic similarity type $\mathcal{L}$ with a binary term $\Rightarrow$ let $\mathsf{K}_{\Rightarrow}^{\mathcal{L}}$ denote the variety axiomatized by the Hilbert equations (E1)-(E4).

Theorem 12. *For every algebraic similarity type and every one of its binary terms $\Rightarrow$ there is a dual isomorphism between the set of Hilbert-based deductive systems relative to $\Rightarrow$, ordered by extension, and the set of all subvarieties of the variety $\mathsf{K}_{\Rightarrow}^{\mathcal{L}}$, ordered by inclusion. The isomorphism is given by $\mathcal{S} \mapsto \mathsf{K}_{\mathcal{S}}$.*

Proof. Recall that for a selfextensional deductive system $\mathcal{S}$ the Frege relation determines exactly the equations that hold in the variety $\mathsf{K}_{\mathcal{S}}$, that is, $\langle \varphi, \psi \rangle \in \Lambda(\mathcal{S})$ iff $\varphi \approx \psi$ holds in $\mathsf{K}_{\mathcal{S}}$. Thus if $\mathcal{S}$ and $\mathcal{S}'$ are Hilbert-based relative to $\Rightarrow$ and $\mathsf{K}_{\mathcal{S}} = \mathsf{K}_{\mathcal{S}'}$, then $\Lambda(\mathcal{S}) = \Lambda(\mathcal{S}')$. By Proposition 11, $\mathcal{S} = \mathcal{S}'$. Thus the function $\mathcal{S} \mapsto \mathsf{K}_{\mathcal{S}}$ is injective. Clearly it is onto since by Proposition 10 every Hilbert-based variety K defines a Hilbert-based deductive system whose class of algebras is K.

From Proposition 11 it follows that $\mathcal{S}$ is an extension of $\mathcal{S}'$ iff $\mathsf{K}_{\mathcal{S}}$ is a subvariety of $\mathsf{K}_{\mathcal{S}'}$. Therefore the function $\mathcal{S} \mapsto \mathsf{K}_{\mathcal{S}}$ is a dual isomorphism. $\qquad\square$

We proceed to show that for any selfextensional deductive systems $\mathcal{S}$ with the uniterm deduction-detachment property, its class of algebras $\mathbf{Alg}\mathcal{S}$ is a variety, indeed we will show that it is the intrinsic variety of $\mathcal{S}$. This will give the following reformulation of the theorem above.

Theorem 13. *For every algebraic similarity type and every one of its binary terms $\Rightarrow$ the map $\mathcal{S} \mapsto \mathbf{Alg}\mathcal{S}$ is a dual isomorphism between the set of Hilbert-based deductive systems relative to $\Rightarrow$, ordered by extension, and the set of all subvarieties of the variety $\mathsf{K}_{\Rightarrow}^{\mathcal{L}}$, ordered by inclusion.*

Lemma 14. *Let $\mathcal{S}$ be a Hilbert-based deductive system relative to $\Rightarrow$. Then for every algebra $\mathbf{A} \in \mathsf{K}_{\mathcal{S}}$, the $\mathcal{S}$-filters of $\mathbf{A}$ are the implicative filters of $\mathbf{A}$.*

Proof. Let $\mathcal{S}$ be a deductive system which is Hilbert-based relative to $\Rightarrow$. Let $\mathbf{A} \in \mathsf{K}_{\mathcal{S}}$ and let F be an $\mathcal{S}$-filter of $\mathbf{A}$. Since $\vdash_{\mathcal{S}} p \Rightarrow p$ and $p, p \Rightarrow q \vdash_{\mathcal{S}} q$ it is clear that F is an implicative filter. Conversely, if F is an implicative filter of $\mathbf{A}$, assume that $\varphi_0, \ldots, \varphi_{n-1} \vdash_{\mathcal{S}} \varphi$ and that $v \in \mathrm{Hom}(\mathbf{Fm}, \mathbf{A})$ is such that $v(\varphi_0), \ldots, v(\varphi_{n-1}) \in F$, then we have $v(\varphi_0 \Rightarrow (\ldots \Rightarrow (\varphi_{n-1} \Rightarrow \psi)\ldots)) = 1 \in F$. Thus, we conclude that $v(\psi) \in F$ as well. This shows that F is an $\mathcal{S}$-filter. $\qquad\square$

Lemma 15. *Let $\mathcal{S}$ be a Hilbert-based deductive system. Then for every algebra $\mathbf{A} \in \mathsf{K}_{\mathcal{S}}$, the Frege relation of the g-matrix $\langle \mathbf{A}, \mathrm{Fi}_{\mathcal{S}}\mathbf{A} \rangle$ is the identity and therefore the matrix is reduced and has the congruence property.*

Proof. Let $a, b \in A$ be different elements. Consider the sets $F_a = \{c \in A : a \Rightarrow c = 1\}$ and $F_b = \{c \in A : b \Rightarrow c = 1\}$. It is easy to see that they are implicative filters, hence by Lemma 14 they belong to $\mathrm{Fi}_{\mathcal{S}}\mathbf{A}$. Clearly, if $F_a = F_b$, then $a \Rightarrow b = b \Rightarrow a = 1$. Hence, $a = b$. Thus, if $a \neq b$, $a \notin F_b$ or $b \notin F_a$. Hence $\langle a, b \rangle \notin \mathbf{\Lambda_A}(\mathrm{Fi}_{\mathcal{S}}\mathbf{A})$. This shows that the Frege relation of the g-matrix $\langle \mathbf{A}, \mathrm{Fi}_{\mathcal{S}}\mathbf{A} \rangle$ is the identity, which implies that $\langle \mathbf{A}, \mathrm{Fi}_{\mathcal{S}}\mathbf{A} \rangle$ is reduced and has the congruence property. □

Theorem 16. *If $\mathcal{S}$ is a Hilbert-based deductive system then*
1. $\mathbf{Alg}\mathcal{S} = \mathsf{K}_{\mathcal{S}} = \mathsf{V}(\mathcal{S})$.
2. $\mathbf{Alg}\mathcal{S}$ *is a variety.*
3. $\mathcal{S}$ *is Hilbert-based relative to* $\mathbf{Alg}\mathcal{S}$.

Proof. 1. We know that $\mathbf{Alg}\mathcal{S} \subseteq \mathsf{K}_{\mathcal{S}}$ always holds. By the previous lemma we obtain that $\mathsf{K}_{\mathcal{S}} \subseteq \mathbf{Alg}\mathcal{S}$. 2 follows from 1 because $\mathsf{K}_{\mathcal{S}}$ is a variety. 3 follows from 1 and item 3 in Proposition 7. □

In [8] it is proved that every selfextensional deductive system with the u-DDP is fully selfextensional. We are going to give a proof of this fact that does not make use of Gentzen systems.

Theorem 17. *Every selfextensional deductive system with the u-DDP is fully selfextensional.*

Proof. Let $\mathcal{S}$ be a selfextensional deductive system with the u-DDP relative to $\Rightarrow$. By Theorem 9 and the corollary to its proof, it is Hilbert-based relative to $\Rightarrow$ and $\mathsf{K}_{\mathcal{S}}$. By Theorem 16, $\mathsf{K}_{\mathcal{S}} = \mathbf{Alg}\mathcal{S}$. Thus, by Lemma 15, if $\mathbf{A} \in \mathbf{Alg}\mathcal{S}$, $\mathbf{\Lambda_A}(\mathrm{Fi}_{\mathcal{S}}\mathbf{A})$ is the identity relation on A; and therefore it is a congruence. If $\mathcal{A}$ is a full g-model of $\mathcal{S}$, its reduction $\mathcal{A}^*$ is of the form $\langle \mathbf{B}, \mathrm{Fi}_{\mathcal{S}}\mathbf{B} \rangle$ for some $\mathbf{B} \in \mathbf{Alg}\mathcal{S}$. By what we have just proved this g-matrix has the congruence property and by Proposition 2.40 in [8] this property is preserved by surjective strict homomorphisms (bilogical morphisms). Therefore, $\mathcal{A}$ has the congruence property, too. We can conclude that $\mathcal{S}$ is fully selfextensional. □

Given a pointed quasivariety Q with constant term $\top$ the $\top$-*assertional logic* of Q is the deductive system $\mathcal{S}^{ASL}\mathsf{Q} = \langle \mathbf{Fm}, \vdash_{\mathcal{S}^{ASL}\mathsf{Q}} \rangle$ defined by

$$\Gamma \vdash_{\mathcal{S}^{ASL}\mathsf{Q}} \varphi \quad \text{iff} \quad \forall \mathbf{A} \in \mathsf{Q} \; \forall v \in \mathrm{Hom}(\mathbf{Fm}, \mathsf{Q})(v[\Gamma] \subseteq \{1^{\mathbf{A}}\} \Longrightarrow v(\varphi) = 1^{\mathbf{A}}),$$

where $1^{\mathbf{A}}$ is the interpretation of the constant term $\top$ in $\mathbf{A}$. We will characterize the selfextensional deductive systems $\mathcal{S}$ with the deduction-detachment property such that $\mathcal{S}$ is equal to the $\top$-assertional logic of $\mathbf{Alg}\mathcal{S}$, where $\top$ is $x \Rightarrow x$ for the deduction-detachment term $\Rightarrow$ of $\mathcal{S}$.

A pointed quasivariety Q with constant term $\top$ is said to be *relatively point-regular* if for every $\mathbf{A} \in \mathsf{Q}$ and all the congruences θ, θ' of $\mathbf{A}$ such that $\mathbf{A}/\theta, \mathbf{A}/\theta' \in \mathsf{Q}$, $1^{\mathbf{A}}/\theta = 1^{\mathbf{A}}/\theta'$ implies $\theta = \theta'$.

The regularly algebraizable deductive systems are the assertional logics of the pointed quasivarieties that are reletively-point regular. If $\mathcal{S}$ is a regularly algebraizable deductive system, then all theorems of $\mathcal{S}$ are equivalent, so any one can be taken as the designated constant term $\top$ and $\mathcal{S}$ is the $\top$-assertional logic of $\mathbf{Alg}\mathcal{S}$.

Theorem 18 ([6] Thm. 1.34). *A deductive system $\mathcal{S}$ is regularly algebraizable iff $\mathbf{Alg}\mathcal{S}$ is a pointed and relatively-point regular quasivariety and $\mathcal{S} = \mathcal{S}^{ASL}\mathbf{Alg}\mathcal{S}$.*

Theorem 19. *Let $\mathcal{S}$ be a selfextensional deductive system with the uniterm deduction-detachment property. Then, $\mathcal{S}$ is regularly algebraizable iff $\mathcal{S} = \mathcal{S}^{ASL}\mathbf{Alg}\mathcal{S}$.*

Proof. Let $\mathcal{S}$ be a selfextensional deductive system with the uniterm deduction-detachment property for $\Rightarrow$. Then $\mathbf{Alg}\mathcal{S}$ is a pointed variety. By the theorem above, if $\mathcal{S}$ is regularly algebraizable, then $\mathcal{S} = \mathcal{S}^{ASL}\mathbf{Alg}\mathcal{S}$. Assume now that $\mathcal{S} = \mathcal{S}^{ASL}\mathbf{Alg}\mathcal{S}$. We show that $\mathbf{Alg}\mathcal{S}$ is point-regular. Let $\mathbf{A} \in \mathbf{Alg}\mathcal{S}$ and let $\theta, \theta' \in \mathrm{Co}\mathbf{A}$ be such that $1/\theta = 1/\theta'$. Suppose that $\langle a, b \rangle \in \theta$. Then $\langle a \Rightarrow a, a \Rightarrow b \rangle \in \theta$, that is $\langle 1, a \Rightarrow b \rangle \in \theta$. Thus, Thus, $\langle 1, a \Rightarrow b \rangle \in \theta'$. Similarly, $\langle 1, b \Rightarrow a \rangle \in \theta'$. Hence, in $\mathbf{A}/\theta'$, $1 = a/\theta' \Rightarrow b/\theta'$ and $1 = b/\theta' \Rightarrow a/\theta'$. Since $\mathbf{Alg}\mathcal{S}$ is a variety, $\mathbf{A}/\theta' \in \mathbf{Alg}\mathcal{S}$. Therefore, $\langle \mathbf{A}/\theta', \Rightarrow \rangle$ is a Hilbert algebra. Hence $a/\theta' = b/\theta'$. Thus, $\langle a, b \rangle \in \theta'$. By a similar argument we get the other inclusion. Now by the above theorem $\mathcal{S}$ is regularly algebraizable. $\square$

In [8] (Thm. 3.18 and Prop. 3.20) it is shown that for any fully selfextensional deductive system $\mathcal{S}$, $\mathcal{S}$ is a Fregean, protoalgebraic deductive system with theorems iff $\mathcal{S}$ is regularly algebraizable. Thus, since every deductive system with the deduction-detachment property is protoalgebraic and has theorems, Theorem 17 implies that a selfextensional deductive system with the uniterm deduction-detachment property is Fregean iff it is regularly algebraizable. Moreover, Czelakowski and Pigozzi prove in [6] (Corollary 80) that if a deductive system is protoalgebraic and Fregean, then it is fully Fregean. Thus we have the equivalences below:

Theorem 20. *Let $\mathcal{S}$ be a selfextensional deductive system with the uniterm deduction-detachment property. The following statements are equivalent:*

1. *$\mathcal{S}$ is Fregean;*
2. *$\mathcal{S}$ is fully Fregean;*
3. *$\mathcal{S}$ is regularly algebraizable;*
4. *$\mathcal{S} = \mathcal{S}^{ASL}\mathbf{Alg}\mathcal{S}$.*

4. Fregean logics with a deduction-detachment theorem

We will obtain some results on Fregean logics with a uniterm deduction-detachment theorem using our results on selfextensional logics with a deduction-detachment theorem. In particular we give a different proof of the second part of Theorem 66 of Czelakowski and Pigozzi in [6].

Lemma 21 ([6]). *If $\mathcal{S}$ is a deductive system and $\Rightarrow$ is a deduction-detachment term for $\mathcal{S}$, then $\mathcal{S}$ is Fregean iff the set $\{p \Rightarrow q, q \Rightarrow p\}$ is an equivalence set of formulas for $\mathcal{S}$.*

If $\mathcal{S}$ is a selfextensional deductive system with a DDT-term $\Rightarrow$, then for all formulas $\varphi_0, \ldots, \varphi_n, \varphi$ and every permutation π of $\{0, \ldots, n\}$,

$$\varphi_0 \Rightarrow (\varphi_1 \Rightarrow (\ldots \Rightarrow (\varphi_n \Rightarrow \varphi) \ldots)) \dashv_{\mathcal{S}} \vdash \varphi_{\pi(0)} \Rightarrow (\varphi_{\pi(1)} \Rightarrow (\ldots \Rightarrow (\varphi_{\pi(n)} \Rightarrow \varphi) \ldots)).$$

In general, for every full model $\langle \mathbf{A}, \mathrm{Fi}_{\mathcal{S}}\mathbf{A} \rangle$ the analogous result holds, that is for every $a_0, \ldots, a_n, b \in A$ and every permutation π of $\{0, \ldots, n\}$, the sets $\mathrm{Clo}_{\mathrm{Fi}_{\mathcal{S}}\mathbf{A}}(a_0 \Rightarrow (a_1 \Rightarrow (\ldots \Rightarrow (a_n \Rightarrow b) \ldots)))$ and $\mathrm{Clo}_{\mathrm{Fi}_{\mathcal{S}}\mathbf{A}}(a_{\pi(0)} \Rightarrow (a_{\pi(1)} \Rightarrow (\ldots \Rightarrow (a_{\pi(n)} \Rightarrow b) \ldots)))$ are equal.

Given a sequence $\varphi_0, \ldots, \varphi_n$ of formulas and a formula ψ we introduce the notation $\overline{\varphi} \Rightarrow \psi$ to refer to the formula $\varphi_0 \Rightarrow (\varphi_1 \Rightarrow (\ldots \Rightarrow (\varphi_n \Rightarrow \psi) \ldots))$. Similarly, given a sequence $a_0, \ldots, a_n$ of elements of an algebra and an element b, $\overline{a} \Rightarrow b$ is the element $a_0 \Rightarrow (a_1 \Rightarrow (\ldots \Rightarrow (a_n \Rightarrow b) \ldots))$.

Proposition 22. *Let $\mathcal{S}$ be a selfextensional deductive system with a DDT-term $\Rightarrow$. $\mathcal{S}$ is Fregean iff for every n-ary connective $\star$, every k and every different variables $p_0, \ldots, p_k, q_0, \ldots, q_{n-1}, r_0, \ldots, r_{n-1}$ the quasiequations*

$$(\bigwedge_{i<n} \overline{p} \Rightarrow (q_i \Rightarrow r_i) \approx \top \wedge \bigwedge_{i<n} \overline{p} \Rightarrow (r_i \Rightarrow q_i) \approx \top) \longrightarrow \tag{4}$$

$$\overline{p} \Rightarrow (\star(q_0, \ldots, q_{n-1}) \Rightarrow \star(r_0, \ldots, r_{n-1})) \approx \top$$

are valid in $\mathbf{Alg}\mathcal{S}$.

Proof. Let $\mathcal{S}$ be a selfextensional deductive system with a DDT-term $\Rightarrow$. Then $\mathcal{S}$ is protoalgebraic. Suppose $\mathcal{S}$ is Fregean. By Theorem 20 it is fully Fregean. Let $\mathbf{A} \in \mathbf{Alg}\mathcal{S}$. Then, $\mathcal{A} = \langle \mathbf{A}, \mathrm{Fi}_{\mathcal{S}}\mathbf{A} \rangle$ is a Fregean g-matrix. Assume that $v \in \mathrm{Hom}(\mathbf{Fm}, \mathbf{A})$ is such that for every $i < n$, $v(\overline{p} \Rightarrow (q_i \Rightarrow r_i)) = 1$ and $v(\overline{p} \Rightarrow (r_i \Rightarrow q_i)) = 1$. Then, letting $X = \{v(p_0), \ldots, v(p_k)\}$, for every $i < n$,

$$\mathrm{Clo}_{\mathcal{A}}(X, v(q_i)) = \mathrm{Clo}_{\mathcal{A}}(X, v(r_i)).$$

Hence, $\langle v(q_i), v(r_i) \rangle \in \mathbf{\Lambda}_{\mathbf{A}}(\mathrm{Clo}_{\mathcal{A}}(X))$. Therefore,

$$\langle \star(v(q_0), \ldots, v(q_{n-1})), \star(v(r_0), \ldots, v(r_{n-1})) \rangle \in \mathbf{\Lambda}_{\mathbf{A}}(\mathrm{Clo}_{\mathcal{A}}(X)).$$

Thus, since $\mathcal{S}$ is fully Fregean

$$\mathrm{Clo}_{\mathcal{A}}(X, \star(v(q_0), \ldots, v(q_{n-1}))) = \mathrm{Clo}_{\mathcal{A}}(X, \star(v(r_0), \ldots, v(r_{n-1}))).$$

Hence, $\star(v(r_0), \ldots, v(r_{n-1})) \in \mathrm{Clo}_{\mathcal{A}}(X, \star(v(q_0), \ldots, v(q_{n-1})))$. Therefore,

$$v(\overline{p} \Rightarrow (\star(q_0, \ldots, q_{n-1}) \Rightarrow \star(q_0, \ldots, q_{n-1}))) \in \mathrm{Clo}_{\mathcal{A}}(1).$$

This implies that $v(\overline{p} \Rightarrow (\star(q_0, \ldots, q_{n-1}) \Rightarrow \star(q_0, \ldots, q_{n-1}))) = 1$.

Suppose now that the quasiequations (4) of the statement of the proposition hold in $\mathbf{Alg}\mathcal{S}$. Let $\mathcal{A} = \langle \mathbf{A}, \mathrm{Fi}_{\mathcal{S}}\mathbf{A} \rangle$ be a reduced full model of $\mathcal{S}$. Then $\mathbf{A} \in \mathbf{Alg}\mathcal{S}$. Let X be a finite subset of A. We will show that $\mathbf{\Lambda}_{\mathcal{A}}(X)$ is a congruence. Let $\star$ be a n-ary connective. Suppose for every $i < n$, $\langle a_i, b_i \rangle \in \mathbf{\Lambda}_{\mathcal{A}}(X)$. Then, $\mathrm{Clo}_{\mathcal{A}}(X, a_i) = \mathrm{Clo}_{\mathcal{A}}(X, b_i)$, for every $i < n$. We can assume without losing generality that $1 \in X$. Thus, for any sequence $\overline{X}$ of all the elements of X of length the cardinality of X, $\overline{\overline{X}} \Rightarrow (a_i \Rightarrow b_i) \in \mathrm{Clo}_{\mathcal{A}}(1)$ and $\overline{\overline{X}} \Rightarrow (b_i \Rightarrow a_i) \in$

$\text{Clo}_{\mathcal{A}}(1)$. Thus, $\overline{\overline{X}} \Rightarrow (a_i \Rightarrow b_i) = 1$ and $\overline{\overline{X}} \Rightarrow (b_i \Rightarrow a_i) = 1$. Hence, using the quasiequations (4), $\overline{\overline{X}} \Rightarrow (\star(a_0,\ldots,a_{n-1}) \Rightarrow \star(b_0,\ldots,b_{n-1})) = 1$, and similarly, $\overline{\overline{X}} \Rightarrow (\star(b_0,\ldots,b_{n-1}) \Rightarrow \star(a_0,\ldots,a_{n-1})) = 1$. Hence, $\text{Clo}_{\mathcal{A}}(X, \star(a_0,\ldots,a_{n-1})) = \text{Clo}_{\mathcal{A}}(X, \star(b_0,\ldots,b_{n-1}))$. Thus, $\langle \star(a_0,\ldots,a_{n-1}), \star(b_0,\ldots,b_{n-1}) \rangle \in \boldsymbol{\Lambda}_{\mathcal{A}}(X)$. $\qquad \square$

Lemma 23. *Let* $\mathbf{A}$ *be an algebra and* $\Rightarrow$ *a binary term such that* $\langle A, \Rightarrow \rangle$ *is a Hilbert algebra. The quasiequations in (4) are valid in* $\mathbf{A}$ *iff for every n-ary connective* $\star$, *letting* X *be any sequence of all the elements of the set* $\{q_i \Rightarrow r_i, r_i \Rightarrow q_i : i < n\}$, *the equations*

$$\overline{X} \Rightarrow (\star(q_0,\ldots,q_{n-1}) \Rightarrow \star(r_0,\ldots,r_{n-1})) \approx 1 \qquad (5)$$

are valid in $\mathbf{A}$.

Proof. Suppose that the quasiequations in (4) are valid in $\mathbf{A}$. Since $\langle A, \Rightarrow \rangle$ is a Hilbert algebra, the equations $\overline{X} \Rightarrow (q_i \Rightarrow r_i) \approx 1$ and $\overline{X} \Rightarrow (r_i \Rightarrow q_i) \approx 1$ are valid in $\mathbf{A}$. Hence, using the quasiequations (4), the equations

$$\overline{X} \Rightarrow (\star(q_0,\ldots,q_{n-1}) \Rightarrow \star(r_0,\ldots,r_{n-1})) \approx 1$$

and

$$\overline{X} \Rightarrow (\star(r_0,\ldots,r_{n-1}) \Rightarrow \star(q_0,\ldots,q_{n-1})) \approx 1$$

are valid in $\mathbf{A}$.

Suppose now that the equations

$$\overline{X} \Rightarrow (\star(q_0,\ldots,q_{n-1}) \Rightarrow \star(r_0,\ldots,r_{n-1})) \approx 1$$

are valid in $\mathbf{A}$. Then so are the equations

$$\overline{X} \Rightarrow (\star(r_0,\ldots,r_{n-1}) \Rightarrow \star(q_0,\ldots,q_{n-1})) \approx 1.$$

Let $p_0,\ldots,p_k, q_0,\ldots,q_{n-1}, r_0,\ldots,r_{n-1}$ be different variables. Let $v \in \text{Hom}(\mathbf{Fm}, \mathbf{A})$ be such that $v(\overline{p} \Rightarrow (q_i \Rightarrow r_i)) = 1$ and $v(\overline{p} \Rightarrow (r_i \Rightarrow q_i)) = 1$ for every $i < n$. From known facts on Hilbert algebras it follows that $v(\overline{p} \Rightarrow (\star(r_0,\ldots,r_{n-1}) \Rightarrow \star(q_0,\ldots,q_{n-1}))) = 1$. $\qquad \square$

Corollary 24. *A Hilbert-based class* K *of algebras relative to* $\Rightarrow$ *is the variety* $\mathbf{Alg}\mathcal{S}$ *of a Fregean deductive system* $\mathcal{S}$ *with* $\Rightarrow$ *as binary term with the deduction-detachment property iff it is a subvariety of the variety axiomatized by the Hilbert equations and, for every connective* $\star$, *the equations*

$$\overline{X} \Rightarrow (\star(q_0,\ldots,q_{n-1}) \Rightarrow \star(r_0,\ldots,r_{n-1})) \approx 1 \qquad (6)$$

where X *is a sequence of all the elements of the set* $\{q_i \Rightarrow r_i, r_i \Rightarrow q_i : i < n\}$.

Given an algebraic similarity type $\mathcal{L}$ and a binary term $\Rightarrow$, let $\mathsf{HI}_{\mathcal{L}}^{\Rightarrow}$ be the variery axiomatized by the Hilbert equations for $\Rightarrow$ and the above equations in (6). As a corollary we have:

Theorem 25. *For every algebraic similarity type* $\mathcal{L}$ *and every one of its binary terms* $\Rightarrow$ *there is a dual isomorphism between the set of Fregean Hilbert-based deductive systems relative to* $\Rightarrow$, *ordered by extension, and the set of all subvarieties of the variety* $\mathsf{HI}_{\mathcal{L}}^{\Rightarrow}$. *The isomorphism is given by* $\mathcal{S} \mapsto \mathbf{Alg}\mathcal{S}$.

5. Selfextensional logics with a deduction-detachment theorem and Gentzen calculi

Given a similarity type $\mathcal{L}$, in the present paper a *sequent* of type $\mathcal{L}$ will be a pair $\langle \Gamma, \varphi \rangle$ where Γ is a possibly empty finite set of formulas and φ is a formula. We will write $\Gamma \rhd \varphi$ instead of $\langle \Gamma, \varphi \rangle$.

A *Gentzen-style rule* is a pair $\langle X, \Gamma \rhd \varphi \rangle$ where X is a (possibly empty) finite set of sequents and $\Gamma \rhd \varphi$ is a sequent. A *substitution instance of a Gentzen-style rule* $\langle X, \Gamma \rhd \varphi \rangle$ is a Gentzen-style rule of the form $\langle \sigma[X], \sigma[\Gamma] \rhd \sigma(\varphi) \rangle$ for some substitution σ, where $\sigma[X] = \{ \sigma[\Delta] \rhd \sigma(\psi) : \Delta \rhd \psi \in X \}$. A Gentzen-style rule $\langle X, \Gamma \rhd \varphi \rangle$ is *initial* if X is empty. We will use the standard fraction notation for Gentzen-style rules

$$\frac{\Gamma_0 \rhd \varphi_0, \ldots, \Gamma_{n-1} \rhd \varphi_{n-1}}{\Gamma \rhd \varphi}$$

For the purposes of this paper, a *Gentzen calculus* is a set of Gentzen-style rules. Just as for Hilbert style axiom systems there is the notion of proof from an arbitrary set of premises, given a Gentzen calculus **G** we can define the notion of proof from an arbitrary set of sequents in a similar way. A *proof* in a Gentzen calculus **G** from a set of sequents X is a finite succession of sequents each one of whose elements is a substitution instance of an initial rule of **G** or a sequent in X or is obtained by applying a substitution instance of a rule of **G** to previous elements in the sequence. A sequent $\Gamma \rhd \varphi$ is *derivable in* **G** *from* a set of sequents X if there is a proof in **G** from X whose last sequent is $\Gamma \rhd \varphi$; in this situation we write $X \vdash_{\mathbf{G}} \Gamma \rhd \varphi$. If $\Gamma \rhd \varphi$ is derivable from the emptyset of sequents it is said to be a *derivable* sequent of **G**. A rule $\langle X, \Gamma \rhd \varphi \rangle$ is a *derived rule* of a Gentzen calculus **G** if $X \vdash_{\mathbf{G}} \Gamma \rhd \varphi$. Notice that if a rule is a derived rule, so are all its substitution instances, and that, by the definition, every (primitive) rule of **G** is a derived rule.

A *Gentzen system* is a pair $\mathcal{G} = \langle \mathbf{Fm}, \vdash_{\mathcal{G}} \rangle$ where $\mathbf{Fm}$ is the algebra of formulas and $\vdash_{\mathcal{G}}$ is a finitary closure operator on the set of sequents that is *substitution-invariant*. This means, using the notation $X \vdash_{\mathcal{G}} \Gamma \rhd \varphi$, where X is any set of sequents, instead of the notation $\Gamma \rhd \varphi \in \vdash_{\mathcal{G}}(X)$ typical for closure operators, that if

$$\{ \Gamma_i \rhd \psi_i : i < n \} \vdash_{\mathcal{G}} \Gamma \rhd \varphi, \tag{7}$$

then for every substitution $\sigma \in \mathrm{Hom}(\mathbf{Fm}, \mathbf{Fm})$

$$\{ \sigma[\Gamma_i] \rhd \sigma(\psi_i) : i < n \} \vdash_{\mathcal{G}} \sigma[\Gamma] \rhd \sigma(\varphi). \tag{8}$$

We say that a Gentzen system $\mathcal{G} = \langle \mathbf{Fm}, \vdash_{\mathcal{G}} \rangle$ *satisfies a Gentzen-style rule*

$$\frac{\Gamma_i \rhd \psi_i : i < n}{\Gamma \rhd \varphi}$$

if $\{\Gamma_i \triangleright \psi_i : i < n\} \vdash_{\mathcal{G}} \Gamma \triangleright \varphi$; in this situation we also say that the rule is a *sound rule* of $\mathcal{G}$. A Gentzen system $\mathcal{G} = \langle \mathbf{Fm}, \vdash_{\mathcal{G}} \rangle$ is said to be *structural* if it satisfies the structural rules of Weakening and Cut, and the Identity rule $\langle \emptyset, p \triangleright p \rangle$ [1].

A Gentzen calculus $\mathbf{G}$ determines the Gentzen system $\mathcal{G}_{\mathbf{G}} = \langle \mathbf{Fm}, \vdash_{\mathbf{G}} \rangle$. If $\mathbf{G}$ has the structural rules (either as primitive or derived), the Gentzen system $\mathcal{G}_{\mathbf{G}}$ is structural.

Every structural Gentzen system $\mathcal{G}$ defines a deductive system $\mathcal{S}_{\mathcal{G}}$ as follows

$$\Gamma \vdash_{\mathcal{S}_{\mathcal{G}}} \varphi \quad \text{iff} \quad \text{there is a finite } \Delta \subseteq \Gamma \text{ such that } \emptyset \vdash_{\mathcal{G}} \Delta \triangleright \varphi.$$

We will say that a Gentzen system $\mathcal{G}$ is *adequate* for a deductive system $\mathcal{S}$ if $\mathcal{S} = \mathcal{S}_{\mathcal{G}}$.

Generalized matrices can be used as models of Gentzen-style rules, Gentzen calculi and Gentzen systems. The double nature of g-matrices as models of both deductive systems and Gentzen systems allows us to study in a natural way the connections between the algebraic theory of deductive systems and the algebraic theory of Gentzen systems. We explore some of these connections here for selfextensional logics with a deduction-detachment term.

A g-matrix $\boldsymbol{\mathcal{A}} = \langle \mathbf{A}, \mathcal{C} \rangle$ is said to be a *model of a Gentzen-style rule*

$$\frac{\{\Gamma_i \triangleright \psi_i : i < n\}}{\Gamma \triangleright \varphi}$$

if for every homomorphism $h \in \mathrm{Hom}(\mathbf{Fm}, \mathbf{A})$, $h(\varphi) \in \mathrm{Clo}_{\mathcal{C}}(h[\Gamma])$ whenever for all $i < n$ $h(\varphi_i) \in \mathrm{Clo}_{\mathcal{C}}(h[\Gamma_i])$. It is a *model of a Gentzen calculus* if it is a model of all its rules, and it is a *model of a Gentzen system* if it is a model of all its sound rules. The following observations follow immediately from the definitions:

1. if a g-matrix is a model of a Gentzen-style rule, it is also a model of all its substitution instances,
2. if a g-matrix is a model of a Gentzen calculus, then it is a model of the Gentzen system that it defines,
3. if a g-matrix is a model of a Gentzen system $\mathcal{G}$, it is a g-model of the associated deductive system $\mathcal{S}_{\mathcal{G}}$.

The *congruence rules* for an n-ary connective $\star$ are the Gentzen-style rules of the form

$$\frac{\{\varphi_i \triangleright \psi_i, \ \psi_i \triangleright \varphi_i : i \leq n\}}{\star(\varphi_0 \ldots \varphi_{n-1}) \triangleright \star(\psi_0 \ldots \psi_{n-1})}$$

We say that a Gentzen calculus *has the congruence rules* if the congruence rules of every connective are derived rules. A Gentzen system *has the congruence rules* if it satisfies the congruence rules of every connective.

Let $\mathcal{S}$ be from now on a selfextensional logic with the deduction-detachment property for $\Rightarrow$. Recall that then on every $\mathbf{A} \in \mathbf{Alg}\mathcal{S}$ the operation $\Rightarrow^{\mathbf{A}}$ defines

---

[1] We do not need to consider the other structural rules - exchange and contraction - because we consider sets of premises in our sequents and not successions.

by condition (1) an order that we denote by $\leq^{\mathbf{A}}$. We say that a Gentzen calculus $\mathbf{G}$ adequate for $\mathcal{S}$ is $\mathbf{Alg}\mathcal{S}$-*order-sound* if whenever

$$\{\varphi_i \triangleright \psi_i : I \in I\} \vdash_{\mathbf{G}} \varphi \triangleright \psi,$$

then for every $\mathbf{A} \in \mathbf{Alg}\mathcal{S}$ and every valuation $v \in \mathrm{Hom}(\mathbf{Fm}, \mathbf{A})$

$$\text{if for all } i \in I,\ v(\varphi_i) \leq^{\mathbf{A}} v(\psi_i), \text{ then } v(\varphi) \leq^{\mathbf{A}} v(\psi).$$

We say that it is $\mathbf{Alg}\mathcal{S}$-*order-complete* if the converse of the main implication in the above statement holds.

Lemma 26. *If $\mathbf{G}$ is a Gentzen calculus adequate for $\mathcal{S}$ which is $\mathbf{Alg}\mathcal{S}$-order-complete, then it has the congruence rules.*

Proof. Let $\star$ be an n-ary connective. Let $\mathbf{A} \in \mathbf{Alg}\mathcal{S}$ and $v \in \mathrm{Hom}(\mathbf{Fm}, \mathbf{A})$. Assume that for φ_i, ψ_i with $i < n$, $v(\varphi_i) \leq^{\mathbf{A}} v(\psi_i)$ and $v(\psi_i) \leq^{\mathbf{A}} v(\varphi_i)$. Thus, $v(\varphi_i) = v(\psi_i)$. Therefore, $v(\star \varphi_0 \ldots \varphi_{n-1}) = v(\star \psi_0 \ldots \psi_{n-1})$. By $\mathbf{Alg}\mathcal{S}$-order-completeness it follows that the congruence rules for $\star$ are derived rules of $\mathbf{G}$. $\qquad\square$

We say that a Gentzen calculus $\mathbf{G}$ *has the DDT rules* if the rules of the forms

$$\frac{\Gamma, \varphi \triangleright \psi}{\Gamma \triangleright \varphi \Rightarrow \psi} \qquad\qquad \frac{\Gamma \triangleright \varphi \Rightarrow \psi}{\Gamma, \varphi \triangleright \psi}$$

are derived rules.

Remark 27. *If $\mathbf{G}$ has the DDT rules then for every sequent $\Gamma \triangleright \varphi$ and every finite sequence $\overline{\Gamma}$ of all the elements of Γ*

$$\Gamma \triangleright \varphi \dashv_{\mathbf{G}}\vdash \top \triangleright \overline{\overline{\Gamma}} \Rightarrow \varphi.$$

The remark implies the lemma below.

Lemma 28. *Let $\mathbf{G}$ be a Gentzen calculus with the DDT rules, then for every family of sequents $\{\Gamma_i \triangleright \varphi_i : i \in I\}$ and every sequent $\Gamma \triangleright \varphi$ the following statements are equivalent:*

1. $\{\Gamma_i \triangleright \varphi_i : i \in I\} \vdash_{\mathbf{G}} \Gamma \triangleright \varphi$
2. $\{\top \triangleright \overline{\overline{\Gamma_i}} \Rightarrow \varphi_i : i \in I\} \vdash_{\mathbf{G}} \top \triangleright \overline{\overline{\Gamma}} \Rightarrow \varphi,$

where $\overline{\Gamma}$ is any finite sequence of all the formulas in Γ.

Lemma 29. *If $\mathbf{G}$ is a structural Gentzen calculus adequate for $\mathcal{S}$ which is $\mathbf{Alg}\mathcal{S}$-order-sound, has the DDT rules and has the congruence rules, then it is $\mathbf{Alg}\mathcal{S}$-order-complete.*

Proof. Assume that the family of sequents with elements $\varphi_i \triangleright \psi_i$ with $i \in I$ and $\varphi \triangleright \psi$ is such that for every $\mathbf{A} \in \mathbf{Alg}\mathcal{S}$ and every valuation $v \in \mathrm{Hom}(\mathbf{Fm}, \mathbf{A})$, if for all $i \in I$, $v(\varphi_i) \leq^{\mathbf{A}} v(\psi_i)$, then $v(\varphi) \leq^{\mathbf{A}} v(\psi)$. Then, setting $\top := p \Rightarrow p$ for some fixed variable p,

$$\{\varphi_i \Rightarrow \psi_i \approx \top : i \in I\} \models_{\mathbf{Alg}\mathcal{S}} \varphi \Rightarrow \psi \approx \top.$$

Thus by completeness of the quasiequational logic of $\mathbf{Alg}\mathcal{S}$ and the fact that $\mathbf{Alg}\mathcal{S}$ is a variety, there is a proof of the equation $\varphi \Rightarrow \psi \approx \top$ from the equations in $\{\varphi_i \Rightarrow \psi_i \approx \top : i \in I\}$ and the equations which are valid in $\mathbf{Alg}\mathcal{S}$, which are the equations $\delta \approx \varepsilon$ such that $\delta \dashv\vdash_{\mathcal{S}} \varepsilon$. An easy inductive argument will show that for every equation $\gamma \approx \delta$ in such a proof

$$\{\varphi_i \triangleright \psi_i : i \in I\} \vdash_{\mathbf{G}} \gamma \triangleright \delta, \delta \triangleright \gamma.$$

If $\gamma \approx \delta$ is $\varphi_i \Rightarrow \psi_i \approx \top$ with $i \in I$, then the above remark and the fact that (using Identity, the DDT rules and Weakening) $\vdash_{\mathbf{G}} \varphi_i \Rightarrow \psi_i \triangleright \top$ gives the result. If $\gamma \approx \delta$ is valid in $\mathbf{Alg}\mathcal{S}$, then $\gamma \dashv\vdash_{\mathcal{S}} \delta$; therefore the sequents $\gamma \triangleright \delta, \delta \triangleright \gamma$ are derivable in $\mathbf{G}$ and we have the result. Now if $\gamma \approx \delta$ follows by symmetry of the equality from previous equations in the proof, then it is clear. If it follows by transitivity of equality, then Cut gives the desired result. Finally, if it follows by replacement from previous equations in the proof, applying the congruence rules and Cut we obtain the result. Hence, by the induccion principle we obtain that $\top \triangleright \varphi \Rightarrow \psi$ is derivable in $\mathbf{G}$ from $\{\varphi_i \triangleright \psi_i : i \in I\}$. Thus $\varphi \triangleright \psi$ is also derivable from this set. $\qquad\square$

Corollary 30. *If $\mathbf{G}$ is a structural Gentzen calculus with the DDT rules which is adequate for the deductive system $\mathcal{S}$ and which is $\mathbf{Alg}\mathcal{S}$-order-sound, then $\mathbf{G}$ is $\mathbf{Alg}\mathcal{S}$-order-complete iff it has the congruence rules.*

From the corollary follows that there is always a structural Gentzen calculus which is adequate for the deductive system $\mathcal{S}$ and is $\mathbf{Alg}\mathcal{S}$-order-sound and $\mathbf{Alg}\mathcal{S}$-order-complete. It is the Gentzen calculus $\mathbf{G}_{\mathcal{S}}$ defined by the following rules:

1. the structural rules of identity, weakening and cut,
2. the congruence rules for the connectives,
3. the DDT rules
4. for every finite Γ and every φ such that $\Gamma \vdash_{\mathcal{S}} \varphi$, the initial rule

$$\overline{\Gamma \triangleright \varphi}$$

That this calculus is $\mathbf{Alg}\mathcal{S}$-order-sound follows easily from Lemma 28 and the fact that $\mathbf{Alg}\mathcal{S}$ is Hilbert-based with respect to $\Rightarrow$. The $\mathbf{Alg}\mathcal{S}$-order-completeness follows from the corollary. We will denote by $\mathcal{G}(\mathcal{S})$ the Gentzen system of the calculus $\mathbf{G}_{\mathcal{S}}$.

If we have a nice Hilbert style axiomatization of $\mathcal{S}$ we can consider the Gentzen calculus like the one described above except that, instead of the rules in (4), it has the Gentzen-style rules that naturally correspond to the axioms and rules of the Hilbert style axiomatization.

Remark 31. From Remark 27 it follows that any two structural Gentzen calculi adequate for $\mathcal{S}$ and with the DDT rules which are $\mathbf{Alg}\mathcal{S}$-order-sound and $\mathbf{Alg}\mathcal{S}$-order-complete define the same Gentzen system.

We say that a Gentzen system adequate for a deductive system $\mathcal{S}$ with theorems is *fully-adequate* if the g-matrix models of the Gentzen system are the full g-models of $\mathcal{S}$. This notion is introduced in [8] under the name 'strongly adequate'. We give a different, simpler proof of Proposition 4.47 (iii) in [8]. We will use the lemma below.

Lemma 32. *Let $\mathcal{S}$ be a Hilbert-based deductive system. If $\mathcal{A} = \langle \mathbf{A}, \mathcal{C} \rangle$ is g-matrix model of $\mathcal{S}$ such that its Frege relation is the identity, then $\mathcal{C} = \mathrm{Fi}_{\mathcal{S}}\mathbf{A}$ and $\mathbf{A} \in \mathsf{K}_{\mathcal{S}}$*

Proof. Since the Frege relation of $\mathcal{A} = \langle \mathbf{A}, \mathcal{C} \rangle$ is the identity, this g-matrix is reduced, so $\mathbf{A} \in \mathbf{Alg}\mathcal{S} \subseteq \mathsf{K}_{\mathcal{S}}$. To prove that $\mathcal{C} = \mathrm{Fi}_{\mathcal{S}}\mathbf{A}$ assume that $F \in \mathrm{Fi}_{\mathcal{S}}\mathbf{A}$. By Lemma 14, F is an implicative filter. We show that $F = \mathrm{Clo}_{\mathcal{A}}(F)$. If $b \in \mathrm{Clo}_{\mathcal{A}}(F)$, then let $a_0, \ldots, a_n \in F$ such that $a \in \mathrm{Clo}_{\mathcal{A}}(\{a_0, \ldots, a_n\})$. Since $\mathcal{A}$ is a model of the DDT rules, $\overline{\overline{a}} \Rightarrow b \in \mathrm{Clo}_{\mathcal{A}}(\top)$. Therefore, $\mathrm{Clo}_{\mathcal{A}}(\top) = \mathrm{Clo}_{\mathcal{A}}(\overline{\overline{a}} \Rightarrow b)$. Thus, $\overline{\overline{a}} \Rightarrow b = \top$. Since $\top \in F$ because it is an implicative filter, $\overline{\overline{a}} \Rightarrow b \in F$. Hence, since $a_0, \ldots, a_n \in F$ and F is an implicative filter, $b \in F$. $\qquad\square$

Theorem 33. *Let $\mathcal{S}$ be a deductive system with the deduction-detachment property. Then, $\mathcal{S}$ is selfextensional iff the Gentzen system $\mathcal{G}(\mathcal{S})$ is fully adequate for $\mathcal{S}$.*

Proof. If the Gentzen system $\mathcal{G}(\mathcal{S})$ for $\mathcal{S}$ is fully adequate, then the full model $\langle \mathbf{Fm}, \mathbf{Th}\mathcal{S} \rangle$ is a model of $\mathcal{G}(\mathcal{S})$, thus of the congruence rules. This implies that $\mathcal{S}$ is selfextensional. Assume now that $\mathcal{S}$ is selfextensional. Then, by Theorem 17, $\mathcal{S}$ is fully selfextensional. Thus, if $\langle \mathbf{A}, \mathcal{C} \rangle$ is a full g-model of $\mathcal{S}$, then it is a model of the congruence rules of every connective and of the sequents $\Gamma \rhd \varphi$ such that $\Gamma \vdash_{\mathcal{S}} \varphi$. Clearly it is also a model of the structural rules. Moreover, by Theorem 2.48 of [8] $\langle \mathbf{A}, \mathcal{C} \rangle$ has the deduction-detachment property. Thus it is a model of the DDT rules. Therefore, it is a model of the Gentzen system $\mathcal{G}(\mathcal{S})$. To finish the proof it is enough to show that if $\mathcal{A} = \langle \mathbf{A}, \mathcal{C} \rangle$ is a reduced g-matrix model of $\mathcal{G}(\mathcal{S})$, then $\mathcal{C} = \mathrm{Fi}_{\mathcal{S}}\mathbf{A}$. If $\mathcal{A}$ is a reduced g-matrix model of $\mathcal{G}(\mathcal{S})$ it is a model of $\mathcal{S}$ and of the congruence rules. This implies that its Frege relation is a congruence; therefore since $\mathcal{A}$ is reduced, its Frege relation is the identity. Moreover, $\mathcal{A}$ is a model of the DDT rules. Lemma 32 implies that $\mathcal{C} = \mathrm{Fi}_{\mathcal{S}}\mathbf{A}$. $\qquad\square$

The notion of $\mathbf{Alg}\mathcal{S}$-order-sound and $\mathbf{Alg}\mathcal{S}$-order-complete structural Gentzen system is strongly related to the notion of algebraizable Gentzen system. We recall this notion here.

A *structural translation t of sequents into equations* is a mapping that maps every sequent to a finite set of equations and satisfies the following structurality property: for every sequent $\Gamma \rhd \varphi$ and every substitution σ,

$$\text{if } t(\Gamma \rhd \varphi) = \{\varepsilon_i \approx \delta_i : i < n\}, \text{ then } t(\sigma[\Gamma] \rhd \sigma(\varphi)) = \{\sigma(\varepsilon_i) \approx \sigma(\delta_i) : i < n\}.$$

A *structural translation s from equations into sequents* is a mapping that maps every equation to a finite set of sequents and has the corresponding structurality property, that is, for every equation $\varepsilon \approx \delta$ and every substitution σ,

$$\text{if } s(\varepsilon \approx \delta) = \{\Gamma_i \rhd \varphi_i : i < n\}, \text{ then } s(\sigma(\varepsilon) \approx \sigma(\delta)) = \{\sigma[\Gamma_i] \rhd \sigma[\varphi_i] : i < n\}.$$

If t is a translation of sequents into equations and X is a set of sequents, the set of equations $t(X)$ is defined by

$$t(X) = \bigcup \{t(\Gamma \rhd \varphi) : \Gamma \rhd \varphi \in X\}.$$

If s is a translation from equations into sequents and E is a set of equations, the set of sequents $s(E)$ is defined by

$$s(E) = \bigcup \{s(\varphi \approx \psi) : \varphi \approx \psi \in E\}.$$

A Gentzen system $\mathcal{G}$ is said to be *algebraizable* if there is a class of algebras K and a structural translation t from sequents into equations and a structural translation s from equations into sequents such that the following two conditions hold

$$\{\Gamma_i \rhd \varphi_i : i \in I\} \vdash_{\mathcal{G}} \Gamma \rhd \varphi \quad \text{iff} \quad t(\{\Gamma_i \rhd \varphi_i : i \in I\}) \models_{\mathsf{K}} t(\Gamma \rhd \varphi). \qquad (9)$$

$$\varphi \approx \psi \models_{\mathsf{K}} t(s(\varphi \approx \psi)) \quad \text{and} \quad t(s(\varphi \approx \psi)) \models_{\mathsf{K}} \varphi \approx \psi, \qquad (10)$$

where $\models_{\mathsf{K}}$ denotes the equational consequence defined by the class of algebras K as follows:

$$\{\varphi_i \approx \psi_i : i \in I\} \models_{\mathsf{K}} \varphi \approx \psi \quad \text{iff} \quad \forall \mathbf{A} \in \mathsf{K} \text{ and } \forall h \in \mathrm{Hom}(\mathbf{Fm}, \mathbf{A}),$$

$$\text{if } \forall i \in I \; h(\varphi_i) = h(\psi_i), \text{ then } h(\varphi) = h(\psi).$$

When to the right of " $\models_{\mathsf{K}}$ " there is a set it means that every element of the set follows from the equations in the set to the left of " $\models_{\mathsf{K}}$ ".

From [16] it follows that if a Gentzen system $\mathcal{G}$ is algebraizable, there is always a quasivariety K, which is unique, such that (9) and (10) are satisfied for some structural translations t and s. This quasivariety is called the *equivalent algebraic semantics* of $\mathcal{G}$.

The notion of algebraizable Gentzen system was introduced in [15] and its theory developed in [16]. The theory of algebraizable Gentzen systems is an extension to these objects of the theory of algebraizable deductive systems developed by W. Blok and D. Pigozzi in [2]. It is also a particular case of the notion of equivalence between Gentzen systems introduced and studied in [16].

Let us fix a selfextensional deductive system $\mathcal{S}$ with the deduction-detachment property for $\Rightarrow$. We define the structural translation t from sequents to equations and the structural translation sq from equations to sequents respectively by

$$t(\Gamma \rhd \varphi) = \top \approx \overline{\overline{\Gamma}} \Rightarrow \varphi \qquad sq(\varphi \approx \psi) = \{\varphi \rhd \psi, \psi \rhd \varphi\}.$$

Theorem 34. *For every selfextensional logic $\mathcal{S}$ with the deduction-detachment property for $\Rightarrow$, the Gentzen system defined by the Gentzen calculus $\mathbf{G}_{\mathcal{S}}$ is algebraizable with equivalent algebraic semantics $\mathbf{Alg}\mathcal{S}$ and translations t and sq.*

Proof. By the results above $\mathbf{G}_{\mathcal{S}}$ is $\mathbf{Alg}\mathcal{S}$-order-sound and $\mathbf{Alg}\mathcal{S}$-order-complete, thus Lemma 28 gives condition (7) of the definition of algebraizable Gentzen system holds. Condition (8) holds because in any Hilbert algebra, $a = b$ iff $a \Rightarrow b = 1$ and $b \Rightarrow a = 1$. $\square$

References

[1] S.V. Babyonishev, *Fully Fregean logics*, Reports on Mathematical Logic 37 (2003), 59-78.

[2] W. Blok, D. Pigozzi, *Algebraizable Logics* (Memoires of the AMS, vol. 396), The American Mathematical Society, Providence 1986.

[3] S. Celani and R. Jansana, *A closer look at some subintuitionistic logics*. Notre Dame Journal of Formal Logic 42 (2001), 225-255.

[4] J. Czelakowski, *Protoalgebraic logics*, Kluwer, 2001.

[5] J. Czelakowski, *The Suszko operator. Part I*. Studia Logica, Special Issue on Abstract Algebraic Logic II, 74 (2003), 181-231.

[6] J. Czelakowski and D. Pigozzi, *Fregean logics*. Annals of Pure and Applied Logic 127 (2004), 17-76.

[7] J.M. Font, *Belnap's four-valued logic and De Morgan lattices*. Logic Journal of the I.G.P.L. 5 (1997), 413–440.

[8] J.M. Font, R. Jansana, *A General Algebraic Semantics for Sentential Logics*, vol. 7 of Lecture Notes in Logic, Springer-Verlag, 1996.

[9] J.M. Font, R. Jansana, D. Pigozzi, *Fully adequate Gentzen systems and closure properties of the class of full g-models*, Studia Logica 83 (2006), 215-278.

[10] J.M. Font, R. Jansana, D. Pigozzi, *A Survey of Abstract Algebraic Logic*. Studia Logica, Special Issue on Algebraic Logic II, 74 (2003), 13-97.

[11] J.M. Font, V. Verdú, *Algebraic Logic for Classical Conjunction and Disjunction*. Studia Logica, Special Issue on Abstract Algebraic Logic, 50 (1991), 391-419.

[12] R. Jansana, Full g-models for Positive Modal Logic. Mathematical Logic Quarterly 48 (2002), 427-445.

[13] A. Pynko *Definitional equivalence and algebraizability of generalized logical systems*. Annals of Pure and Applied Logic 98 (1999), 1-68.

[14] H. Rasiowa, *An Algebraic Approach to non-classical logics*, North-Holland, 1974.

[15] J. Rebagliato, V. Verdú, *On the algebraization of some Gentzen sytems*. Fundamenta Informaticae, Special Issue on Algebraic Logic and its Applications, 18 (1993), 319-338.

[16] J. Rebagliato, V. Verdú, *Algebraizable Gentzen systems and the Deduction Theorem for Gentzen systems*, Mathematics Preprint Series 1975, Univeristat de Barcelona.

[17] R. Wójcicki, *Referential Matrix Semantics for Propositional Calculi*. Bulletin of the Section of Logic 8 (1979), 170-76.

[18] R. Wójcicki, *Theory of Logical Calculi. Basic Theory of Consequence Operations*, (Synthese Library, vol. 199), Reidel, 1988.

Acknowledgment

I would like to express my gratitude to Josep Maria Font for the suggestions that helped to improve the readability of the paper.

Ramon Jansana
Departament de Lògica, Història i Filosofia de la Ciència,
Universitat de Barcelona,
Montalegre 6,
08001 Barcelona
Spain
e-mail: jansana@ub.edu

J.-Y. Beziau (Ed.), *Logica Universalis*, 2nd edition, 87–93

Logic without Self-Deductibility

Pierre Ageron

Abstract. Self-deductibility is the Stoic version of the law of identity : if A, then A. After a discussion on its role, we suggest a natural system of axioms and rules for a logic in which this law is not valid, based on a simple model where proofs are families of strictly injective maps. Finally we develop some general theory of taxonomies (i.e. "categories without identities") and place this particular example into a more general algebraic picture.

Mathematics Subject Classification (2000). 03B22, 03B60, 18A15.

Keywords. law of identity, self-deductibility, taxonomy.

1. The law of self-deductibility

What we call here (following [4]) the "law of self-deductibility" is the Stoic version of the "law of identity", as incidentally cited by Sextus Empiricus: it reads "if A, then A", so that it expresses reflexivity of entailment. This is a law of the logic of propositions, not to be confused with the more well known Peripatetic law of identity which belongs to the logic of terms and is the equation $a = a$.

Self-deductibility has a paradoxical status. It seems so obvious that, unlike the law of excluded middle or the law of contradiction, it was hardly ever a matter of controversy (interesting attempts at challenging it appear in [4]). Also it seems of no use whatsoever in the mathematical practice and totally sterile in terms of deductive power. However in Gentzen's sequent calculus, the law of self-deductibility is the only axiom, apart from those for truth and absurdity (recall that an axiom is a rule with no premiss). It follows that self-deductibility, far from being superfluous, is essential to start any proof. Moreover this axiom is really only one half of the law of identity : the "cut rule" (expressing transitivity of entailment) is in many senses the dual rule. Since self-deductibility is unavoidable in sequent calculus, it is not surprising (although difficult to prove) that the cut rule is redundant. This also means that challenging self-deductibility implies to renounce cut elimination.

One of the first modern attempts to formalize deductive logic is Charles Peirce's *Algebra of logic*, published in 1885 [8]. This paper mentions in passing a quite remarkable interpretation of the law of self-deductibility. Peirce refers to it as the "first icon of algebra" and argues that it "justifies our continuing to hold what we have held, though we may, for instance, forget how we were originally justified in holding it". This clearly suggests that self-deductibility works as an amnesia principle : once some statement is proved, it allows us to forget how it was proved. In mathematical practice, it is obviously a good thing that we can continue to trust our old theorems even if we are unable to reconstruct their proofs. However, to give a a complete account of some new result, we should be able to keep track of our old proofs very precisely. If Peirce's view is correct, this implies some restriction on the scope of self-deductibility.

Now let us see proofs as certain functions between sets (the so-called Heyting paradigm). The analysis we have just done rather naturally leads to the idea that these functions should be injective (in order to trace back the history of a proof unambiguously) and not surjective (at each step we need some space to save the job we accomplished). So we write $A \vdash B$ if there is a non surjective injective map from A into B. From the constructive point of view we shall try and stick to, it is in fact stronger and better to consider *strict* injective maps, i.e. those injective maps whose coimage is inhabited. In other terms, we end up with the following : a proof of $A \vdash B$ consists of an injective map $f : A \to B$ together with some element $x_0 \in B \setminus f[A]$ witnessing that f is strict. Assuming from now on that $A \notin A$ for every set A, it is convenient to encode these data as an injective map from $A \cup \{A\}$ into B.

In this interpretation, self-deductible propositions correspond precisely to reflexive sets, also known as Dedekind-infinite sets : recall that reflexive sets are not finite, and that the converse holds if excluded middle and countable choice are assumed. Also, assuming excluded middle, mutually deductible propositions (i.e. $A \vdash B$ and $B \vdash A$) correspond to equipotent reflexive sets (via the Cantor-Bernstein theorem). It should by now be clear that the idea of a logic without self-deductibility is a perfectly consistent and natural one: such a logic really underlies Dedekind's theory of "infinite systems" [3]. Our task in the next section will be to make that hidden logic explicit. A basic knowledge of category theory will be assumed from the reader, but proofs will be omitted.

2. A model for a logic without self-deductibility

To find out axioms and rules for a logic without self-deductibility, our strategy consists in describing a particular categorical structure: our selection of axioms and rules will then reflect the kind of structure that this model is equipped with. Of course we build on the ideas of section **1**. But in our model, propositions will not be just sets, but arbitrary families of sets : this extension allows for a nice interpretation of conjunction and disjunction. We certainly do not pretend this model is what we want ultimately : in particular it is too simple to account for implication or negation.

The model is as follows. We denote by $\mathbf{A}$ the category with objects all families $A = (A_i)_{i \in I}$ of sets, such that an arrow from $A = (A_i)_{i \in I}$ to $B = (B_j)_{j \in J}$ is a map $\phi : J \to I$ together with a family of *injective* maps $A_{\phi(j)} \to B_j$. Recall that we assume $A \notin A$ for every set A throughout the paper. Then we have at our disposal the endofunctor $? : \mathbf{A} \to \mathbf{A}$ defined on objects by

$$?A = (A_i \cup \{A_i\})_{i \in I}$$

and on arrows in a straightforward manner. Also it can be checked that $\mathbf{A}$ has a terminal object $\top$ and binary products $A \wedge B$, as well as an initial object $\perp$ and binary sums $A \vee B$. For instance one has $\top = (\ )$ and $\perp = (\emptyset)$. It is the case that $?$ preserves $\top$ and $\wedge$, but not $\perp$ or $\vee$. Now write $A \vdash B$ if there is some arrow from $?A$ to B in $\mathbf{A}$. It can be checked that the following hold:

$$\frac{}{A \vdash ?A} \quad (1) \qquad\qquad \frac{C \vdash A \quad C \vdash B}{C \vdash A \wedge B} \quad (9)$$

$$\frac{}{A \vdash ??A} \quad (2) \qquad\qquad \frac{C \vdash A \quad D \vdash B}{C \wedge D \vdash A \wedge B} \quad (10)$$

$$\frac{A \vdash B}{?A \vdash ?B} \quad (3) \qquad\qquad \frac{}{\perp \vdash ?C} \quad (11)$$

$$\frac{A \vdash ?B \quad B \vdash C}{A \vdash C} \quad (4) \qquad\qquad \frac{}{A \vdash ?(A \vee B)} \quad (12)$$

$$\frac{A \vdash A \quad A \vdash B}{B \vdash B} \quad (5) \qquad\qquad \frac{}{B \vdash ?(A \vee B)} \quad (13)$$

$$\frac{}{A \vdash \top} \quad (6) \qquad\qquad \frac{}{A \vdash ?A \vee B} \quad (14)$$

$$\frac{}{A \wedge B \vdash ?A} \quad (7) \qquad\qquad \frac{}{B \vdash A \vee ?B} \quad (15)$$

$$\frac{}{A \wedge B \vdash ?B} \quad (8) \qquad\qquad \frac{A \vdash C \quad B \vdash C}{A \vee B \vdash ?C} \quad (16)$$

We can read this as a system of logical axioms and rules in the style of categorical sequents, i.e. sequents with exactly one hypothesis and one conclusion, as they appear in [5]. The calculus of categorical sequents is sometimes criticized by proof theorists because it does not enjoy cut elimination. On the other hand, precisely this shortcoming makes it possible to weaken self-deductibility. The weakening is performed by a modality "?", involved in all axioms except (6). This modality is very different from those of modal or linear logics. Note that if we erase "?" from axioms and rules (1) to (16), we get only valid axioms and rules of usual intuitionistic logic.

Although self-deductibility is not derivable in this system, some of its strength can be recovered from rule (4), which can be seen as a strong form of the cut rule. For instance (4) combined with (7) yields the usual weakening rule

$$\frac{A \vdash C}{A \wedge B \vdash C}$$

in spite of the fact that the axiom

$$\frac{}{A \wedge B \vdash A}$$

does not hold. It follows from (1), (3) and (4) that a proposition A is self-deductible if and only if A and $?A$ are mutually deductible. And ordinary cut (i.e. transitivity of entailment) can be derived from (2), (3) and (4).

The most unexpected rule is certainly (5). In the basic case where A and B consist of just one set, it exactly says that a set containing a reflexive subset is itself reflexive. As this result, theorem 68 in [3], crucially depends on excluded middle, rule (5) gives a definite Boolean flavour to our system, even though it does not include negation or implication. In the case of an arbitrary family, it relies on the fact that $A \vdash A$ holds if and only if each A_i is reflexive, a consequence of theorem 72 in [3]: reflexive sets are precisely those sets containing a subset equipotent to the set of natural numbers. If we were to accept *infinitary* rules, we should consider adding this one where the premisses consist in an infinite regression of causes:

$$\frac{\ldots \; A_3 \vdash A_2 \quad A_2 \vdash A_1 \quad A_1 \vdash A}{A \vdash A}$$

In the next section, we shall develop a general framework in order to elucidate the features of our model from an algebraic point of view.

3. Category-theoretic aspects of the subject

There is a link between category theory and universal propositional logic, first explored by Joachim Lambek in a series of papers starting with [5]. The idea is to view the objects of a category as propositions and its arrows as proofs. In fact each notion of "category equipped with some extra algebraic structure of a certain kind" can be seen to describe a set of logical axioms and rules, together with a "natural" algebra of proofs. E.g. it is well known that bicartesian closed categories (i.e. categories with terminal object, binary products, initial object, binary sums and exponentials) are the algebraic counterpart of intuitionistic logic. Similarly the so-called ∗-autonomous categories correspond to linear logic. Using the language of Ehresmann's sketches, a general logic is nothing but a limit sketch containing the sketch for categories as a subsketch and built according to certain constraints : see [1] for details. What matters for us here is just this obvious observation : since abstract "identity arrows" $\mathrm{id}_A : A \to A$ are part of the definition of a category, such a logic automatically satisfies self-deductibility.

Now suppose we want to drop the principle of self-deductibility, while retaining the cut rule and remaining in the categorical spirit. Then we have to consider structures defined similarly to categories except that identities need not exist. Such structures do appear but sporadically in the categorical literature. They have been given various names: semicategories, semigroupoids, multiplicative graphs, taxonomies. Here we will retain the latter terminology, suggested by Robert Paré and Richard Wood in a paper that was never published. The theory of taxonomies is not much developed since it was considered by many as a blind alley. As put down by the influential category theorist William Lawvere in 1991: "Even today there are many who think one could usefully 'generalize' categories by omitting the requirement that there must be identity maps [...] However [how] useless would be an intricate network of speeding buses without bus-stops" [6]. Moreover it is often objected that taxonomies are nothing but (two-sided) ideals in categories: this is certainly true, but of little use in practice, because there is in general no natural category in which a given taxonomy is an ideal. In particular, the formal process of freely adding identity arrows to a taxonomy destroys the identities that might originally exist for some objects.

Recently taxonomies have attracted a renewal of interest. E.g. Isar Stubbe's thesis [9] demonstrates nicely that they cannot be avoided even in a context that does not involve them a priori. Note that his taxonomies are "regular" in the sense of [7]: every arrow $f : A \to B$ can be parsed as $f = h \circ g$, where moreover $g : A \to C$ and $h : C \to B$ are unique "up to homotopy". This "interpolation" (or "density") property happens to make regular taxonomies rather close to categories. Looking at them with a logical glance would probably be worthwhile.

A completely different example of a taxonomy is the effective taxonomy of a pointed endofunctor, which was introduced in [3]. Given a category $\mathbf{C}$, we start with a functor $F : \mathbf{C} \to \mathbf{C}$ together with a natural transformation $\eta : \mathrm{id}_{\mathbf{C}} \Rightarrow F$. Recall that the latter means that for every object A, we are given an arrow $\eta_A : A \to F(A)$ in such a way that we have $\eta_B \circ f = F(f) \circ \eta_A$ for every arrow $f : A \to B$ in $\mathbf{C}$. Now consider the graph whose objects are juste those of $\mathbf{C}$ while its arrows from A to B are the arrows from $F(A)$ to B in $\mathbf{C}$. Consecutive arrows of that graph compose in a canonical way : if $f : A \to B$ and $g : B \to C$, let

$$g \bullet f = g \circ \eta_B \circ f.$$

It is easy to check that $\bullet$ is associative, so that we have in fact defined a taxonomy that we will denote by $\mathrm{Tax}(\mathbf{C}, F, \eta)$. In general, $\mathrm{Tax}(\mathbf{C}, F, \eta)$ fails to be a category because of the lack of identities. More precisely an object A has an identity in $\mathrm{Tax}(\mathbf{C}, F, \eta)$ if and only if the arrow η_A is invertible in $\mathbf{C}$.

Of course an effective taxonomy $\mathrm{Tax}(\mathbf{C}, F, \eta)$ inherits from F and η a much richer structure than merely being a taxononomy : the rather exotic kind of structure that arises has been characterized in [2] under the name "supertaxonomy".

Further enrichments of the taxonomy structure (and of the underlying logic) appear if, e.g., we assume that $\mathbf{C}$ has and F preserves all products (empty, finite or infinite). A typical situation is that where $\mathbf{C}$ is the free completion under products of some multicocomplete category $\mathbf{A}$ in sense of Yves Diers. Let us first recall that a *cocomplete* category $\mathbf{A}$ is one in which every diagram δ has a colimit, i.e. a cocone over δ through which every cocone over δ factorizes uniquely. A *multicocomplete* category is one in which every diagram δ has a multicolimit, i.e. a family of cocones over δ such that every cocone over δ factorizes uniquely through precisely one of these cocones. Equivalently a category $\mathbf{A}$ is multicocomplete if and only if its completion under products $\mathbf{C} = \Pi\mathbf{A}$ is cocomplete. Note that the completion $\Pi\mathbf{A}$ can be described very explicitly : its objects are all families $A = (A_i)_{i \in I}$ of objects of $\mathbf{A}$, and an an arrow from $A = (A_i)_{i \in I}$ to $B = (B_j)_{j \in J}$ is a map $\phi : J \to I$ together with a family of arrows $A_{\phi(j)} \to B_j$. In this situation any endofunctor $F : \mathbf{A} \to \mathbf{A}$ uniquely extends to an endofunctor $\Pi F : \Pi\mathbf{A} \to \Pi\mathbf{A}$ preserving products (but not, in general, colimits); furthermore any natural transformation $\eta : \mathrm{id}_\mathbf{A} \Rightarrow F$ extends to a natural transformation $\Pi\eta : \mathrm{id}_{\Pi\mathbf{A}} \Rightarrow \Pi F$.

Now a typical example of a multicocomplete category which is not cocomplete is the category $\mathbf{A}$ of sets and injective maps between them. It should be clear that the category $\mathbf{C}$ described in section **2** is just $\Pi\mathbf{A}$ for that particular $\mathbf{A}$; similarly the endofunctor $? : \mathbf{C} \to \mathbf{C}$ is just ΠF with $F : \mathbf{A} \to \mathbf{A}$ the obvious pointed endofunctor such that $F(A) = A \cup \{A\}$. What we considered to be proofs in our logic are precisely arrows in $\mathbf{T} = \mathrm{Tax}(\Pi\mathbf{A}, \Pi F, \Pi\eta)$ where $\eta : \mathrm{id}_\mathbf{A} \Rightarrow F$ is the obvious natural transformation. In fact, all of our axioms and rules except for rule (5) hold in $\mathbf{T} = \mathrm{Tax}(\Pi\mathbf{A}, \Pi F, \Pi\eta)$ for $\mathbf{A}$ an arbitrary multicocomplete category, $F : \mathbf{A} \to \mathbf{A}$ an arbitrary endofunctor and $\eta : \mathrm{id}_\mathbf{A} \Rightarrow F$ an arbitrary natural transformation. Rules (1) to (4) above reflect the supertaxonomy structure of $\mathbf{T}$. Rules (6) to (10) reflect the structure inherited by $\mathbf{T}$ from the fact that $\Pi\mathbf{A}$ has and ΠF preserves products. Rules (11) to (16) reflect the structure inherited by $\mathbf{T}$ from the fact that $\Pi\mathbf{A}$ has sums (that are not necessarily preserved by ΠF).

As can be suspected from the fact that it is some kind of Booleanness (see section **2**), rule (5) is more specific and the underlying algebra is more subtle. Namely, if $\mathbf{A}$ is the category of sets and injective maps, it happens that $\mathrm{Tax}(\mathbf{A}, F, \eta) \to \mathbf{A}$ is a so-called "crossed taxonomy" : see [2]. As for the extension of rule (5) to $\mathrm{Tax}(\Pi\mathbf{A}, \Pi F, \Pi\eta)$, it is not fully understood yet.

References

[1] Pierre Ageron, *Structure des logiques et logique des structures : logiques, catégories, esquisses*, thèse de doctorat, Université Paris 7, 1991

[2] Pierre Ageron, *Effective taxonomies and crossed taxonomies*, Cahiers de topologie et de géométrie différentielle catégoriques **XXXVII** (1996), 82–90

[3] Richard Dedekind, *Was sind und was sollen die Zahlen ?*, Vieweg, Braunschweig, 1888

[4] Décio Krause and Jean-Yves Béziau, *Relativizations of the principe of identity*, Logic Journal of the Interest Group in Pure and Applied Logic **5** (1997), 327–338

[5] Joachim Lambek, *Deductive systems and categories. I: Syntactic calculus and residuated categories*, Mathematical System Theory **2** (1968), 287–318

[6] F. William Lawvere, *Some thoughts on the future of category theory*, in: Category Theory, Proceedings of the International Conference (Como, 1990), Lectures Notes in Mathematics **1488**, Springer, Heidelberg-New York-Berlin, 1991, 1–13

[7] Marie-Anne Moens, Ugo Berni-Catani and Francis Borceux, *On regular presheaves and regular semicategories*, Cahiers de topologie et de géométrie différentielle catégoriques **XLIII** (2002) 163–190

[8] Charles Peirce, *On the algebra of logic: a contribution to the philosophy of notation*, American Journal of Mathematics **7** (1885), 180–202

[9] Isar Stubbe, *Categorical structures enriched in a quantaloid: categories and semicategories*, dissertation doctorale, Université catholique de Louvain-la-Neuve, 2003

Pierre Ageron
Département de mathématiques et mécanique
Université de Caen, campus II
BP 5186
14032 Caen cedex
France
e-mail: `ageron@math.unicaen.fr`

Part II
Identity and Nature
of Logical Structures

J.-Y. Beziau (Ed.), *Logica Universalis, 2nd edition*, 97–109
© 2007 Birkhäuser Verlag Basel/Switzerland

Equipollent Logical Systems

Carlos Caleiro and Ricardo Gonçalves

Abstract. When can we say that two distinct logical systems are, nevertheless, essentially the "same"? In this paper we discuss the notion of "sameness" between logical systems, bearing in mind the expressive power of their associated spaces of theories, but without neglecting their syntactical dimension. Departing from a categorial analysis of the question, we introduce the new notion of *equipollence* between logical systems. We use several examples to illustrate our proposal and to support its comparison to other proposals in the literature, namely homeomorphisms [7], and translational equivalence (or synonymity) [6].

Keywords. Logical system, theory space, equipollence.

1. Introduction

When we talk about classical propositional logic (CPL), for example, we are most often not referring to just a particular entity but rather to a family of (possibly very) different logical systems that do all present essentially the "same" CPL. But what do we mean when we say that two logical systems are the "same"? Our goal is to find a satisfactory answer to this question and to show how our proposal, *equipollence*, relates to earlier proposals in the literature [7, 6, 1].

Certainly there is no point in even discussing this question without first agreeing on what a logical system is. Still, we do not wish to dwell on such a delicate subject here. The interested reader can find a very complete discussion of this theme in [4], along with a myriad of different viewpoints and possible definitions. For what we are concerned in this paper, we will restrict our attention to Tarski-style consequence operators over a structured language. This choice is certainly not without controversy, but most logicians should at least agree that it covers a wide range of well known examples.

This work was partially supported by FCT and EU FEDER, namely, via the Projects POCTI/2001/MAT/37239 FibLog and POCTI/MAT/55796/2004 QuantLog of CLC. The second author was also supported by FCT under the PhD grant SFRH/BD/18345/2004/SV7T.

Methodologically, we shall adopt a category-theoretical perspective, where logical systems will constitute our main category of interest. An adequate notion of morphism between logical systems will however be essential, since we are not only interested in logical systems as objects by themselves, but we are particularly interested in the way different logical systems relate to each other. Morphisms between logical systems will be uniform translations between the structured languages that preserve the corresponding consequence operators. We will also take into account that logical systems generate spaces of theories, which in their turn can also be endowed with a suitable categorial structure, in a functorial way. This categorial setup is directly inspired by [8, 2].

As a first attempt to attack the problem, we shall analyze the most obvious idea, that is, to try and characterize "sameness" between logical systems using the built-in notion of isomorphism in the corresponding category. It turns out that isomorphisms between logical systems are very closely related to Pollard's homeomorphisms, as presented in [7], and thus suffer from very similar merits and shortcomings. In fact, with the support of suitable examples, we can claim that, although meaningful, this notion is syntactically too strict as a definition of "sameness". To overcome this strictness, we shall then take a brief look at the theory spaces generated by these logical systems, thus abstracting away from syntactical fine details. At the level of logical systems, individual formulas play a fundamental role, while at the level of their associated theory spaces, the expressive power of isolated formulas is not so much important. It turns out, however, that the built-in notion of isomorphism between theory spaces is now too broad, as we also illustrate. The correct notion of "sameness" seems therefore to lie somewhere in between these two notions of isomorphism. By capturing the right amount of interplay between the two, and taking into account the functorial relationship between the categories of logical systems and theory spaces, we finally manage to isolate our notion of *equipollence*, in a way similar to the one developed in [2]. It is interesting to note that the ideas underlying *equipollence* are very closely related to those that stand behind Pelletier and Urquhart's proposal [6] of translational equivalence (or synonymity). *Equipollence* can nevertheless be shown to be more widely applicable, although the two notions coincide under mild assumptions, that we make explicit.

We begin by introducing, in section 2, the categories **Log** of logical systems and **Tsp** of theory spaces, that will be used throughout the paper. In section 3, we analyze the notions of isomorphism in these categories, and we provide examples that help us to conclude that none of them, per se, is satisfactory as a definition of "sameness" between logical systems. We also show how isomorphisms in **Log** relate to the homeomorphisms of [7]. Our notion of *equipollence* is then introduced and analyzed in section 4. We illustrate our proposal with a few examples, that will also enhance its comparison with the notion of synonymity of [6]. We conclude, in section 5, with an overview of our proposal and a discussion of its adequation.

2. From logical systems to theory spaces

In this section we introduce the precise definitions of logical systems and theory spaces, along with their associated categorial structure. We should however start with the more syntactical details. We will consider logical languages that are freely generated from a given signature including constructors of different arities, as is most often the case.

Definition 2.1. A *signature* is an indexed set $\Sigma = \{\Sigma^n\}_{n \in \mathbb{N}}$, where each Σ^n is the set of n-ary constructors.

We consider that the set of *propositional variables* is included in Σ^0.

Definition 2.2. The *language* over a given a signature Σ, which we denote by L_Σ, is build inductively in the usual way:

- $\Sigma^0 \subseteq L_\Sigma$;
- If $n \in \mathbb{N}$, $\varphi_1, \ldots, \varphi_n \in L_\Sigma$ and $c \in \Sigma^n$ then $c(\varphi_1, \ldots, \varphi_n) \in L_\Sigma$.

We call Σ-*formulas* to the elements of L_Σ, or simply *formulas* when Σ is clear from the context.

Definition 2.3. A *logical system* is a pair $\mathcal{L} = \langle \Sigma, \vdash \rangle$, where Σ is a signature and $\vdash$ is a consequence operator on L_Σ (in the sense of Tarski, cf. eg. [9]), that is, $\vdash : 2^{L_\Sigma} \to 2^{L_\Sigma}$ is a function that satisfies the following properties, for every $\Gamma, \Phi \subseteq L_\Sigma$:

Extensiveness: $\Gamma \subseteq \Gamma^\vdash$;
Monotonicity: If $\Gamma \subseteq \Phi$ then $\Gamma^\vdash \subseteq \Phi^\vdash$;
Idempotence: $(\Gamma^\vdash)^\vdash \subseteq \Gamma^\vdash$.

For the sake of generality, we do not require here the consequence operator to be finitary, or even structural.

Since we will need to talk about the expressive power of the language of a given logical system, we will need to refer to its connectives (primitive or derived). For the purpose, we consider fixed once and for all a set $\Xi = \{\xi_i\}_{i \in \mathbb{N}^+}$ of *metavariables*. Then, given a signature Σ and $k \in \mathbb{N}$, we can consider the set L_Σ^k defined inductively by:

- $\{\xi_1, \ldots, \xi_k\} \subseteq L_\Sigma^k$;
- $\Sigma^0 \subseteq L_\Sigma^k$;
- If $n \in \mathbb{N}$, $\varphi_1, \ldots, \varphi_n \in L_\Sigma^k$ and $c \in \Sigma^n$ then $c(\varphi_1, \ldots, \varphi_n) \in L_\Sigma^k$.

Clearly, we have that $L_\Sigma = L_\Sigma^0$. We can also consider the set $L_\Sigma^\omega = \bigcup_{n \in \mathbb{N}} L_\Sigma^n$. Given $\varphi \in L_\Sigma^k$ we will write $\varphi(\xi_1 \setminus \psi_1, \ldots, \xi_k \setminus \psi_k)$ to denote the formula that is obtained from φ by simultaneously replacing each occurrence of ξ_i in φ by ψ_i, for every $i \leq k$.

A *derived connective* of arity $k \in \mathbb{N}$ is a λ-term $d = \lambda \xi_1 \ldots \xi_k.\varphi$ where $\varphi \in L_\Sigma^k$. We denote by DC_Σ^k the set of all derived connectives of arity k over Σ. Note that, if $c \in \Sigma_k$ is a primitive connective, it can also be considered as the derived connective $c = \lambda\xi_1 \ldots \xi_k.c(\xi_1, \ldots, \xi_k)$. Given a derived connective $d = \lambda\xi_1 \ldots \xi_n.\varphi$ we will often write $d(\psi_1, \ldots, \psi_n)$ instead of $\varphi(\xi_1 \setminus \psi_1, \ldots, \xi_n \setminus \psi_n)$.

Different languages generated from different signatures can be translated according to the following notion of morphism, where primitive connectives from one signature are mapped to derived connectives from another signature, while preserving the corresponding arities.

Definition 2.4. Given signatures Σ_1 and Σ_2, a *signature morphism* $h : \Sigma_1 \to \Sigma_2$ is an $\mathbb{N}$-indexed family of functions $h = \{h^n : \Sigma_1^n \to DC_{\Sigma_2}^n\}_{n \in \mathbb{N}}$.

Given a signature morphism $h : \Sigma_1 \to \Sigma_2$, we can define its free extensions $h : L_{\Sigma_1}^k \to L_{\Sigma_2}^k$ for $k \in \mathbb{N}$, and $h : L_{\Sigma_1}^\omega \to L_{\Sigma_2}^\omega$ inductively, as follows:

- $h(\xi_i) = \xi_i$ if $\xi_i \in \Xi$;
- $h(c) = h^0(c)$ if $c \in \Sigma_1^0$;
- $h(c(\varphi_1, \ldots, \varphi_n)) = h^n(c)(h(\varphi_1), \ldots, h(\varphi_n))$ if $c \in \Sigma_1^n$.

A translation function h that satisfies the above requirements will be dubbed *uniform*.

Signatures and their morphisms constitute a category **Sig** with identities $id_\Sigma : \Sigma \to \Sigma$ such that $id_\Sigma^n(c) = \lambda\xi_1 \ldots \xi_n.c(\xi_1, \ldots, \xi_n)$ for every $n \in \mathbb{N}$ and $c \in \Sigma^n$, and the composition of signature morphisms $f : \Sigma_1 \to \Sigma_2$ and $g : \Sigma_2 \to \Sigma_3$ defined to be $g \circ f : \Sigma_1 \to \Sigma_3$ such that $(g \circ f)^n(c) = \lambda\xi_1 \ldots \xi_n.g(\varphi)$, assuming that $f^n(c) = \lambda\xi_1 \ldots \xi_n.\varphi$.

We can now take advantage of uniform translations to put forth the notion of morphism between logical systems. Given a function $h : L_{\Sigma_1} \to L_{\Sigma_2}$ and $\Phi \subseteq L_{\Sigma_1}$ we can consider the set $h[\Phi] = \{h(\varphi) : \varphi \in \Phi\}$.

Definition 2.5. Let $\mathcal{L}_1 = \langle \Sigma_1, \vdash_1 \rangle$ and $\mathcal{L}_2 = \langle \Sigma_2, \vdash_2 \rangle$ be logical systems. A *logical system morphism* $h : \mathcal{L}_1 \to \mathcal{L}_2$ is a signature morphism $h : \Sigma_1 \to \Sigma_2$ such that $h[\Phi^{\vdash_1}] \subseteq h[\Phi]^{\vdash_2}$ for every $\Phi \subseteq L_{\Sigma_1}$.

Logical systems and their morphisms constitute a concrete category **Log**, over **Sig**. The following is a well known useful lemma.

Lemma 2.6. *Let $\mathcal{L}_1 = \langle \Sigma_1, \vdash_1 \rangle$ and $\mathcal{L}_2 = \langle \Sigma_2, \vdash_2 \rangle$ be logical systems, and $h : \mathcal{L}_1 \to \mathcal{L}_2$ a **Log**-morphism. Then, $h[\Phi^{\vdash_1}]^{\vdash_2} = h[\Phi]^{\vdash_2}$ for every $\Phi \subseteq L_{\Sigma_1}$.*

Proof. Clearly, by the extensiveness of $\vdash_1$, $\Phi \subseteq \Phi^{\vdash_1}$. Therefore, $h[\Phi] \subseteq h[\Phi^{\vdash_1}]$ and by the monotonicity of $\vdash_2$ we get that $h[\Phi]^{\vdash_2} \subseteq h[\Phi^{\vdash_1}]^{\vdash_2}$. On the other hand, since h is a morphism, we have that $h[\Phi^{\vdash_1}] \subseteq h[\Phi]^{\vdash_2}$. Thus, by the monotonicity of $\vdash_2$ it follows that $h[\Phi^{\vdash_1}]^{\vdash_2} \subseteq (h[\Phi]^{\vdash_2})^{\vdash_2}$. Now, by the idempotence of $\vdash_2$ we get that $(h[\Phi]^{\vdash_2})^{\vdash_2} \subseteq h[\Phi]^{\vdash_2}$. Therefore, we have $h[\Phi^{\vdash_1}]^{\vdash_2} \subseteq h[\Phi]^{\vdash_2}$. $\square$

As usual, a *theory* of a logical system $\mathcal{L} = \langle \Sigma, \vdash \rangle$ is a set $\Phi \subseteq L_\Sigma$ such that $\Phi^\vdash = \Phi$. We denote by $Th(\mathcal{L})$ the set of all theories of $\mathcal{L}$. It is well known that the structure of the set $Th(\mathcal{L})$ under the inclusion ordering is very important. Namely, it is always a complete lattice.

Definition 2.7. A *theory space* is a complete lattice $tsp = \langle Th, \leq \rangle$, that is, a partial order $\leq$ on the set Th such that every $T \subseteq Th$ has a least upper-bound (or join) $\bigvee T$.

In particular, given a logical system $\mathcal{L} = \langle \Sigma, \vdash \rangle$, $tsp_{\mathcal{L}} = \langle Th(\mathcal{L}), \subseteq \rangle$ is always a theory space (cf. eg. [9]). Moreover, the language translations associated to logical system morphisms always act on the consequence operators in such a way that joins are preserved in the corresponding theory spaces.

Definition 2.8. Let $tsp_1 = \langle Th_1, \leq_1 \rangle$ and $tsp_2 = \langle Th_2, \leq_2 \rangle$ be theory spaces. A *theory spaces morphism* $h : tsp_1 \to tsp_2$ is a function $h : Th_1 \to Th_2$ such that $h(\bigvee_1 T) = \bigvee_2 h[T]$ for every $T \subseteq Th_1$.

We prove now a straightforward but useful property of theory spaces morphisms.

Lemma 2.9. *Let* $tsp_1 = \langle Th_1, \leq_1 \rangle$ *and* $tsp_2 = \langle Th_2, \leq_2 \rangle$ *be theory spaces and* $h : tsp_1 \to tsp_2$ *a theory spaces morphism. Then* h *is order preserving, that is, for every* $\Phi, \Gamma \in Th_1$, *if* $\Phi \leq_1 \Gamma$ *then* $h(\Phi) \leq_2 h(\Gamma)$.

Proof. Clearly, if $\Phi \leq_1 \Gamma$ then $\bigvee_1 \{\Phi, \Gamma\} = \Gamma$. Therefore, since h preserves joins, $h(\Gamma) = h(\bigvee_1 \{\Phi, \Gamma\}) = \bigvee_2 \{h(\Phi), h(\Gamma)\}$. Consequently, $h(\Phi) \leq_2 h(\Gamma)$. $\square$

Theory spaces and their morphisms constitute the category **Tsp**, with the usual identity and composition of functions. What is more, the definition of the space of theories induced by a logical system can be extended to a functor.

Definition 2.10. The maps

- $Th(\mathcal{L}) = tsp_{\mathcal{L}}$;
- $Th(h : \mathcal{L}_1 \to \mathcal{L}_2) : tsp_{\mathcal{L}_1} \to tsp_{\mathcal{L}_2}$, with $Th(h)(\Phi) = h[\Phi]^{\vdash_2}$ if $\mathcal{L}_2 = \langle \Sigma_2, \vdash_2 \rangle$, for every $\Phi \in Th(\mathcal{L}_1)$,

constitute a functor $Th : \mathbf{Log} \to \mathbf{Tsp}$.

Indeed, it can be shown that Th is an adjoint functor, although we shall not need to use this fact here. What is important, however, is that Th does not reflect isomorphisms from **Tsp** to **Log**. Recall however that, as a simple consequence of the functoriality of Th, isomorphisms in **Log** are preserved along Th to isomorphisms in **Tsp**.

3. Isomorphisms

In this section we analyze the notions of isomorphism in the categories **Log** and **Tsp**. These would certainly be the first obvious ways of measuring the degree of "sameness" of two given logical systems. We will see, however, with the help of some examples, that none of these notions is fully satisfactory.

First of all let us recall that an *isomorphism* in an arbitrary category $\mathbf{C}$ is a morphism $f : C_1 \to C_2$ for which there exists a morphism $g : C_2 \to C_1$ such that $g \circ f = id_{C_1}$ and $f \circ g = id_{C_2}$. In this case, the morphism g is also an isomorphism, and is usually referred to as the *inverse* of f, and denoted by f^{-1} due to its uniqueness. Of course, it is also the case that $f = g^{-1}$. Two $\mathbf{C}$-objects C_1 and C_2 are *isomorphic* provided that there exists an isomorphism between them.

Since it will be useful, we first present a characterization of isomorphism in the category $\mathbf{Sig}$ of signatures.

Proposition 3.1. *Two signatures Σ_1 and Σ_2 are isomorphic if and only if there exists a family of bijections $h = \{h^n : \Sigma_1^n \to \Sigma_2^n\}_{n \in \mathbb{N}}$.*

Proof. Suppose first that Σ_1 and Σ_2 are isomorphic in $\mathbf{Sig}$. Then there exist signature morphisms $f : \Sigma_1 \to \Sigma_2$ and $g : \Sigma_2 \to \Sigma_1$ such that $g \circ f = id_{\Sigma_1}$ and $f \circ g = id_{\Sigma_2}$. Given $n \in \mathbb{N}$ and $c \in \Sigma_1^n$, let us first prove that $f^n(c) = \lambda \xi_1 \ldots \xi_n.c'(\xi_1, \ldots, \xi_n)$, where $c' \in \Sigma_2^n$. Assume by absurd that $f^n(c) = \lambda \xi_1 \ldots \xi_n.\varphi$ where φ had more than one constructor. Then, clearly, by the uniformity condition, $g(f^n(c))$ would also have more than one connective, and so, $g(f^n(c))$ could not be $\lambda \xi_1 \ldots \xi_n.c(\xi_1, \ldots, \xi_n)$, which would contradict the fact that $g \circ f = id_{\Sigma_1}$. Thus, we can define $h^n(c) = c'$. The fact that, for each $n \in \mathbb{N}$, the function h^n must then be a bijection follows immediately.

Suppose now that there exists a family of bijections $h = \{h^n : \Sigma_1^n \to \Sigma_2^n\}_{n \in \mathbb{N}}$. Then we can build signature morphisms $f : \Sigma_1 \to \Sigma_2$ and $g : \Sigma_2 \to \Sigma_1$ defined by $h^n(c) = \lambda \xi_1 \ldots \xi_n.f^n(c)(\xi_1, \ldots, \xi_n)$ for every $c \in \Sigma_1^n$, and $g^n(c') = \lambda \xi_1 \ldots \xi_n.c(\xi_1, \ldots, \xi_n)$ for every $c' = f^n(c) \in \Sigma_2^n$, respectively. It is straightforward that $g \circ f = id_{\Sigma_1}$ and $f \circ g = id_{\Sigma_2}$. $\square$

It should be clear that signature isomorphisms induce bijective translations between the generated languages.

We can now characterize isomorphisms in $\mathbf{Log}$.

Proposition 3.2. *Two logical systems $\mathcal{L}_1 = \langle \Sigma_1, \vdash_1 \rangle$ and $\mathcal{L}_2 = \langle \Sigma_2, \vdash_2 \rangle$ are isomorphic if and only if there exists a signature isomorphism $h : \Sigma_1 \to \Sigma_2$ such that $h[\Phi^{\vdash_1}] = h[\Phi]^{\vdash_2}$ for every $\Phi \subseteq L_{\Sigma_1}$.*

Proof. Suppose first that $h : \mathcal{L}_1 \to \mathcal{L}_2$ is an isomorphism in $\mathbf{Log}$. That is, there exists a $\mathbf{Log}$-morphism $g : \mathcal{L}_2 \to \mathcal{L}_1$ such that $g \circ h = id_{\mathcal{L}_1}$ and $h \circ g = id_{\mathcal{L}_2}$. Both h and g are also signature isomorphisms $h : \Sigma_1 \to \Sigma_2$ and $g : \Sigma_2 \to \Sigma_1$ that also satisfy $h[\Phi^{\vdash_1}] \subseteq h[\Phi]^{\vdash_2}$ and $g[\Gamma^{\vdash_2}] \subseteq g[\Gamma]^{\vdash_1}$, for every $\Phi \subseteq L_{\Sigma_1}$ and every $\Gamma \subseteq L_{\Sigma_2}$. But then we have that $g[h[\Phi]^{\vdash_2}] \subseteq g[h[\Phi]]^{\vdash_1} = \Phi^{\vdash_1}$, and so it follows that $h[\Phi]^{\vdash_2} = h[g[h[\Phi]^{\vdash_2}]] \subseteq h[\Phi^{\vdash_1}]$, thus rendering $h[\Phi^{\vdash_1}] = h[\Phi]^{\vdash_2}$.

Suppose now that there exists a signature isomorphism $h : \Sigma_1 \to \Sigma_2$ such that $h[\Phi^{\vdash_1}] = h[\Phi]^{\vdash_2}$ for every $\Phi \subseteq L_{\Sigma_1}$. Then there exists a $\mathbf{Sig}$-morphism $g : \Sigma_2 \to \Sigma_1$ such that $g \circ h = id_{\Sigma_1}$ and $h \circ g = id_{\Sigma_2}$. Clearly $h : \mathcal{L}_1 \to \mathcal{L}_2$ is also a $\mathbf{Log}$-morphism, but so is $g : \mathcal{L}_2 \to \mathcal{L}_1$. In fact, $h[g[\Gamma]^{\vdash_1}] = h[g[\Gamma]]^{\vdash_2} = \Gamma^{\vdash_2}$, and so $g[\Gamma^{\vdash_1}] = g[h[g[\Gamma]^{\vdash_1}]] = g[\Gamma^{\vdash_2}]$. Therefore, $\mathcal{L}_1$ and $\mathcal{L}_2$ are isomorphic. $\square$

This notion of isomorphism in the category **Log** is very closely related to Pollard's notion of *homeomorphism*, as introduced in [7]. Indeed, the only difference is that homeomorphism does only require the language translation function $h : L_{\Sigma_1} \to L_{\Sigma_2}$ to be a bijection, but not necessarily one that is uniform with respect to the structure of formulas. The same applies, of course, to its inverse $h^{-1} : L_{\Sigma_2} \to L_{\Sigma_1}$. In this respect, we should make clear that we find Pollard's notion slightly odd. Either one does not require the language to bear any structure at all, in which case a simple translation function would make perfect sense, or else one should not neglect this structure when translating formulas across logical systems. Still, if we choose to simply ignore the way L_Σ is build from the signature Σ, then we can as well assume that the language is build from a new signature Ω where all the relevant formulas come now without any structure whatsoever, that is, $\Omega^0 = L_\Sigma$ and $\Omega^n = \emptyset$ for every $n > 0$. In doing this transformation we would get $L_\Omega = L_\Sigma$, and it is clear that isomorphisms in **Log** and homeomorphisms would now coincide. Nevertheless, we maintain that translations across logical systems should be effective, to a certain extent, and therefore some kind of uniformity must be required (even if in a more general form than the one we are considering here).

Isomorphisms in **Log** are, however, too strict as a definition of "sameness". Many times logical systems differ just in the number of equivalent sentences they possess, while still exhibiting essentially the same closure properties. The following example illustrates this fact, and applies also to homeomorphisms.

Example. Consider the following logical systems:

- $\mathcal{L}_1 = \langle \Sigma_1, \vdash_1 \rangle$ where $\Sigma_1^0 = \{\top\}$ and $\Sigma_1^n = \emptyset$ for $n > 0$; and $\vdash_1$ is such that $\emptyset^{\vdash_1} = \{\top\}$; and
- $\mathcal{L}_2 = \langle \Sigma_2, \vdash_2 \rangle$ where $\Sigma_2^0 = \{\top_1, \top_2\}$ and $\Sigma_2^n = \emptyset$ for $n > 0$; and $\vdash_2$ is such that $\emptyset^{\vdash_2} = \{\top_1, \top_2\}$.

Clearly, $\mathcal{L}_1$ and $\mathcal{L}_2$ have exactly the same expressive power and there is absolutely no reason why they should not be considered the "same". But it is also clear that they are not isomorphic, nor homeomorphic.

Even if the set of formulas of each logical system is not finite, isomorphism remains a too strong condition. We present one example where the two given logical systems should clearly be the "same", but there cannot exist an isomorphism between them. They are homeomorphic, though, although in a non-uniform way (even if we allow a more general definition of uniformness).

Example. Consider the following logical systems:

- $\mathcal{L}_1 = \langle \Sigma_1, \vdash_1 \rangle$ where $\Sigma_1^0 = \{p\}$, $\Sigma_1^1 = \{\neg\}$ and $\Sigma_1^n = \emptyset$ for $n > 1$; and $\vdash_1$ is such that $\neg$ behaves like classical negation; and
- $\mathcal{L}_2 = \langle \Sigma_2, \vdash_2 \rangle$ where $\Sigma_2^0 = \{p\}$, $\Sigma_2^1 = \{\neg_1, \neg_2\}$ and $\Sigma_2^n = \emptyset$ if $n > 1$; and $\vdash_2$ is such that both $\neg_1$ and $\neg_2$ behave like classical negation.

Clearly these two logics must be the "same", since $\mathcal{L}_2$ is just $\mathcal{L}_1$ with two copies of $\neg$. However, it is obvious that they are not isomorphic.

Despite the existence of a homeomorphism between $\mathcal{L}_1$ and $\mathcal{L}_2$, it cannot be made uniform (in our sense nor in any broader sense). It is clear that $Th(\mathcal{L}_1) = \{\{\neg^{2n}p : n \in \mathbb{N}\}, \{\neg^{2n+1}p : n \in \mathbb{N}\}\}$, and that $Th(\mathcal{L}_2) = \{\{\{\neg_1, \neg_2\}^{2n}p : n \in \mathbb{N}\}, \{\{\neg_1, \neg_2\}^{2n+1}p : n \in \mathbb{N}\}\}$. Assume by absurd that there existed a uniform homeomorphism $h : \mathcal{L}_2 \to \mathcal{L}_1$. Being uniform, h would have to be presented inductively. So, for some $i \in \mathbb{N}$, there would exist distinct sequences $u, v \in \{\neg_1, \neg_2\}^i$ such that $h(u\varphi) = \neg^k h(\varphi)$ and $h(v\varphi) = \neg^j h(\varphi)$, for some $k, j \in \mathbb{N}$ and every formula φ. Then, $h(uv\varphi) = \neg^k h(v\varphi) = \neg^{k+j} h(\varphi)$ and $h(vu\varphi) = \neg^j h(u\varphi) = \neg^{j+k} h(\varphi)$. So we would have $h(uv\varphi) = h(vu\varphi)$ which, together with the bijectivity of h, would contradict the fact that $uv\varphi \neq vu\varphi$.

In both these examples, the one fact that stands out in support of the "sameness" of the logical systems involved is the fact that their theory spaces have exactly the same structure. This fact should certainly be, at least, a necessary condition for considering the logical systems to be the "same". Of course, an isomorphism in **Log** is always mapped by the functor Th to an isomorphism in **Tsp**. Let us then present a characterization of isomorphisms in **Tsp**, which should be interesting to analyze, even if we have reasons to believe that having isomorphic theory spaces may not be enough as a criterion for dubbing two logical systems the "same".

Proposition 3.3. *Two theory spaces $tsp_1 = \langle Th_1, \leq_1 \rangle$ and $tsp_2 = \langle Th_2, \leq_2 \rangle$ are isomorphic if and only if there exists a bijection $h : Th_1 \to Th_2$ such that, for every $\Phi, \Gamma \in Th_1$, $\Phi \leq_1 \Gamma$ if and only if $h(\Phi) \leq_2 h(\Gamma)$.*

Proof. Suppose that $h : tsp_1 \to tsp_2$ is an isomorphism in **Tsp**, that is, there exists a **Tsp**-morphism $g : tsp_2 \to tsp_1$ such that $g \circ h = id_{tsp_1}$ and $h \circ g = id_{tsp_2}$. Clearly $h : Th_1 \to Th_2$ must be a bijection and $g = h^{-1}$. If $\Phi \leq_1 \Gamma$, since h is a **Tsp**-morphism, lemma 2.9 implies that $h(\Phi) \leq_2 h(\Gamma)$. On the other hand, since g is also a **Tsp**-morphism, if $h(\Phi) \leq_2 h(\Gamma)$, lemma 2.9 implies that $\Phi = h^{-1}(h(\Phi)) \leq_1 h^{-1}(h(\Gamma)) = \Gamma$.

Assume now that $h : Th_1 \to Th_2$ is a bijection, and $\Phi \leq_1 \Gamma$ if and only if $h(\Phi) \leq_2 h(\Gamma)$, for every $\Phi, \Gamma \in Th_1$. We first show that $h : tsp_1 \to tsp_2$ is a **Tsp**-morphism. If $T \subseteq Th_1$ and $\Phi \in T$, then $\Phi \leq_1 \bigvee_1 T$. Therefore $h(\Phi) \leq_2 h(\bigvee_1 T)$, and consequently $\bigvee_2 h[T] \leq_2 h(\bigvee_1 T)$. On the other hand, clearly $h(\Phi) \leq_2 \bigvee_2 h[T]$. Since h is a bijection, we can rewrite this to $h(\Phi) \leq_2 h(h^{-1}(\bigvee_2 h[T]))$. Therefore, $\Phi \leq_1 h^{-1}(\bigvee_2 h[T])$ and we conclude that $\bigvee_1 T \leq_1 h^{-1}(\bigvee_2 h[T])$. Thus $h(\bigvee_1 T) \leq_2 \bigvee_2 h[T]$ and h is indeed a **Tsp**-morphism. Analogously, we can show that $h^{-1} : tsp_2 \to tsp_1$ is also a **Tsp**-morphism. $\square$

Saying that two logical systems, $\mathcal{L}_1$ and $\mathcal{L}_2$, are the "same" if $tsp_{\mathcal{L}_1}$ and $tsp_{\mathcal{L}_2}$ are isomorphic in **Tsp** is expectedly not very satisfactory. Namely, this is due to the fact that theory space morphisms are not guided by syntax, that is, they neglect the expressive power of isolated formulas, by translating directly theories to theories. Although a necessary condition, too many logical systems that we would not want to consider the "same" end up having isomorphic theory spaces. Below, we present two such examples.

Example. Consider the following logical systems:

- $\mathcal{L}_1 = \langle \Sigma_1, \vdash_1 \rangle$ where $\Sigma_1^0 = \{a, b\}$ and $\Sigma_1^n = \emptyset$ for $n > 0$; and $\vdash_1$ is the identity; and
- $\mathcal{L}_2 = \langle \Sigma_2, \vdash_2 \rangle$ where $\Sigma_2^0 = \{a, b, ab\}$ and $\Sigma_2^n = \emptyset$ for $n > 0$; and $\vdash_2$ extends $\vdash_1$ by letting $ab \in \Phi^{\vdash_2}$ if and only if $\{a, b\} \subseteq \Phi^{\vdash_2}$.

Clearly, $Th(\mathcal{L}_1) = \{\emptyset, \{a\}, \{b\}, \{a, b\}\}$ and $Th(\mathcal{L}_2) = \{\emptyset, \{a\}, \{b\}, \{a, b, ab\}\}$ are isomorphic. However, $\mathcal{L}_1$ and $\mathcal{L}_2$ should not be considered the "same". In particular, $\mathcal{L}_2$ contains the formula ab that can be seen as a bottom particle since $\{ab\}^{\vdash_2} = L_{\Sigma_2}$. It is also clear that, in $\mathcal{L}_1$, no such formula exists.

A more interesting example is that of linear temporal logic.

Example. Let P be a set of propositional variables. Consider the following two fragments of discrete linear temporal logic (LTL), eg. as in [5], where X stands for "in the next instant" and G for "always in the future":

- $\mathcal{L}_1 = \langle \Sigma_1, \vdash_1 \rangle$ where $\Sigma_1^0 = P$, $\Sigma_1^1 = \{\neg, X\}$, $\Sigma_1^2 = \{\Rightarrow\}$, and $\Sigma_1^n = \emptyset$ for $n > 2$; and
- $\mathcal{L}_2 = \langle \Sigma_2, \vdash_2 \rangle$ where $\Sigma_2^0 = P$, $\Sigma_2^1 = \{\neg, X, G\}$, $\Sigma_2^2 = \{\Rightarrow\}$, and $\Sigma_2^n = \emptyset$ for $n > 2$; and

both $\vdash_1$ and $\vdash_2$ are the corresponding fragments of the consequence operator $\vdash$ of full LTL.

Most notably, it turns out that $\{G\varphi\}^{\vdash} = \{X^n\varphi : n > 0\}^{\vdash}$, which is of course also true in $\mathcal{L}_2$. Therefore, it is straightforward to verify that $tsp_{\mathcal{L}_1}$ and $tsp_{\mathcal{L}_2}$ are isomorphic. However, in $\mathcal{L}_1$ there is no single formula with the same expressive power of $G\varphi$. This is certainly a very good reason not to dub these two logical systems the "same".

4. Equipollence

At this point, it seems clear that the notion of "sameness" between logical systems must lie somewhere in between the notions of isomorphism in **Log** and **Tsp**. For the reasons already discussed, two logical systems should indeed have isomorphic theory spaces whenever they are to be called the "same". This isomorphism should however be based on a formula by formula translation, which must also be uniform on the structure of formulas. Isomorphisms in **Log** do satisfy this constraint, although they seem to be too sensitive to differences in the cardinality of the sets of formulas and constructors of the two logical systems. In the end, what we seem to need is a back and forth translation between the logical systems, that may perhaps not constitute an isomorphism in **Log**, but which induces an isomorphism between the corresponding theory spaces. Figure 1 depicts the idea behind our notion of *equipollence*, that is rigorously formulated below.

Definition 4.1. Two logical systems $\mathcal{L}_1 = \langle \Sigma_1, \vdash_1 \rangle$ and $\mathcal{L}_2 = \langle \Sigma_2, \vdash_2 \rangle$ are *equipollent* if there exist **Log**-morphisms $h : \mathcal{L}_1 \to \mathcal{L}_2$ and $g : \mathcal{L}_2 \to \mathcal{L}_1$ such that $Th(h)$ and $Th(g)$ establish an isomorphism of $tsp_{\mathcal{L}_1}$ and $tsp_{\mathcal{L}_2}$ with $Th(h) = Th(g)^{-1}$.

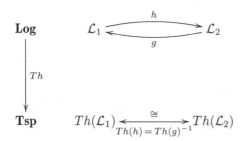

FIGURE 1. Equipollent logical systems.

As we intended, it is trivial to check that isomorphisms in **Log** constitute a very special case of equipollence. Furthermore, as expected, equipollent logical systems are always required to have isomorphic theory spaces.

We shall now provide an alternative, more appealing, characterization of equipollence in terms of the internal notion of logical equivalence provided by each logical system $\mathcal{L} = \langle \Sigma, \vdash \rangle$. Recall that two formulas $\varphi, \gamma \in L_\Sigma$ are said to be *logically equivalent* in $\mathcal{L}$ if both $\varphi \in \{\gamma\}^\vdash$ and $\gamma \in \{\varphi\}^\vdash$, or equivalently if $\{\varphi\}^\vdash = \{\gamma\}^\vdash$. We denote this fact by $\varphi \equiv_\mathcal{L} \gamma$. The following lemma shows that the theories of $\mathcal{L}$ are in fact independent, modulo logically equivalent formulas, of the way they are presented.

Lemma 4.2. *Let* $\Phi, \Gamma \subseteq L_\Sigma$. *Then* $\Phi^\vdash = \Gamma^\vdash$ *whenever the following two conditions are satisfied:*

- *for every* $\varphi \in \Phi$ *there exists* $\varphi' \in \Gamma$ *such that* $\varphi \equiv_\mathcal{L} \varphi'$;
- *for every* $\gamma \in \Gamma$ *there exists* $\gamma' \in \Phi$ *such that* $\gamma \equiv_\mathcal{L} \gamma'$.

Proof. Let us assume that both conditions hold. If $\varphi \in \Phi$ then $\varphi' \in \Gamma$. But $\varphi \equiv_\mathcal{L} \varphi'$ and therefore $\varphi \in \{\varphi'\}^\vdash \subseteq \Gamma^\vdash$, using also the monotonicity of $\vdash$. Analogously, we can show that $\Gamma \subseteq \Phi^\vdash$, and conclude that $\Phi^\vdash = \Gamma^\vdash$ by using the monotonicity and idempotence of $\vdash$. $\qquad\square$

Next we state the aimed alternative characterization of the notion of equipollence. Besides its simplicity, the characterization will also be useful in order to compare equipollence with the notion of translational equivalence (or synonymity) due to Pelletier and Urquhart [6].

Proposition 4.3. *Let* $\mathcal{L}_1 = \langle \Sigma_1, \vdash_1 \rangle$ *and* $\mathcal{L}_2 = \langle \Sigma_2, \vdash_2 \rangle$ *be logical systems. Then* $\mathcal{L}_1$ *and* $\mathcal{L}_2$ *are equipollent if and only if there exist* **Log***-morphisms* $h : \mathcal{L}_1 \to \mathcal{L}_2$ *and* $g : \mathcal{L}_2 \to \mathcal{L}_1$ *such that the following two conditions hold:*

- $\varphi \equiv_{\mathcal{L}_1} g(h(\varphi))$ *for every* $\varphi \in L_{\Sigma_1}$;
- $\gamma \equiv_{\mathcal{L}_2} h(g(\gamma))$ *for every* $\gamma \in L_{\Sigma_2}$.

Proof. Assuming that $\mathcal{L}_1$ and $\mathcal{L}_2$ are equipollent, let $h : \mathcal{L}_1 \to \mathcal{L}_2$ and $g : \mathcal{L}_2 \to \mathcal{L}_1$ be **Log**-morphisms such that $Th(h)$ and $Th(g)$ are isomorphisms, inverse of each other. Hence, given $\varphi \in L_{\Sigma_1}$, it must be the case that $\{\varphi\}^{\vdash_1} =$

$Th(g)(Th(h)(\{\varphi\}^{\vdash_1}))$. However, $Th(h)(\{\varphi\}^{\vdash_1}) = h[\{\varphi\}^{\vdash_1}]^{\vdash_2} = \{h(\varphi)\}^{\vdash_2}$, just using lemma 2.6. Similarly, $Th(g)(\{h(\varphi)\}^{\vdash_2}) = g[\{h(\varphi)\}^{\vdash_2}]^{\vdash_1} = \{g(h(\varphi))\}^{\vdash_1}$, thus implying that $\{\varphi\}^{\vdash_1} = \{g(h(\varphi))\}^{\vdash_1}$, or equivalently that $\varphi \equiv_{\mathcal{L}_1} g(h(\varphi))$. Analogously, we can prove that $\gamma \equiv_{\mathcal{L}_2} h(g(\gamma))$ for every $\gamma \in L_{\Sigma_2}$.

Assume now that $h : \mathcal{L}_1 \rightarrow \mathcal{L}_2$ and $g : \mathcal{L}_2 \rightarrow \mathcal{L}_1$ are **Log**-morphisms satisfying the two conditions stated above. If we have that $\Phi \in Th(\mathcal{L}_1)$, then $Th(g)(Th(h)(\Phi)) = Th(g \circ h)(\Phi) = g[h[\Phi]]^{\vdash_1}$. Hence, $\Phi = \Phi^{\vdash_1}$ and $g[h[\Phi]]^{\vdash_1}$ are in the conditions of proposition 4.2 and we can conclude that $g[h[\Phi]]^{\vdash_1} = \Phi$. Using an analogous argument, we can conclude that also $Th(h)(Th(g)(\Gamma)) = \Gamma$ for every $\Gamma \in Th(\mathcal{L}_2)$. $\square$

Let us illustrate the notion of equipollence with a meaningful example.

Example. Let P be a set of propositional variables. Consider the following two fragments of CPL:

- $\mathcal{L}_1 = \langle \Sigma_1, \vdash_1 \rangle$ where $\Sigma_1^0 = P$, $\Sigma_1^1 = \{\neg\}$, $\Sigma_1^2 = \{\Rightarrow\}$ and $\Sigma_1^n = \emptyset$ for $n > 2$; and
- $\mathcal{L}_2 = \langle \Sigma_2, \vdash_2 \rangle$ where $\Sigma_2^0 = P$, $\Sigma_2^1 = \{\neg\}$, $\Sigma_2^2 = \{\vee, \wedge\}$ and $\Sigma_1^n = \emptyset$ for $n > 2$.

Clearly we can define the following **Log**-morphisms $h : \mathcal{L}_1 \rightarrow \mathcal{L}_2$ and $g : \mathcal{L}_2 \rightarrow \mathcal{L}_1$:

- $h^0(p) = p$ for every $p \in P$, $h^1(\neg) = \lambda\xi_1.\neg\xi_1$, and $h^2(\Rightarrow) = \lambda\xi_1\xi_2.(\neg\xi_1) \vee \xi_2$; and
- $g^0(p) = p$ for every $p \in P$, $g^1(\neg) = \lambda\xi_1.\neg\xi_1$, $g^2(\vee) = \lambda\xi_1\xi_2.((\neg\xi_1) \Rightarrow \xi_2)$, and $g^2(\wedge) = \lambda\xi_1\xi_2.\neg(\xi_1 \Rightarrow (\neg\xi_2))$.

Using proposition 4.3, it is an immediate consequence of well known facts about CPL that the logical systems $\mathcal{L}_1$ and $\mathcal{L}_2$ are equipollent. It is also easy to see that $\mathcal{L}_1$ and $\mathcal{L}_2$ are not isomorphic.

In [6], Pelletier and Urquhart have proposed to capture "sameness" of logical systems using a notion of translational equivalence, that turns out to be very closely related to our notion of equipollence. Indeed, translational equivalence is stated exactly as our alternative characterization of equipollence in proposition 4.3, but using a biconditional connective instead of logical equivalence. According to [6], in order to be translationally equivalent, the two logical systems must both be equivalential in the sense of [3], and with respect to precisely the same biconditional connective, which must therefore be expressible in both. Pelletier and Urquhart also show that this notion of translational equivalence turns out to be equivalent to requiring that both logical systems share a common definitional extension, a notion to which they call synonymity, and which has been considered before by several other logicians (cf. [6]).

One immediate observation that we can make is that equipollence is certainly more widely applicable than translational equivalence, since the logical systems at hand are not required to be equivalential, even less with the same biconditional. Still, equipollence and translational equivalence will obviously coincide if the two logical systems at hand are equivalential with a shared biconditional, that further-more satisfies the deduction theorem in both. If that is not the case, then, transla-tional equivalence always implies equipollence but nothing can be said about the converse.

The conditions underlying translational equivalence seem therefore very re-strictive, since many logical systems fail to be equivalential. The next example illustrates equipollence between two logical systems where no reasonable bicondi-tional connective can even be defined.

Example. Let P be a set of propositional variables. Consider the following two logical systems:

- $\mathcal{L}_1 = \langle \Sigma_1, \vdash_1 \rangle$ where $\Sigma_1^0 = P$, $\Sigma_1^1 = \emptyset$, $\Sigma_1^2 = \{\vee\}$ and $\Sigma_1^n = \emptyset$ for $n > 2$; and $\vdash_1$ is the corresponding restriction of the CPL consequence operator; and
- $\mathcal{L}_2 = \langle \Sigma_2, \vdash_2 \rangle$ where $\Sigma_2^0 = P$, $\Sigma_2^1 = \Sigma_2^2 = \emptyset$, $\Sigma_2^3 = \{\mathbb{W}\}$ and $\Sigma_1^n = \emptyset$ for $n > 3$; and $\vdash_2$ behaves classically with respect to $\mathbb{W}$, understood as ternary disjunction.

Clearly we can define the following **Log**-morphisms $h : \mathcal{L}_1 \to \mathcal{L}_2$ and $g : \mathcal{L}_2 \to \mathcal{L}_1$:

- $h^0(p) = p$ for every $p \in P$, and $h^2(\vee) = \lambda \xi_1 \xi_2.\mathbb{W}(\xi_1, \xi_2, \xi_2)$; and
- $g^0(p) = p$ for every $p \in P$, and $g^3(\mathbb{W}) = \lambda \xi_1 \xi_2 \xi_3.(\xi_1 \vee \xi_2) \vee \xi_3$.

Using proposition 4.3, it is an immediate consequence of well known facts about CPL that the logical systems $\mathcal{L}_1$ and $\mathcal{L}_2$ are equipollent.

5. Conclusion

In this paper we have discussed the notion of "sameness" between logics. By adopt-ing a categorial approach to the problem, and keeping an eye on previous propos-als [6, 7, 1], we end up proposing the definition of equipollence. Two logical systems are equipollent whenever there exist uniform translations between the two logical languages that induce an isomorphism on the corresponding theory spaces. Sev-eral examples of equipollence and non-equipollence are presented along with the exposition.

We have shown, and illustrated with examples, that, as a notion of "same-ness", equipollence is more accurate than Pollard's notion of homeomorphism [7]. Indeed, contrarily to our proposal, homeomorphisms are not even required to pre-serve the structure of formulas. Moreover, even if we ignore this fact, homeo-morphisms (just like logical system isomorphisms) are too sensitive to cardinality issues, and end up distinguishing logical systems that are equipollent, and should in our opinion be considered the "same".

Equipollence is also comparable, with advantage, to Pelletier and Urquhart's notion of translational equivalence (or synonymity) [6]. Although very similar in spirit, equipollence is first of all much more widely applicable than translational equivalence, since the logical systems at hand are not bound to being equivalential, even less regarding the same shared biconditional connective. Still, once these very strong conditions are fulfilled, the two notions simply coincide if we further require the deduction theorem to hold in both systems. Nevertheless, there is a gap in this relationship that we have not been able to fill in. Under the conditions for the applicability of the notion of translational equivalence, if the deduction theorem fails in some of the logical systems, it is still the case that translational equivalence implies equipollence, as an immediate consequence of detachment. The converse, however, may not hold, but we were unable to find any meaningful example where two logical systems would, under these conditions, be equipollent but fail to be translational equivalent.

We conclude with a remark on the notion of postmodern equivalence put forth by Béziau, de Freitas and Viana in [1], which is indeed too lax as a proposal to capture the "sameness" of logical systems. As suspected by its very authors, postmodern equivalence indeed seems to mix "bananas with tomatoes", simply because it tries to solve "too many problems". Perhaps with the exception of any postmodernist joker, there is certainly no one willing to defend that propositional and first-order classical logic are the "same".

References

[1] J.-Y. Béziau, R.P. de Freitas, J.P. Viana. What is Classical Propositional Logic? (A Study in Universal Logic), Logica Studies 7, 2001.

[2] C. Caleiro. Combining Logics. PhD thesis, IST, TU Lisbon, Portugal, 2000.

[3] J. Czelakowski. Equivalential logics I, II, Studia Logica 40, pp. 227–236, 335–372.

[4] D. Gabbay. What is a Logical System? Oxford University Press, 1994.

[5] R. Goldblatt. Logics of Time and Computation. CSLI, 1992. Second edition.

[6] F.J. Pelletier and A. Urquhart. Synonymous Logics, Journal of Philosophical Logic, 32, 2003, pp. 259–285.

[7] S. Pollard. Homeomorphism and the Equivalence of Logical Systems, Notre Dame Journal of Formal Logic, 39, 1998, pp. 422–435.

[8] A. Sernadas, C. Sernadas, and C. Caleiro. Synchronization of logics, Studia Logica, 59(2):217–247, 1997.

[9] R. Wójcicki, Theory of Logical Calculi. Synthese Library, vol. 199, Kluwer Academic Publishers, 1988.

Carlos Caleiro and Ricardo Gonçalves
Center for Logic and Computation
Department of Mathematics
Instituto Superior Técnico
TU Lisbon
Portugal
e-mail: ccal@math.ist.utl.pt
        rgon@math.ist.utl.pt

J.-Y. Beziau (Ed.), *Logica Universalis, 2nd edition*, 111–133

What is a Logic?

In memoriam Joseph Goguen

Till Mossakowski, Joseph Goguen, Răzvan Diaconescu and
Andrzej Tarlecki

Abstract. This paper builds on the theory of institutions, a version of abstract
model theory that emerged in computer science studies of software specification and semantics. To handle proof theory, our institutions use an extension
of traditional categorical logic with sets of sentences as objects instead of single sentences, and with morphisms representing proofs as usual. A natural
equivalence relation on institutions is defined such that its equivalence classes
are logics. Several invariants are defined for this equivalence, including a Lindenbaum algebra construction, its generalization to a Lindenbaum category
construction that includes proofs, and model cardinality spectra; these are
used in some examples to show logics inequivalent. Generalizations of familiar results from first order to arbitrary logics are also discussed, including
Craig interpolation and Beth definability.

Mathematics Subject Classification (2000). Primary 03C95; Secondary 18A15,
03G30, 18C10, 03B22.

Keywords. Universal logic, institution theory, category theory, abstract model
theory, categorical logic.

1. Introduction

Logic is often informally described as the study of *sound reasoning*. As such, it
plays a crucial role in several areas of mathematics (especially foundations) and
of computer science (especially formal methods), as well as in other fields, such as
analytic philosophy and formal linguistics. In an enormous development beginning
in the late 19th century, it has been found that a wide variety of different principles are needed for sound reasoning in different domains, and "a logic" has come
to mean a set of principles for some form of sound reasoning. But in a subject
the essence of which is formalization, it is embarrassing that there is no widely
acceptable formal definition of "a logic". It is clear that two key problems here are

to define what it means for two presentations of a logic to be equivalent, and to provide effective means to demonstrate equivalence and inequivalence.

This paper addresses these problems using the notion of "institution", which arose within computer science in response to the population explosion among the logics in use there, with the ambition of doing as much as possible at a level of abstraction independent of commitment to any particular logic [19, 36, 21]. The *soundness* aspect of sound reasoning is addressed by axiomatizing the notion of satisfaction, and the *reasoning* aspect is addressed by calling on categorical logic, which applies category theory to proof theory by viewing proofs as morphisms. Thus, institutions provide a balanced approach, in which both syntax and semantics play key roles. However, much of the institutional literature considers sentences without proofs and models without (homo)morphisms, and a great deal can be done just with satisfaction, such as giving general foundations for modularization of specifications and programs, which in turn has inspired aspects of the module systems of programming languages including C++, ML, and Ada.

Richer variants of the institution notion consider entailment relations on sentences and/or morphisms of models, so that they form categories; using proof terms as sentence morphisms provides a richer variant which fully supports proof theory. We call these the set/set, set/cat, cat/set, and cat/cat variants (where the first term refers to sentences, and the second to models); the table in Thm. 5.20 summarizes many of their properties. See [21] for a general treatment of the variant notions of institution, and [38, 40, 11, 12, 14] for some non-trivial results in abstract model theory done institutionally.

This paper adds to the literature on institutions a notion of equivalence, such that a logic is an equivalence class of institutions. To support this thesis, we consider a number of logical properties, model and proof theoretical, that are, and that are not, preserved under equivalence, and apply them to a number of examples. Perhaps the most interesting invariants are versions of the Lindenbaum algebra; some others concern cardinality of models. We also develop a normal form for institutions under our notion of equivalence, by extending the categorical notion of "skeleton".

We extend the Lindenbaum algebra construction to a *Lindenbaum category* construction, defined on any institution with proofs, by identifying not only equivalent sentences, but also equivalent proofs. We show that this construction is an invariant, i.e., preserved up to isomorphism by our equivalence on institutions. This construction extends the usual approach of categorical logic by having sets of sentences as objects, rather than just single sentences, and thus allows treating a much larger class of logics in a uniform way.

A perhaps unfamiliar feature of institutions is that satisfaction is not a dyadic relation, but rather a relation among sentence, model, and "signature", where signatures form a category the objects of which are thought of as vocabularies over which the sentences are constructed. In concrete cases, these may be propositional variables, relation symbols, function symbols, and so on. Since these form a category, it is natural that the constructions of sentences (or formulae) and models appear as functors on this category, and it is also natural to have an axiom expressing the invariance of "truth" (i.e., satisfaction) under change of notation. See Def. 2.1 below. When the vocabulary is fixed, the category of signatures is the one-object category **1**. (Another device can be used to eliminate models, giving pure proof theory as a special case, if desired.) If $\sigma\colon \Sigma \longrightarrow \Sigma'$ is an inclusion of signatures, then its application to models (via the model functor) is "reduct". The institutional triadic satisfaction can be motivated philosophically by arguments like those given by Peirce [30] for his "interpretants", which allow for context dependency of denotation in his semiotics, as opposed to Tarski's dyadic satisfaction. We also use this feature to resolve a problem about cardinality raised in [3]; see Example 2.2.

Joseph Goguen. Our co-author Joseph Goguen died on July 3rd, 2006. The scientific community has lost a great scientist doing pioneering research in many diverse areas, and we also have lost a close friend and teacher. Shortly before his death, we were privileged to take part in the Festschrift colloquium for his 65th birthday. His most important message to the Festschrift participants was a commitment to solidarity and cooperation.

2. Institutions and Logics

We assume the reader is familiar with basic notions from category theory; e.g., see [1, 25] for introductions to this subject. By way of notation, $|\mathbb{C}|$ denotes the class of objects of a category $\mathbb{C}$, and composition is denoted by "∘". Categories are assumed by convention to be locally small (i.e., to have a small set of morphisms between any two objects) unless stated otherwise. The basic concept of this paper in its set/cat variant is as follows[1]:

Definition 2.1.   An *institution* $\mathcal{I} = (\mathbb{S}ign^{\mathcal{I}}, \mathsf{Sen}^{\mathcal{I}}, \mathsf{Mod}^{\mathcal{I}}, \models^{\mathcal{I}})$ consists of

1. a category $\mathbb{S}ign^{\mathcal{I}}$, whose objects are called *signatures*,
2. a functor $\mathsf{Sen}^{\mathcal{I}}\colon \mathbb{S}ign^{\mathcal{I}} \to \mathbb{S}et$, giving for each signature a set whose elements are called *sentences* over that signature,
3. a functor $\mathsf{Mod}^{\mathcal{I}}\colon (\mathbb{S}ign^{\mathcal{I}})^{\mathrm{op}} \to \mathbb{C}\mathbb{A}\mathbb{T}$ giving for each signature Σ a category whose objects are called Σ-*models*, and whose arrows are called Σ-*(model) morphisms*[2] and
4. a relation $\models_{\Sigma}^{\mathcal{I}} \subseteq |\mathsf{Mod}^{\mathcal{I}}(\Sigma)| \times \mathsf{Sen}^{\mathcal{I}}(\Sigma)$ for each $\Sigma \in |\mathbb{S}ign^{\mathcal{I}}|$, called Σ-*satisfaction*,

---

[1] A more concrete definition is given in [22], which avoids category theory by spelling out the conditions for functoriality, and assuming a set theoretic construction for signatures. Though less general, this definition is sufficient for everything in this paper; however, it would greatly complicate our exposition. Our use of category theory is modest, oriented towards providing easy proofs for very general results, which is precisely what is needed for the goals of this paper.
[2] $\mathbb{C}\mathbb{A}\mathbb{T}$ is the category of all categories; strictly speaking, it is only a quasi-category living in a higher set-theoretic universe. See [25] for a discussion of foundations.

114   T. Mossakowski, J. Goguen, R. Diaconescu and A. Tarlecki

such that for each morphism $\sigma : \Sigma \to \Sigma'$ in $\mathbb{S}ign^{\mathcal{I}}$, the *satisfaction condition*

$$M' \models^{\mathcal{I}}_{\Sigma'} \mathsf{Sen}^{\mathcal{I}}(\sigma)(\varphi) \quad \text{iff} \quad \mathsf{Mod}^{\mathcal{I}}(\sigma)(M') \models^{\mathcal{I}}_{\Sigma} \varphi$$

holds for each $M' \in |\mathsf{Mod}^{\mathcal{I}}(\Sigma')|$ and $\varphi \in \mathsf{Sen}^{\mathcal{I}}(\Sigma)$. We denote the *reduct* functor $\mathsf{Mod}^{\mathcal{I}}(\sigma)$ by $\_\!\restriction_\sigma$ and the sentence translation $\mathsf{Sen}^{\mathcal{I}}(\sigma)$ by $\sigma(\_)$. When $M = M'\!\restriction_\sigma$ we say that M' is a σ-*expansion of* M.

A *set/set institution* is an institution where each model category is discrete; this means that the model functor actually becomes a functor $\mathsf{Mod}^{\mathcal{I}} : (\mathbb{S}ign^{\mathcal{I}})^{\mathrm{op}} \to \mathbb{C}lass$ into the quasi-category of classes and functions.

General assumption: We assume that all institutions are such that satisfaction is invariant under model isomorphism, i.e., if Σ-models M, M' are isomorphic, then $M \models_\Sigma \varphi$ iff $M' \models_\Sigma \varphi$ for all Σ-sentences φ. $\square$

We now consider classical propositional logic, perhaps the simplest non-trivial example (see the extensive discussion in [3]), and also introduce some concepts from the theory of institutions:

Example 2.2. Fix a countably infinite[3] set $\mathcal{X}$ of variable symbols, and let $\mathbb{S}ign$ be the category with finite subsets Σ of $\mathcal{X}$ as objects, and with inclusions as morphisms (or all set maps, if preferred, it matters little). Let $\mathsf{Mod}(\Sigma)$ have $[\Sigma \to \{0,1\}]$ (the set of functions from Σ to $\{0,1\}$) as its set of objects; these models are the row labels of truth tables. Let a (unique) Σ-model morphism $h \colon M \longrightarrow M'$ exist iff for all $p \in \Sigma$, $M(p) = 1$ implies $M'(p) = 1$. Let $\mathsf{Mod}(\Sigma' \hookrightarrow \Sigma)$ be the restriction map $[\Sigma \to \{0,1\}] \to [\Sigma' \to \{0,1\}]$. Let $\mathsf{Sen}(\Sigma)$ be the (absolutely) free algebra generated by Σ over the propositional connectives (we soon consider different choices), with $\mathsf{Sen}(\Sigma \hookrightarrow \Sigma')$ the evident inclusion. Finally, let $M \models_\Sigma \varphi$ mean that φ evaluates to true (i.e., 1) under the assignment M. It is easy to verify the satisfaction condition, and to see that φ is a tautology iff $M \models_\Sigma \varphi$ for all $M \in |\mathsf{Mod}(\Sigma)|$. Let **CPL** denote this institution of propositional logic, with the connectives conjunction, disjunction, negation, implication, true and false. Let **CPL**$^{\neg, \wedge, false}$ denote propositional logic with negation, conjunction and false, and **CPL**$^{\neg, \vee, true}$ with propositional logic negation, disjunction and true[4].

This arrangement puts truth tables on the side of semantics, and formulas on the side of syntax, each where it belongs, instead of trying to treat them the same way. It also solves the problem raised in [3] that the cardinality of $\mathcal{L}(\Sigma)$ varies with that of Σ, where $\mathcal{L}(\Sigma)$ is the quotient of $\mathsf{Sen}(\Sigma)$ by the semantic equivalence $\models|_\Sigma$, defined by $\varphi \models|_\Sigma \varphi'$ iff ($M \models_\Sigma \varphi$ iff $M \models_\Sigma \varphi'$, for all $M \in |\mathsf{Mod}(\Sigma)|$); it is the Lindenbaum algebra, in this case, the free Boolean algebra over Σ, and its cardinality is 2^{2^n} where n is the cardinality of Σ. Hence this cardinality cannot be considered an invariant of **CPL** without the parameterization by Σ (see also Def. 4.13 below). $\square$

---

[3]The definition also works for finite or uncountable $\mathcal{X}$.
[4]The truth constant avoids the empty signature having no sentences at all.

The moral of the above example is that everything should be parameterized by signature. Although the construction of the underlying set of the Lindenbaum algebra above works for any institution, its algebraic structure depends on how sentences are defined. However, Section 4 shows how to obtain at least part of this structure for any institution.

Example 2.3. The institution **FOLR** of unsorted first-order logic with relations has signatures Σ that are families Σ_n of sets of relation symbols of arity $n \in I\!N$, and **FOLR** signature morphisms $\sigma \colon \Sigma \longrightarrow \Sigma'$ that are families $\sigma_n \colon \Sigma_n \longrightarrow \Sigma'_n$ of arity-preserving functions on relation symbols. An **FOLR** Σ-sentence is a closed first-order formula using relation symbols in Σ, and sentence translation is relation symbol substitution. A **FOLR** Σ-model is a set M and a subset $R_M \subseteq M^n$ for each $R \in \Sigma_n$. Model translation is reduct with relation translation. A Σ-model morphism is a function $h \colon M \longrightarrow M'$ such that $h(R_M) \subseteq R_{M'}$ for all R in Σ. Satisfaction is as usual. The institution **FOL** adds function symbols to **FOLR** in the usual way, and **MSFOL** is its many sorted variant.   □

Example 2.4. In the institution **EQ** of many sorted equational logic, a signature consists of a set of sorts with a set of function symbols, each with a string of argument sorts and a result sort. Signature morphisms map sorts and function symbols in a compatible way. Models are many sorted algebras, i.e., each sort is interpreted as a carrier set, and each function symbol names a function among carrier sets specified by its argument and result sorts. Model translation is reduct, sentences are universally quantified equations between terms of the same sort, sentence translation replaces translated symbols (assuming that variables of distinct sorts never coincide in an equation), and satisfaction is the usual satisfaction of an equation in an algebra.   □

Example 2.5. **K** is propositional modal logic with □ and ◇. Its models are Kripke structures, and satisfaction is defined using possible-world semantics in the usual way. **IPL** is intuitionistic propositional logic, differing from **CPL** in having Kripke structures as models, and possible-world satisfaction. The proof theory of **IPL** (which is favored over the model theory by intuitionists) is discussed in Section 5.
                    □

Both intuitionistic and modal logic in their first-order variants, with both constant and varying domains, form institutions, as do other modal logics restricting **K** by further axioms, such as **S4** or **S5**, as well as substructural logics, like linear logic, where judgements of the form $\varphi_1 \ldots \varphi_n \vdash \psi$ are sentences. Higher-order [7], polymorphic [37], temporal [18], process [18], behavioural [4], coalgebraic [9] and object-oriented [20] logics also form institutions. Many familiar basic concepts can be defined over any institution:

Definition 2.6. Given a set of Σ-sentences Γ and a Σ-sentence φ, then φ is a *semantic consequence* of Γ, written $\Gamma \models_\Sigma \varphi$ iff for all Σ-models M, we have $M \models_\Sigma \Gamma$ implies $M \models_\Sigma \varphi$, where $M \models_\Sigma \Gamma$ means $M \models_\Sigma \psi$ for each $\psi \in \Gamma$. Two sentences are *semantically equivalent*, written $\varphi_1 \models\mid \varphi_2$, if they are satisfied by the same models. Two models are *elementary equivalent*, written $M_1 \equiv M_2$, if they satisfy the same sentences. An institution is *compact* iff $\Gamma \models_\Sigma \varphi$ implies $\Gamma' \models_\Sigma \varphi$ for some finite subset Γ' of Γ. A *theory* is a pair (Σ, Γ) where Γ is a set of Σ-sentences, and is *consistent* iff it has at least one model. $\qquad \square$

Cardinality properties associate cardinalities to objects in a category. It is natural to do this using *concrete categories* [1], which have a faithful *forgetful* or *carrier functor* to $\mathbb{S}et$. Since we also treat many sorted logics, we generalize from $\mathbb{S}et$ to categories of many sorted sets $\mathbb{S}et^S$, where the sets S range over sort sets of an institution's signatures. The following enriches institutions with carrier sets for models [5]:

Definition 2.7. A *concrete institution* is an institution $\mathcal{I}$ together with a functor $sorts^{\mathcal{I}} : \mathbb{S}ign^{\mathcal{I}} \to \mathbb{S}et$ and a natural transformation $|\_|^{\mathcal{I}} : \mathbf{Mod}^{\mathcal{I}} \to \mathbb{S}et^{(sorts^{\mathcal{I}})^{op}(\_)}$ between functors from $\mathbb{S}ign^{op}$ to $\mathbb{CAT}$ such that for each signature Σ, the *carrier functor* $|\_|^{\mathcal{I}}_\Sigma : \mathbf{Mod}^{\mathcal{I}}(\Sigma) \to \mathbb{S}et^{sorts^{\mathcal{I}}(\Sigma)}$ is faithful (that is, $\mathbf{Mod}^{\mathcal{I}}(\Sigma)$ is a concrete category, with carrier functors $|\_|^{\mathcal{I}}_\Sigma : \mathbf{Mod}^{\mathcal{I}}(\Sigma) \to \mathbb{S}et^{sorts^{\mathcal{I}}(\Sigma)}$ natural in Σ). Here, $\mathbb{S}et^{(sorts^{\mathcal{I}})^{op}(\_)}$ stands for the functor that maps each signature $\Sigma \in |\mathbb{S}ign^{\mathcal{I}}|$ to the category of $sorts^{\mathcal{I}}(\Sigma)$-sorted sets. A concrete institution has the *finite model property* if each satisfiable theory has a finite model (i.e., a model M with the carrier $|M|$ being a family of finite sets). A concrete institution *admits free models* if all carrier functors for model categories have left adjoints. $\qquad \square$

The following notion from [29] also provides signatures with underlying sets of symbols, by extending $sorts^{\mathcal{I}}$; essentially all institutions that arise in practice have this structure:

Definition 2.8. A *concrete institution with symbols* is a concrete institution $\mathcal{I}$ together with a faithful functor $Symb^{\mathcal{I}} : \mathbb{S}ign^{\mathcal{I}} \to \mathbb{S}et$ that naturally extends $sorts^{\mathcal{I}}$, that is, such that for each signature Σ, $sorts^{I}(\Sigma) \subseteq Symb^{\mathcal{I}}(\Sigma)$, and for each σ in $\mathbb{S}ign^{\mathcal{I}}$, $Symb^{\mathcal{I}}(\sigma)$ extends $sorts^{\mathcal{I}}(\sigma)$. A concrete institution with symbols *admits free models* if all the forgetful functors for model categories have left adjoints. $\qquad \square$

3. Equivalence of Institutions

Relationships between institutions are captured mathematically by 'institution morphisms', of which there are several variants, each yielding a category under a canonical composition. For the purposes of this paper, institution comorphisms [21] are technically more convenient, though the definition of institution equivalence below is independent of this choice. The original notion, from [19] in the set/cat form, works well for 'forgetful' morphisms from one institution to another having less structure:

Definition 3.1. Given institutions $\mathcal{I}$ and $\mathcal{J}$, then an *institution morphism* $(\Phi, \alpha, \beta) : \mathcal{I} \to \mathcal{J}$ consists of

1. a functor $\Phi : \mathbb{S}ign^{\mathcal{I}} \to \mathbb{S}ign^{\mathcal{J}}$,

2. a natural transformation $\alpha: \mathsf{Sen}^{\mathcal{J}} \circ \Phi \Rightarrow \mathsf{Sen}^{\mathcal{I}}$, and
3. a natural transformation $\beta: \mathsf{Mod}^{\mathcal{I}} \Rightarrow \mathsf{Mod}^{\mathcal{J}} \circ \Phi^{\mathrm{op}}$

such that the following *satisfaction condition* holds

$$M \models^{\mathcal{I}}_{\Sigma} \alpha_{\Sigma}(\varphi) \quad \text{iff} \quad \beta_{\Sigma}(M) \models^{\mathcal{J}}_{\Phi(\Sigma)} \varphi$$

for each signature $\Sigma \in |\mathbb{S}ign^{\mathcal{I}}|$, each Σ-model M and each $\Phi(\Sigma)$-sentence φ. $\quad\square$

Institution morphisms form a category $\mathbb{I}ns$ under the natural composition.

Definition 3.2. Given institutions $\mathcal{I}$ and $\mathcal{J}$, then an *institution comorphism* $(\Phi, \alpha, \beta): \mathcal{I} \longrightarrow \mathcal{J}$ consists of

- a functor $\Phi: \mathbb{S}ign^{\mathcal{I}} \longrightarrow \mathbb{S}ign^{\mathcal{J}}$,
- a natural transformation $\alpha: \mathsf{Sen}^{\mathcal{I}} \Rightarrow \mathsf{Sen}^{\mathcal{J}} \circ \Phi$,
- a natural transformation $\beta: \mathsf{Mod}^{\mathcal{J}} \circ \Phi^{op} \Rightarrow \mathsf{Mod}^{\mathcal{I}}$

such that the following *satisfaction condition* is satisfied for all $\Sigma \in |\mathbb{S}ign^{\mathcal{I}}|$, $M' \in |\mathsf{Mod}^{\mathcal{J}}(\Phi(\Sigma))|$ and $\varphi \in \mathsf{Sen}^{\mathcal{I}}(\Sigma)$:

$$M' \models^{\mathcal{J}}_{\Phi(\Sigma)} \alpha_{\Sigma}(\varphi) \quad \text{iff} \quad \beta_{\Sigma}(M') \models^{\mathcal{I}}_{\Sigma} \varphi .$$

With the natural compositions and identities, this gives a category $\mathbb{C}oIns$ of institutions and institution comorphisms.

A *set/set institution comorphism* is like a set/cat comorphism, except that β_{Σ} is just a function on the objects of model categories; the model morphisms are ignored.

Given concrete institutions $\mathcal{I}, \mathcal{J}$, then a *concrete comorphism* from $\mathcal{I}$ to $\mathcal{J}$ is an institution comorphism $(\Phi, \alpha, \beta): \mathcal{I} \longrightarrow \mathcal{J}$ plus a natural transformation $\delta: sorts^{\mathcal{I}} \Rightarrow sorts^{\mathcal{J}} \circ \Phi$ and a natural in $\mathcal{I}$-signatures Σ family of natural transformations $\mu_{\Sigma}: |\beta_{\Sigma}(\_)|^{\mathcal{I}}_{\Sigma} \Rightarrow (|\_|^{\mathcal{J}}_{\Phi(\Sigma)})\!\upharpoonright_{\delta_{\Sigma}}$ between functors from $\mathsf{Mod}^{\mathcal{J}}(\Phi(\Sigma))$ to $sorts^{\mathcal{I}}(\Sigma)$-sorted sets, so that for each $\mathcal{I}$-signature Σ, $\Phi(\Sigma)$-model M' in $\mathcal{J}$ and sort $s \in sorts^{\mathcal{I}}(\Sigma)$, we have a function $\mu_{\Sigma,M',s}: (|\beta_{\Sigma}(M')|^{\mathcal{I}}_{\Sigma})_s \to (|M'|^{\mathcal{J}}_{\Phi(\Sigma)})_{\delta_{\Sigma}(s)}$.

Given concrete institutions with symbols $\mathcal{I}$ and $\mathcal{J}$, a *concrete comorphism with symbols* from $\mathcal{I}$ to $\mathcal{J}$ extends an institution comorphism $(\Phi, \alpha, \beta): \mathcal{I} \longrightarrow \mathcal{J}$ by a natural transformation $\delta: Symb^{\mathcal{I}} \Rightarrow Symb^{\mathcal{J}} \circ \Phi$ that restricts to $\delta': sorts^{\mathcal{I}} \Rightarrow sorts^{\mathcal{J}} \circ \Phi$, and a family of functions $\mu_{\Sigma}: (|\beta_{\Sigma}(\_)|^{\mathcal{I}}_{\Sigma})_s \to (|\_|^{\mathcal{J}}_{\Phi(\Sigma)})_{\delta'_{\Sigma}(s)}$, required to be natural in $\mathcal{I}$-signatures Σ. Notice that then $(\Phi, \alpha, \beta, \delta', \mu)$ is a concrete comorphism. $\quad\square$

Fact 3.3. An institution comorphism is an isomorphism in $\mathbb{C}oIns$ iff all its components are isomorphisms. $\quad\square$

Unfortunately, institution isomorphism is too strong to capture the notion of "a logic," since it can fail to identify logics that differ only in irrelevant details:

Example 3.4. Let **CPL$'$** be **CPL** with arbitrary finite sets as signatures. Then **CPL$'$** has a proper class of signatures, while **CPL** only has countably many. Hence, **CPL** and **CPL$'$** cannot be isomorphic. $\quad\square$

However, **CPL** and **CPL$'$** are essentially the same logic. We now give a notion of institution *equivalence* that is weaker than that of institution isomorphism, very much in the spirit of equivalences of categories. The latter weakens isomorphism of categories: two categories are equivalent iff they have isomorphic *skeletons*. A subcategory $\mathbb{S} \hookrightarrow \mathbb{C}$ is a *skeleton* of $\mathbb{C}$ if it is full and each object of $\mathbb{C}$ is isomorphic (in $\mathbb{C}$) to exactly one object in $\mathbb{S}$. In this case, the inclusion $\mathbb{S} \hookrightarrow \mathbb{C}$ has a left inverse (i.e., a retraction) $\mathbb{C} \to \mathbb{S}$ mapping each object to the unique representative of its isomorphism class (see [25]).

Definition 3.5. A (set/cat) institution comorphism (Φ, α, β) is a (set/cat) *institution equivalence* iff

- Φ is an equivalences of categories,
- α_Σ has an inverse up to semantic equivalence α'_Σ, (i.e., $\alpha_\Sigma(\alpha'_\Sigma(\varphi)) \models|_\Sigma \varphi$ and $\alpha'_\Sigma(\alpha_\Sigma(\psi)) \models|_{\Phi(\Sigma)} \psi$) which is natural in Σ, and
- β_Σ is an equivalence of categories, such that its inverse up to isomorphism and the corresponding isomorphism natural transformations are natural in Σ.

$\mathcal{I}$ is *equivalent* to $\mathcal{J}$ if there is an institution equivalence from $\mathcal{I}$ to $\mathcal{J}$. □

This definition is very natural; it is 2-categorical equivalence in the appropriate 2-category of institutions [10]. The requirement for a set/set institution comorphism to be a *set/set equivalence* is weaker: each β_Σ need only have an inverse up to elementary equivalence β'_Σ.

Definition 3.6. A concrete institution comorphism is a *concrete equivalence* if the underlying institution comorphism is an equivalence and all δ_Σ and $\mu_{\Sigma,M',s}$ are bijective, for each $\Sigma \in |\mathbb{S}ign^\mathcal{I}|$, $M' \in |\mathsf{Mod}^\mathcal{J}(\Phi(\Sigma))|$ and $s \in sorts^\mathcal{I}(\Sigma)$.

A concrete comorphism with symbols is a *concrete equivalence with symbols* if the underlying institution comorphism is an equivalence and δ_Σ is bijective for each signature Σ. □

Proposition 3.7. Both set/cat and set/set equivalence of institutions are equivalence relations, and set/cat equivalence implies set/set equivalence. □

The following is important for studying invariance properties of institutions under equivalence:

Lemma 3.8. If $(\Phi, \alpha, \beta) \colon \mathcal{I} \longrightarrow \mathcal{J}$ is a set/cat or set/set institution equivalence, $\Gamma \models^\mathcal{I}_\Sigma \varphi$ iff $\alpha_\Sigma(\Gamma) \models^\mathcal{J}_{\Phi(\Sigma)} \alpha_\Sigma(\varphi)$ for any signature Σ in $\mathcal{I}$ and $\Gamma \cup \{\varphi\} \subseteq \mathsf{Sen}^\mathcal{I}(\Sigma)$; also $M_1 \equiv M_2$ iff $\beta_\Sigma(M_1) \equiv \beta_\Sigma(M_2)$, for any $M_1, M_2 \in \mathsf{Mod}^\mathcal{J}(\Phi(\Sigma))$. □

Example 3.9. **CPL** and **CPL$'$** are set/cat equivalent. So are **CPL**$^{\neg,\vee,true}$ and **CPL**$^{\neg,\wedge,false}$: signatures and models are translated identically, while sentences are translated using de Morgan's laws. Indeed, **CPL**$^{\neg,\vee,true}$ and **CPL**$^{\neg,\wedge,false}$ are isomorphic, but the isomorphism is far more complicated than the equivalence. □

Definition 3.10. Given a set/cat institution $\mathcal{I}$, an institution $\mathcal{J}$ is a set/cat *skeleton* of $\mathcal{I}$, if

- $\mathbb{S}ign^{\mathcal{J}}$ is a skeleton of $\mathbb{S}ign^{\mathcal{I}}$,
- $\mathsf{Sen}^{\mathcal{J}}(\Sigma) \cong \mathsf{Sen}^{\mathcal{I}}(\Sigma)/\mathord{=}\!|$ for $\Sigma \in |\mathbb{S}ign^{\mathcal{J}}|$ (the bijection being natural in Σ), and $\mathsf{Sen}^{\mathcal{J}}(\sigma)$ is the induced mapping between the equivalence classes,
- $\mathsf{Mod}^{\mathcal{J}}(\Sigma)$ is a skeleton of $\mathsf{Mod}^{\mathcal{I}}(\Sigma)$, and $\mathsf{Mod}^{\mathcal{J}}(\sigma)$ is the restriction of $\mathsf{Mod}^{\mathcal{I}}(\sigma)$,
- $M \models^{\mathcal{J}}_{\Sigma} [\varphi]$ iff $M \models^{\mathcal{I}}_{\Sigma} \varphi$.

Set/set skeletons are defined similarly, except that $\mathsf{Mod}^{\mathcal{J}}(\Sigma)$ is in bijective correspondence with $\mathsf{Mod}^{\mathcal{I}}(\Sigma)/\mathord{\equiv}$. $\square$

Theorem 3.11. Assuming the axiom of choice, every institution has a skeleton. Every institution is equivalent to any of its skeletons. Any two skeletons of an institution are isomorphic. Institutions are equivalent iff they have isomorphic skeletons. $\square$

We have now reached a central point, where we can claim

> *The identity of a logic is the isomorphism type of its skeleton institution.*

This isomorphism type even gives a normal form for equivalent logics. It follows that a *property of a logic* must be a property of institutions that is invariant under equivalence, and the following sections explore a number of such properties.

4. Model-Theoretic Invariants of Institutions

This section discusses some model-theoretic invariants of institutions; the table in Thm. 5.20 summarizes the results on this topic in this paper.

Every institution has a Galois connection between its sets Γ of Σ-sentences and its classes $\mathcal{M}$ of Σ-models, defined by $\Gamma^{\bullet} = \{M \in \mathsf{Mod}(\Sigma) \mid M \models_{\Sigma} \Gamma\}$ and $\mathcal{M}^{\bullet} = \{\varphi \in \mathsf{Sen}(\Sigma) \mid \mathcal{M} \models_{\Sigma} \varphi\}$. A Σ-theory Γ is *closed* if $(\Gamma^{\bullet})^{\bullet} = \Gamma$.[5] Closed Σ-theories are closed under arbitrary intersections; hence they form a complete lattice. This leads to a functor $\mathcal{C}^{\models} \colon \mathbb{S}ign \longrightarrow \mathbb{C}Lat$. Although $\mathcal{C}^{\models}$ is essentially preserved under equivalence, the closure operator $({}_{-}{}^{\bullet})^{\bullet}$ on theories is not. This means it makes too fine-grained distinctions; for example, in **FOL**, $(true^{\bullet})^{\bullet}$ is infinite, while in a skeleton of **FOL**, $([true]^{\bullet})^{\bullet}$ is just the singleton $\{[true]\}$. As already noted in [33], the closure operator at the same time is too coarse for determining the identity of a logic: while e.g., proof theoretic falsum in a sound and complete logic (see Section 5) is preserved by homeomorphisms of closure operators in the sense of [33], external semantic falsum (see Dfn. 4.2) is not. Because the theory closure operator is not preserved under equivalence, we do not study it further, but instead use the closed theory lattice functor $\mathcal{C}^{\models}$ and the Lindenbaum functor $\mathcal{L}$ defined below. (We note in passing that this Galois connection generalizes some results considered important in the study of ontologies in the computer science sense.)

The category of theories of an institution is often more useful than its lattice of theories, where a theory morphism $(\Sigma, \Gamma) \longrightarrow (\Sigma', \Gamma')$ is a signature morphism $\sigma \colon \Sigma \longrightarrow \Sigma$ such that $\Gamma' \models_{\Sigma'} \sigma(\Gamma)$. Let $\mathbb{T}h(\mathcal{I})$ denote this category (it should

---

[5]The closed theories can serve as models in institutions lacking (non-trivial) models.

120   T. Mossakowski, J. Goguen, R. Diaconescu and A. Tarlecki

be skeletized to become an invariant). The following result is basic for combining theories, and has important applications to both specification and programming languages [19]:

Theorem 4.1. The category of theories of an institution has whatever colimits its category of signatures has. □

Definition 4.2. An institution has *external semantic conjunction* [39] if for any pair of sentences φ_1, φ_2 over the same signature, there is a sentence ψ such that ψ holds in a model iff both φ_1 and φ_2 hold in it. ψ will also be denoted $\varphi_1 \bigcirc\!\!\!\wedge \varphi_2$, a meta-notation which may not agree with the syntax for sentences in the institution. Similarly, one can define what it means for an institution to have external semantic disjunction, negation, implication, equivalence, true, false, and we will use similar circle notations for these. An institution is *truth functionally complete*, if any Boolean combination of sentences is equivalent to a single sentence. □

Example 4.3. **FOL** is truth functionally complete, while **EQ** has no external semantic connectives. □

The Lindenbaum construction of Example 2.2 works for any institution $\mathcal{I}$:

Definition 4.4. Let $\Xi^{\mathcal{I}}$ be the single sorted algebraic signature having that subset of the operations $\{\bigcirc\!\!\!\wedge, \bigcirc\!\!\!\vee, \ominus, \ominus, \ominus, \textcircled{t}, \textcircled{f}\}$ (with standard arities) that are external semantic for $\mathcal{I}$; $\Xi^{\mathcal{I}}$ may include connectives not provided by the institution $\mathcal{I}$, or provided by $\mathcal{I}$ with a different syntax. We later prove that $\Xi^{\mathcal{I}}$ is invariant under equivalence[6]. For any signature in $\mathcal{I}$, let $\mathcal{L}(\Sigma)$ have as carrier set the quotient $\mathsf{Sen}(\Sigma)/\!\models\!|$, as in Example 2.2. Every external semantic operation of $\mathcal{I}$ has a corresponding operation $\mathcal{L}(\Sigma)$, so $\mathcal{L}(\Sigma)$ can be given a $\Xi^{\mathcal{I}}$-algebra structure. Any subsignature of $\Xi^{\mathcal{I}}$ can also be used (indicated with superscript notation as in Example 2.2), in which case crypto-isomorphisms[7] can provide Lindenbaum algebra equivalence. Moreover, $\mathcal{L}$ is a functor $\mathbb{S}ign \longrightarrow \mathsf{Alg}(\Xi^{\mathcal{I}})$ because $\models\!|$ is preserved by translation along signature morphisms[8]. If $\mathcal{I}$ is truth functionally complete, then $\mathcal{L}(\Sigma)$ is a Boolean algebra. A proof theoretic variant of $\mathcal{L}(\Sigma)$ is considered in Section 5 below. □

Definition 4.5. An institution has *external semantic universal $\mathcal{D}$-quantification* [40] for a class $\mathcal{D}$ of signature morphisms iff for each $\sigma : \Sigma \to \Sigma'$ in $\mathcal{D}$ and each Σ'-sentence, there is a Σ-sentence $\forall \sigma.\varphi$ such that $M \models_\Sigma \forall \sigma.\varphi$ iff $M' \models \varphi$ for each σ-expansion M' of M. External semantic existential quantification is defined similarly. □

---

[6]By determining $\Xi^{\mathcal{I}}$ in a purely model-theoretic way, we avoid the need to deal with different signatures of Lindenbaum algebras of equivalent logics, as it is necessary in the framework of [31].

[7]A *cryptomorphism* is a homomorphism between algebras of different signatures linked by a signature morphism; the homomorphism goes from the source algebra into the reduct of the target algebra.

[8]$\mathcal{L}$ is also functorial in the institution, though the details are rather complex.

This definition accommodates quantification over any entities which are part of the relevant concept of signature. For conventional model theory, this includes second order quantification by taking $\mathcal{D}$ to be all extensions of signatures by operation and relation symbols. First order quantification is modeled with $\mathcal{D}$ the *representable signature morphisms* [11, 13] defined below, building on the observation that an assignment for a set of (first order) variables corresponds to a model morphism from the free (term) model over that set of variables:

Definition 4.6. A signature morphism $\chi \colon \Sigma \to \Sigma'$ is *representable* iff there are a Σ-model M_χ called the *representation of* χ and a category isomorphism i_χ such that the diagram below commutes, where $(M_\chi/\mathsf{Mod}(\Sigma))$ is a comma category and U is the forgetful functor. $\qquad\square$

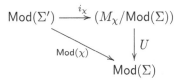

It seems likely that if external semantic universal quantification over representable quantifiers is included in the signature $\Xi^{\mathbf{FOL}}$, then our Lindenbaum algebra functor yields cylindric algebras, though not all details have been checked as of this writing.

Theorem 4.7. Let $(\Phi, \alpha, \beta)\colon \mathcal{I} \longrightarrow \mathcal{J}$ be an institution equivalence. Then $\mathcal{I}$ has universal (or existential) representable quantification iff $\mathcal{J}$ also has universal (or existential) representable quantification.

It follows that the set of the external semantic connectives an institution has is preserved under institution equivalence.

Example 4.8. Horn clause logic is not equivalent to **FOL**, because it does not have negation (nor implication etc.). Horn clause logic with predicates and without predicates are not equivalent: in the latter logic, model categories of theories have (regular epi, mono)-factorizations, which is not true for the former logic. $\qquad\square$

Example 4.9. Propositional logic **CPL** and propositional modal logic **K** are not equivalent: the former has external semantic disjunction, the latter does not: $(M \models_\Sigma p)$ or $(M \models_\Sigma \neg p)$ means that p is interpreted homogeneously in all worlds of M, which is not expressible by a modal formula. Indeed, the Lindenbaum algebra signature for **CPL** is $\{\oslash, \oslash, \ominus, \ominus, \ominus, \oslash, \oslash\}$, while that for **K** is $\{\oslash, \ominus, \oslash, \oslash\}$. Likewise, first-order logic and first-order modal logic are not equivalent. These assertions also hold replacing "modal" by "intuitionistic". $\qquad\square$

Definition 4.10. An institution is liberal iff for any theory morphism $\sigma \colon T_1 \longrightarrow T_2$, $\mathsf{Mod}(\sigma)\colon \mathsf{Mod}(T_2) \longrightarrow \mathsf{Mod}(T_1)$ has a left adjoint. An institution has initial (terminal) models if $\mathsf{Mod}(T)$ has so for each theory T. $\qquad\square$

Definition 4.11. For any classes $\mathcal{L}$ and $\mathcal{R}$ of signature morphisms in an institution $\mathcal{I}$, the institution has the *semantic Craig $\langle \mathcal{L}, \mathcal{R} \rangle$-interpolation property*[9] [39] if for any pushout

$$
\begin{array}{ccc}
\Sigma & \xrightarrow{\sigma_1} & \Sigma_1 \\
\downarrow{\sigma_2} & & \downarrow{\theta_2} \\
\Sigma_2 & \xrightarrow{\theta_1} & \Sigma'
\end{array}
$$

in $\mathbb{S}ign$ such that $\sigma_1 \in \mathcal{L}$ and $\sigma_2 \in \mathcal{R}$, any set of Σ_1-sentences Γ_1 and any set of Σ_2-sentences Γ_2 with $\theta_2(\Gamma_1) \models \theta_1(\Gamma_2)$, there exists a set of Σ-sentences Γ (called the *interpolant*) such that $\Gamma_1 \models \sigma_1(\Gamma)$ and $\sigma_2(\Gamma) \models \Gamma_2$. □

This generalizes the conventional formulation of interpolation from intersection/union squares of signatures to arbitrary classes of pushout squares. While **FOL** has interpolation for all pushout squares [17], many sorted first order logic has it only for those where one component is injective on sorts [8, 6, 23], and **EQ** and Horn clause logic only have it for pushout squares where $\mathcal{R}$ consists of injective morphisms [35, 14], or where L consists of sort-injective morphisms that encapsulate the operation symbols, i.e., no new operation symbol has an old result sort [34]. Using sets of sentences rather than single sentences accommodates interpolation results for equational logic [35] as well as for other institutions having Birkhoff-style axiomatizability properties [14].

Definition 4.12. An institution is *(semi-)exact* if Mod maps finite colimits (pushouts) to limits (pullbacks). □

Semi-exactness is important because many model theoretic results depend on it. It is also important for instantiating parameterized specifications. It means that given a pushout as in Def. 4.11 above, any pair $(M_1, M_2) \in \mathsf{Mod}(\Sigma_1) \times \mathsf{Mod}(\Sigma_2)$ that is *compatible* in the sense that M_1 and M_2 reduce to the same Σ-model can be *amalgamated* to a unique Σ'-model M (i.e., there exists a unique $M \in \mathsf{Mod}(\Sigma')$ that reduces to M_1 and M_2, respectively), and similarly for model morphisms. *Elementary amalgamation* [14] is like semi-exactness but considers the model reducts up to elementary equivalence.

It is also known how to define reduced products, Łoś sentences (i.e., sentences preserved by both ultraproducts and ultrafactors) and Łoś institutions [11], elementary diagrams of models [12], and (Beth) definability[10], all in an institution independent way, such that the expected theorems hold under reasonable assumptions. All this is very much in the spirit of "abstract model theory," in the sense advocated by Jon Barwise [2], but it goes much further, including even some new results for known logics, such as many sorted first order logic [14, 23].

---

[9] Recently, it has been noticed that a slightly stronger property, the Craig-Robinson interpolation property, seems to be more appropriate in many contexts; see [16] for details.

[10] Some important results on Beth definability (e.g., the Beth theorem, which asserts the equivalence of explicit and implicit definability) have been stimulated by the first version of the present paper and now have been published, see [32].

The faithful functors to $\mathbb{S}et$ make it possible to consider cardinalities for signatures and models in a concrete institutions with symbols. By restricting signature morphisms to a subcategory, it is often possible to view these cardinality functions as functors.

Definition 4.13. The *Lindenbaum cardinality spectrum* of a concrete institution with symbols maps a cardinal number κ to the maximum number of non-equivalent sentences for a signature of cardinality κ. The *model cardinality spectrum* of a concrete institution with symbols maps each pair of a theory T and a cardinal number κ to the number of non-isomorphic models of T of cardinality κ. □

Theorem 4.14. Sentence and model cardinality spectra, and the finite model property, are preserved under concrete equivalence. □

5. Proof Theoretic Invariants

Proof theoretic institutions include both proofs and sentences. Categorical logic usually works with categories of sentences, where morphisms are (equivalence classes of) proof terms [24]. But this only captures provability between single sentences, whereas logic traditionally studies provability from a *set* of sentences. The following overcomes this limitation by considering categories of sets of sentences:

Definition 5.1. A *cat/cat institution* is like a set/cat institution, except that now Sen: $\mathbb{S}ign \longrightarrow \mathbb{S}et$ comes with an additional categorical structure on sets of sentences, which is carried by a functor Pr: $\mathbb{S}ign \longrightarrow \mathbb{C}at$ such that $(\_)^{op} \circ \mathcal{P} \circ$ Sen is a subfunctor of Pr, and the inclusion $\mathcal{P}(\text{Sen}(\Sigma))^{op} \hookrightarrow \text{Pr}(\Sigma)$ is broad and preserves products of disjoint sets of sentences[11]. Here $\mathcal{P}$: $\mathbb{S}et \longrightarrow \mathbb{C}at$ is the functor taking each set to its powerset, ordered by inclusion, construed as a thin category[12]. Preservation of products implies that proofs of $\Gamma \to \Psi$ are in bijective correspondence with families of proofs $(\Gamma \to \psi)_{\psi \in \Psi}$, and that there are monotonicity proofs $\Gamma \to \Psi$ whenever $\Psi \subseteq \Gamma$.

A *cat/cat institution comorphism* between cat/cat institutions $\mathcal{I}$ and $\mathcal{J}$ consists of a set/cat institution comorphism $(\Phi, \alpha, \beta)\colon \mathcal{I} \longrightarrow \mathcal{J}$ and a natural transformation $\gamma\colon \text{Pr}^{\mathcal{I}} \longrightarrow \text{Pr}^{\mathcal{J}} \circ \Phi$ such that translation of sentence sets is compatible with translation of single sentences: $\gamma_\Sigma \circ \iota_\Sigma = \iota'_\Sigma \circ \mathcal{P}(\alpha_\Sigma)^{op}$, where ι_Σ and ι'_Σ are the appropriate inclusions. A cat/cat institution comorphism $(\Phi, \alpha, \beta, \gamma)$ is a *cat/cat equivalence* if Φ is an equivalence of categories, β is a family of equivalences natural in Σ, and so is γ. Note that there is no requirement on α. As before, all this also extends to the case of omitting of model morphisms, i.e., the cat/set case. Henceforth, the term *proof theoretic institution* will refer to both the cat/cat and the cat/set cases. □

---

[11]Instead of having two functors Pr and Sen, it is also possible to have one functor into a comma category.

[12]A category is *thin* if between two given objects, there is at most one morphism, i.e., the category is a pre-ordered class.

Given an arbitrary but fixed proof theoretic institution, we can define an entailment relation $\vdash_\Sigma$ between sets of Σ-sentences as follows: $\Gamma \vdash_\Sigma \Psi$ if there exists a morphism $\Gamma \to \Psi$ in $\mathsf{Pr}(\Sigma)$. A proof theoretic institution is *sound* if $\Gamma \vdash_\Sigma \Psi$ implies $\Gamma \models_\Sigma \Psi$; it is *complete* if the converse implication holds. In the sequel, we will assume that all proof theoretic institutions are sound, which in particular implies the following:

Proposition 5.2. Any cat/cat equivalence is a set/cat equivalence. $\square$

Proposition 5.3. $\vdash$ satisfies the properties of an *entailment system* [28], i.e., it is reflexive, transitive, monotonic and stable under translation along signature morphisms. In fact, entailment systems are in bijective correspondence with proof theoretic institutions having trivial model theory (i.e., $\mathsf{Mod}(\Sigma) = \emptyset$) and thin categories of proofs. $\square$

The requirement for sentence translation in proof theoretic institution equivalences is very close to the notion of translational equivalence introduced in [31]. A set/set institution equivalence basically requires that the back-and-forth translation of sentence is semantically equivalent to the original sentence (i.e., $\alpha'_\Sigma(\alpha_\Sigma(\varphi)) \models\mid \varphi$); a similar notion would arise when using $\vdash$. Note, however, that this does not work well for modal logics, since e.g., in **S5**, $\varphi \vdash \Box\varphi$. Therefore, [31] require $\vdash \alpha'_\Sigma(\alpha_\Sigma(\varphi)) \leftrightarrow \varphi$. However, this is based upon the presence of equivalence as a proof theoretic connective, which is not present in all institutions. Our solution to this problem comes naturally out of the above definition of proof theoretic (i.e., cat/cat or cat/set) equivalence: $\alpha'_\Sigma(\alpha_\Sigma(\varphi))$ and φ have to be isomorphic in the category of proofs. We thus neither identify φ and $\Box\varphi$ in modal logics, nor rely on the presence of a connective $\leftrightarrow$.

Definition 5.4. A proof theoretic institution is *finitary* if $\Gamma \vdash_\Sigma \varphi$ implies $\Gamma' \vdash_\Sigma \varphi$ for some finite $\Gamma' \subseteq \Gamma$.

A proof theoretic institution has *proof theoretic conjunction* if each category $\mathsf{Pr}(\Sigma)$ has distinguished products of singletons, which are singletons again and which are preserved by the proof translations $\mathsf{Pr}(\varphi)$. In terms of derivability, this implies that for φ_1, φ_2 Σ-sentences, there is a product sentence $\varphi_1 \boxtimes \varphi_2$, and two "projection" proof terms $\pi_1 : \varphi_1 \boxtimes \varphi_2 \longrightarrow \varphi_1$ and $\pi_2 : \varphi_1 \boxtimes \varphi_2 \longrightarrow \varphi_2$, such that for any ψ with $\psi \vdash_\Sigma \varphi_1$ and $\psi \vdash_\Sigma \varphi_2$, then $\psi \vdash_\Sigma \varphi_1 \boxtimes \varphi_2$.

Similarly, a proof theoretic institution has *proof theoretic disjunction (true, false)* if each proof category has distinguished coproducts of singletons that are singletons (a distinguished singleton terminal object, a distinguished singleton initial object) which are preserved by the proof translations.

For each set Γ of Σ-sentences, there is a canonical homomorphism of graphs $\_\cup\Gamma \colon \mathsf{Pr}(\Sigma) \longrightarrow \mathsf{Pr}(\Sigma)$ as defined by the following commutative diagram of proofs:

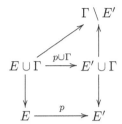

(where the monotonicity proofs are not labelled). In general, the graph homomorphism $\_\cup\Gamma$ is *not* functorial!

A proof theoretic institution has *proof theoretic implication* if each graph homomorphism $\_\cup\{\varphi\} \colon \mathsf{Pr}(\Sigma) \longrightarrow \mathsf{Pr}(\Sigma)$ has a distinguished "right adjoint", denoted by $\varphi\boxed{\to}\_$, such that $\varphi\boxed{\to}\_$ maps singletons to singletons and $\varphi\boxed{\to}\Gamma = \{\varphi\boxed{\to}\psi | \psi \in \Gamma\}$ and such that it commutes with the proof translations. This means there exists a bijective correspondence, called the 'Deduction Theorem' in classical logic, between $\mathsf{Pr}(\Sigma)(\Gamma \cup \{\rho\}, E)$ and $\mathsf{Pr}(\Sigma)(\Gamma, \rho \to E)$ natural in Γ and E, and such that the following diagram commutes for all signature morphisms $\sigma \colon \Sigma \to \Sigma'$:

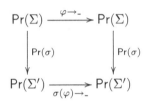

In case $\boxed{f}$ is present, it has *proof theoretic negation* if each sentence ψ has a distinguished negation $\boxed{\neg}\psi$ which is preserved by the proof translations $\mathsf{Pr}(\varphi)$ and such that $\mathsf{Pr}(\Sigma)(\Gamma \cup \{\psi\}, \boxed{f})$ is in natural bijective correspondence to $\mathsf{Pr}(\Sigma)(\Gamma, \{\boxed{\neg}\psi\})$.

A proof theoretic institution is *propositional* if it has proof theoretic conjunction, disjunction, implication, negation, true and false. $\qquad\square$

Definition 5.5. A proof theoretic institution with proof theoretic negation has $\neg\neg$-*elimination* if for each Σ-sentence φ, $\boxed{\neg}\boxed{\neg}\varphi \vdash_\Sigma \varphi$ (the converse relation easily follows from the definition). $\qquad\square$

For example, **CPL** and **FOL** have $\neg\neg$-elimination, while **IPL** has not. Clearly, any complete proof theoretic institution with external semantic and proof theoretic negation has $\neg\neg$-elimination.

Proposition 5.6. A proof theoretic institution having proof theoretic implication enjoys the deduction theorem and modus ponens for $\vdash_\Sigma$. A complete proof theoretic institution is finitary iff it is compact. $\qquad\square$

Example 5.7.   The modal logic **K** does not have proof theoretic implication, nor negation, and this is a difference from intuitionistic logic **IPL**, showing that the two logics are not equivalent. (See [24] for the proof category of **IPL**.)   □

While **K** does not have proof theoretic implication, it still has a form of *local* implication, which does not satisfy the deduction theorem. This can be axiomatized as follows:

Definition 5.8.   A proof theoretic institution has *Hilbert implication* if for each signature Σ, there is a unique binary operator $\boxminus$ on Σ-sentences satisfying the Hilbert axioms for implication, i.e.,

$$(K) \quad \emptyset \vdash_\Sigma \{\varphi \boxminus \psi \boxminus \varphi\}$$
$$(S) \quad \emptyset \vdash_\Sigma \{(\varphi \boxminus \psi \boxminus \chi) \boxminus (\varphi \boxminus \psi) \boxminus \varphi \boxminus \chi\}$$
$$(MP) \quad \{\varphi \boxminus \psi, \varphi\} \vdash_\Sigma \{\psi\}$$

There is a proof theoretic variant of the Lindenbaum algebra of Def. 4.4: □

Definition 5.9.   Let $\Psi^{\mathcal{I}}$ be the single sorted algebraic signature having a subset of the operations $\{\boxwedge, \boxvee, \boxminus, \boxminus, \boxleftrightarrow, \boxed{t}, \boxed{f}\}$ (with their standard arities), chosen according to whether $\mathcal{I}$ has proof theoretic conjunction, disjunction, negation etc., and Hilbert implication for implication. Note that like the signature $\Xi^{\mathcal{I}}$ introduced in Def. 4.4, $\Psi^{\mathcal{I}}$ may include connectives not provided by the institution $\mathcal{I}$, or provided by $\mathcal{I}$ with a different syntax. By Thm. 5.11, $\Xi^{\mathcal{I}}$ has a canonical embedding into $\Psi^{\mathcal{I}}$. Consider $\mathcal{L}^{\mathsf{H}}(\Sigma) = \mathsf{Sen}(\Sigma)/\!\cong$, where $\cong$ is isomorphism in $\mathsf{Pr}(\Sigma)$. Since products etc. are unique up to isomorphism, it is straightforward to make this a $\Psi^{\mathcal{I}}$-algebra.

The Lindenbaum algebra is the basis for the *Lindenbaum category* $\mathcal{LC}^{\mathsf{H}}(\Sigma)$, which has object set $\mathcal{P}(\mathcal{L}^{\mathsf{H}}(\Sigma))$. By choosing a system of canonical representatives for $\mathsf{Sen}(\Sigma)/\!\cong$, this object set can be embedded into $|\mathsf{Pr}(\Sigma)|$; hence we obtain an induced full subcategory, which we denote by $\mathcal{LC}^{\mathsf{H}}(\Sigma)$. Different choices of canonical representatives may lead to different but isomorphic Lindenbaum categories. While the Lindenbaum category construction is functorial, the proof theoretic Lindenbaum algebra construction is generally not. Also, the closed theory functor $\mathcal{C}^{\models}$ has a proof theoretic counterpart $\mathcal{C}^{\vdash}$ taking theories closed under $\vdash$.   □

Definition 5.10.   A proof theoretic institution is *compatible* if for each circled (i.e., external semantic) operator in $\Xi^{\mathcal{I}}$, the corresponding boxed (i.e., proof theoretic) operator in $\Psi^{\mathcal{I}}$ is present. It is *bicompatible* if also the converse holds.   □

Theorem 5.11.   A complete proof theoretic institution with thin proof categories is compatible, but not necessarily bicompatible.   □

Proposition 5.12.   Assume a proof theoretic institution with thin proof categories. If deduction is complete, then $\mathcal{L}(\Sigma)$ and $\mathcal{L}^{\mathsf{H}}(\Sigma)|_{\Xi^{\mathcal{I}}}$ are isomorphic; just soundness gives a surjective cryptomorphism $\mathcal{L}^{\mathsf{H}}(\Sigma)|_{\Xi^{\mathcal{I}}} \to \mathcal{L}(\Sigma)$, and just completeness gives one in the opposite direction.   □

Example 5.13. Intuitionistic propositional logic shows that proof theoretic disjunction does not imply external semantic disjunction.   □

Definition 5.14. A proof theoretic institution is *classical modal* if its Lindenbaum algebras $\mathcal{L}^{\mathsf{H}}(\Sigma)$ are Boolean algebras (also having implication) with an operator $\Box$ (congruent with $\boxed{\wedge}$ and $\boxed{t}$). A classical modal proof theoretic institution is *normal* if the operator satisfies the necessitation law: $\varphi \vdash_{\Sigma} \Box\varphi$. (Note that modus ponens already follows from implication being present in $\Psi^{\mathcal{I}}$.)   □

It is clear that equivalences between classical modal proof theoretic institutions need to preserve $\mathcal{L}^{\mathsf{H}}$ (but not necessarily the operator). It should hence be possibly to apply the results of [31].

Example 5.15. **S4** has a non-idempotent operator (congruent with $\boxed{\wedge}$ and $\boxed{t}$) on its Lindenbaum algebra, while **S5** does not have one. Hence, **S4** and **S5** are not equivalent.   □

Definition 5.16. Given cat/cat institutions $\mathcal{I}$ and $\mathcal{J}$, $\mathcal{J}$ is a cat/cat *skeleton* of $\mathcal{I}$ if it is like a set/cat skeleton, but such that $\mathsf{Sen}^{\mathcal{J}}(\Sigma) = \mathsf{Sen}^{\mathcal{I}}(\Sigma)/{\cong}$, and such that $\mathsf{Pr}^{\mathcal{J}}(\Sigma) = \mathcal{L}\mathcal{C}^{\mathsf{H}}(\Sigma)$, the Lindenbaum category.   □

Lawvere [26, 27] defined quantification as adjoint to substitution. Here we define quantification as adjoint to sentence translation along a class $\mathcal{D}$ of signature morphisms, which typically introduce new constants to serve as quantification "variables":

Definition 5.17. A cat/cat institution *has proof theoretic universal (existential) $\mathcal{D}$-quantification* for a class $\mathcal{D}$ of signature morphisms stable under pushouts, if for all signature morphisms $\sigma \in \mathcal{D}$, $\mathsf{Pr}(\sigma)$ has a distinguished right (left) adjoint, denoted by $(\forall\sigma)_{-}$ $((\exists\sigma)_{-})$ and preserved by proof translations along signature morphisms. This means there exists a bijective correspondence between $\mathsf{Pr}(\Sigma)(E, (\forall\sigma)E')$ and $\mathsf{Pr}(\Sigma')(\sigma(E), E')$ natural in E and E', in classical logic known as the 'Generalisation Rule', such that for each signature pushout with $\sigma \in \mathcal{D}$,

the pair $\langle \mathsf{Pr}(\theta), \mathsf{Pr}(\theta') \rangle$ is a morphism of adjunctions.   □

One may define a proof theoretic concept of consistency. A theory (Σ, Γ) is *consistent* when its closure under $\vdash$ is a proper subset of $\mathsf{Sen}(\Gamma)$. Craig interpolation also has a proof theoretic version: for any proof $p:\ \theta_1(E_1) \to \theta_2(E_2)$, there exist proofs $p_1:\ E_1 \to \sigma_1(E)$ and $p_2:\ \sigma_2(E) \to E_2$ such that $p = \theta_2(p_2) \circ \theta_1(p_1)$.

Given a set/cat (or set/set) institution $\mathcal{I}$, we can obtain a complete cat/cat (or cat/set) institution $\mathcal{I}^{\models}$ by letting $\mathsf{Pr}(\Sigma)$ be the pre-order defined by $\Gamma \leq \Psi$ if

$\Gamma \models_\Sigma \Psi$, considered as a category. Some of the proof theoretic notions are useful when interpreted in $\mathcal{I}^\models$:

Definition 5.18. An institution $\mathcal{I}$ has *internal semantic conjunction* if $\mathcal{I}^\models$ has proof theoretic conjunction; similarly for the other connectives. □

Example 5.19. Intuitionistic logic **IPL** has internal, but not external semantic implication. Higher-order intuitionistic logic interpreted in a fixed topos (see [24]) has proof theoretic and Hilbert implication, but neither external nor internal semantic implication. Modal logic **S5** has just Hilbert implication. □

Theorem 5.20. The properties in the table below are invariant under set/set, set/cat, cat/set and cat/cat equivalence, resp.[13] Sect. 5.) Properties in *italics* rely on concrete institutions (as in Def. 2.7).

set/set	set/cat
compactness, (semi-)exactness, elementary amalgamation, semantic Craig interpolation, Beth definability, having external semantic conjunction, disjunction, negation, true, false, being truth functionally complete, Lindenbaum signature $\Xi^{\mathcal{I}}$, Lindenbaum algebra functor $\mathcal{L}$, closed theory lattice functor $\mathcal{C}^\models$, (co)completeness of the signature category, *Lindenbaum cardinality spectrum, finite model property.*	all of set/set, having external semantic universal or existential (representable) quantification, exactness, elementary diagrams, (co-)completeness of model categories, existence of reduced products, preservation for formulæ along reduced products, being a Łoś-institution, *model cardinality spectrum, admission of free models.*
cat/set	cat/cat
all of set/set plus its proof theoretic counterparts where applicable, soundness, completeness, Hilbert implication, $\neg\neg$-elimination.	all of set/cat and cat/set, having proof theoretic universal or existential quantification, compatibility, bicompatibility.

6. $\mathbb{C}/\mathbb{D}$-institutions

In this final section, we take an even more relativistic view on institutions. We already have introduced a number of variants of institutions: set/set, set/cat, cat/set and cat/cat. This can be made more formal in the following way.

---

[13] Functors such as the Lindenbaum algebra functor are preserved in the sense that $\mathcal{L}_{\mathcal{I}}$ is naturally isomorphic to $\mathcal{L}_{\mathcal{J}} \circ \Phi$.

A category $\mathbb{C}$ is *concrete* [1], if it is equipped with a forgetful functor $|\_| : \mathbb{C} \longrightarrow \mathbb{S}et.$[14]

Given concrete categories $\mathbb{C}$ and $\mathbb{D}$, a $\mathbb{C}/\mathbb{D}$-institution consists of

- a category $\mathbb{S}ign$ of signatures,
- a sentence functor $\mathsf{Sen} : \mathbb{S}ign \to \mathbb{C}$,
- a model functor $\mathsf{Mod} : \mathbb{S}ign^{\mathrm{op}} \to \mathbb{D}$, and
- a satisfaction relation $\models_\Sigma \subseteq |\mathsf{Mod}(\Sigma)| \times |\mathsf{Sen}(\Sigma)|$ for each $\Sigma \in |\mathbb{S}ign|$,

subject to the satisfaction condition: $M' \models_{\Sigma'} |\mathsf{Sen}(\sigma)|(\varphi)$ iff $|\mathsf{Mod}(\sigma)|(M') \models_\Sigma \varphi$ for each $M' \in |\mathsf{Mod}(\Sigma')|$ and $\varphi \in |\mathsf{Sen}(\Sigma)|$.

With this terminology, set/set-institutions are just $\mathbb{S}et/\mathbb{S}et$-institutions, and set/cat-institutions are $\mathbb{S}et/\mathbb{C}\mathbb{A}\mathbb{T}$-institutions. What about the cat/set and cat/cat. variants?

Let $\mathcal{P} : \mathbb{S}et \longrightarrow \mathbb{S}et$ be the covariant power set functor, and $|\_| : \mathbb{C}at \longrightarrow \mathbb{S}et$ be the functor sending each small category to its set of objects. Then $\mathbb{P}owerCat$, the comma category $(\mathcal{P} \downarrow |\_|)$, consists of small categories whose set of objects is the power set of a given set (of "sentences"), and functors whose object mapping is induced by a mapping between the sets of sentences. Our cat/cat-institutions are more precisely $\mathbb{P}owerCat/\mathbb{C}\mathbb{A}\mathbb{T}$-institutions; a similar remark holds for cat/set-instititions.

Florian Rabe (personal communication) pointed out the following phenomenon: if a proof $p : \{\varphi, \psi\} \longrightarrow \{\chi\}$ is translated along a signature morphism σ that identifies φ and ψ, we get a proof $\sigma(p) : \{\sigma(\varphi)\} \longrightarrow \{\sigma(\chi)\}$ where the information about the original numbers of premises is lost. In some cases (for example, when generating free proof systems [15, 16]), it is desirable to have the facility to keep this multiplicity information[15], which amounts to work with *multisets* of sentences (instead of just sets) in the premises of proofs. A more algebraic way of formulating this is the following

Proof categories for Lawvere/cat-institutions are many-sorted Lawvere theories

offering a new variant of the Curry-Howard isomorphism.

Recall that a single-sorted finitary Lawvere theory [26, 27] is a product-preserving object-bijective functor from $\mathbb{N}^{op}$ to $\mathbb{P}$, where $\mathbb{N}$ is the category of finite ordinals (considered as a subcategory of $\mathbb{S}et$) and $\mathbb{P}$ is any category (of "term tuples"). Now a many-sorted Lawvere theory (*without rank*) is a product-preserving object-bijective functor from $\mathbb{F}am(S)^{op}$ to $\mathbb{P}$. Here, for a set S, $\mathbb{F}am(S)$ is the comma category $(\mathbb{S}et, S)$ of families $\phi : X \longrightarrow S$ of elements of S (where X is an arbitrary index set) and reindexing morphisms. $\mathbb{F}am(S)$ can be shown to be equivalent to the free category with coproducts over the set S of generating objects.

---

[14]The use of the same symbol $|\_|$ for the forgetful functor here and for the collection of objects of a category should not be confusing. Moreover, in some cases (for example, in the case of $\mathbb{C}\mathbb{A}\mathbb{T}$), we need to replace the category of sets by the quasi-category of classes.

[15]The problem is that otherwise disjointness of sentence sets are not preserved, and hence products (=disjoint unions) are not preserved either

Notice that a family $\phi\colon X \longrightarrow S$ a family is nothing but an S-sorted system of variables; a proof judgement hence is a variable context. Each variable can be thought of as standing for an unknown proof, and the type of the variable being the sentence being witnessed by the proof. This nicely corresponds to the intuition (guided by many examples, and also by the standard interpretation of algebraic theories as Lawvere theories) that the category of proofs has as morphisms tuples of terms with variables, where the variable context is given by the source proof judgement. Further note that for a sentence ψ, the projections $\pi_1, pi_2\colon \psi \times \psi \longrightarrow \psi$ are retractions with one-sided inverse $\langle id, id \rangle$, but they are in general not isomorphisms. This exactly yields the difference between ψ and $\psi \times \psi$ needed for the multiset-character of proof judgements.

Though it has to be elaborated whether the concept of Lawvere/cat institution works well with the examples, the naturalness and elegance of the concepts already sounds very promising.

7. Conclusions

We believe this paper has established four main points: (1) The notion of "a logic" should depend on the purpose; in particular, proof theory and model theory sometimes treat essentially the same issue in different ways. Institutions provide an appropriate framework, having a balance between model theory and proof theory. (2) Every plausible notion of equivalence of logics can be formalized using institutions and various equivalence relations on them. (3) Inequivalence of logics can be established using various constructions on institutions that are invariant under the appropriate equivalence, such as Lindenbaum algebras and cardinality spectra. We have given several examples of such inequivalences. (4) A great deal of classical logic can be generalized to arbitrary institutions, and the generalized formulations are often quite interesting in themselves. Perhaps the fourth point is the most exciting, as there remains a great deal more to be done, particularly in the area of proof theory.

Among the proof theoretic properties that we have not treated: Proof theoretic ordinals, while an important device, would deviate a bit from the subject of this paper, because they are a measure for the proof theoretic strength of a *theory* in a logic, not a measure for the logic itself. But properties like (strong) normal forms for proofs could be argued to contribute to the identity of a logic; treating them would require $Pr(\Sigma)$ to become an order-enriched category or a 2-category of sentences, with proof terms and proof term reductions. A related topic is cut elimination, which would require an even finer structure on $Pr(\Sigma)$, with proof rules of particular format. Another direction is the introduction of numberings in order to study recursiveness of entailment. We hope this paper provides a good starting point for such investigations.

Acknowledgements. Till Mossakowski has been supported by the project MULTIPLE of the *Deutsche Forschungsgemeinschaft* under Grants KR 1191/5-1 and KR

1191/5-2. Thanks to Florian Rabe and Lutz Schröder for helpful discussions that eventually arrived at the connection to Lawvere theories.

References

[1] J. Adámek, H. Herrlich, and G. Strecker. *Abstract and Concrete Categories*. Wiley, New York, 1990.

[2] J. Barwise. Axioms for abstract model theory. *Annals of Mathematical Logic*, 7:221–265, 1974.

[3] J.-Y. Béziau, R.P. de Freitas, and J.P. Viana. What is classical propositional logic? (a study in universal logic). *Logica Studies*, 7, 2001.

[4] M. Bidoit and R. Hennicker. On the integration of observability and reachability concepts. In M. Nielsen and U. Engberg, editors, *FoSSaCS 2002*, volume 2303 of *Lecture Notes in Computer Science*, pages 21–36. Springer, 2002.

[5] M. Bidoit and A. Tarlecki. Behavioural satisfaction and equivalence in concrete model categories. *Lecture Notes in Computer Science*, 1059:241–256, 1996.

[6] T. Borzyszkowski. Generalized interpolation in first-order logic. Technical report, 2003. Submitted to Fundamenta Informaticae.

[7] T. Borzyszkowski. Moving specification structures between logical systems. In J. L. Fiadeiro, editor, *13th WADT'98*, volume 1589 of *Lecture Notes in Computer Science*, pages 16–30. Springer, 1999.

[8] T. Borzyszkowski. Generalized interpolation in CASL. *Information Processing Letters*, 76:19–24, 2000.

[9] C. Cîrstea. Institutionalising many-sorted coalgebraic modal logic. In *CMCS 2002*, Electronic Notes in Theoretical Computer Science. Elsevier Science, 2002.

[10] R. Diaconescu. Grothendieck institutions. *Applied Categorical Structures*, 10(4):383–402, 2002. Preliminary version appeared as IMAR Preprint 2-2000, ISSN 250-3638, February 2000.

[11] R. Diaconescu. Institution-independent ultraproducts. *Fundamenta Informaticæ*, 55:321–348, 2003.

[12] R. Diaconescu. Elementary diagrams in institutions. *Journal of Logic and Computation*, 14(5):651–674, 2004.

[13] R. Diaconescu. Herbrand theorems in arbitrary institutions. *Information Processing Letters*, 90:29–37, 2004.

[14] R. Diaconescu. An institution-independent proof of Craig Interpolation Theorem. *Studia Logica*, 77(1):59–79, 2004.

[15] R. Diaconescu. Proof systems for institutional logic. *Journal of Logic and Computation*, 16:339–357, 2006.

[16] R. Diaconescu. *Institution-independent model theory*. Springer Verlag, to appear, 2007.

[17] T. Dimitrakos and T. Maibaum. On a generalized modularization theorem. *Information Processing Letters*, 74:65–71, 2000.

[18] J. L. Fiadeiro and J. F. Costa. Mirror, mirror in my hand: A duality between specifications and models of process behaviour. *Mathematical Structures in Computer Science*, 6(4):353–373, 1996.

[19] J. A. Goguen and R. M. Burstall. Institutions: Abstract model theory for specification and programming. *Journal of the Association for Computing Machinery*, 39:95–146, 1992. Predecessor in: LNCS 164, 221–256, 1984.

[20] J. A. Goguen and R. Diaconescu. Towards an algebraic semantics for the object paradigm. In *WADT*, number 785 in Lecture Notes in Computer Science. Springer Verlag, Berlin, Germany, 1994.

[21] J. A. Goguen and G. Roşu. Institution morphisms. *Formal Aspects of Computing*, 13:274–307, 2002.

[22] J. A. Goguen and W. Tracz. An Implementation-Oriented Semantics for Module Composition. In Gary Leavens and Murali Sitaraman, editors, *Foundations of Component-based Systems*, pages 231–263. Cambridge, 2000.

[23] D. Găină and A. Popescu. An institution-independent proof of Robinson consistency theorem. Studia Logica, to appear.

[24] J. Lambek and P. J. Scott. *Introduction to Higher Order Categorical Logic*. Cambridge University Press, 1986.

[25] S. Mac Lane. *Categories for the working mathematician. Second Edition.* Springer, 1998.

[26] F. W. Lawvere. *Functional Semantics of Algebraic Theories.* PhD thesis, Columbia University, 1963.

[27] F. W. Lawvere. Functorial semantics of elementary theories. *Journal of Symbolic Logic*, 31:294–295, 1966.

[28] J. Meseguer. General logics. In *Logic Colloquium 87*, pages 275–329. North Holland, 1989.

[29] T. Mossakowski. Specification in an arbitrary institution with symbols. In C. Choppy, D. Bert, and P. Mosses, editors, *14th WADT*, volume 1827 of *Lecture Notes in Computer Science*, pages 252–270. Springer-Verlag, 2000.

[30] C. S. Peirce. *Collected Papers.* Harvard, 1965. In 6 volumes; see especially Volume 2: Elements of Logic.

[31] F.J. Pelletier and A. Urquhart. Synonymous logics. *Journal of Philosophical Logic*, 32:259–285, 2003.

[32] M. Petria and R. Diaconescu. Abstract Beth definability in institutions. *Journal of Symbolic Logic*, 71(3):1002–1028, 2006.

[33] S. Pollard. Homeomorphism and the equivalence of logical systems. *Notre Dame Journal of Formal Logic*, 39:422–435, 1998.

[34] A. Popescu, T. Şerbănuţă, and G. Roşu. A semantic approach to interpolation. In Luca Aceto and Anna Ingólfsdóttir, editors, *Foundations of Software Science and Computation Structures, 9th International Conference, FOSSACS 2006*, volume 3921 of *Lecture Notes in Computer Science*, pages 307–321. Springer, 2006. also appeared as Technical Report UIUCDCS-R-2005-2643, May 2005.

[35] P.-H. Rodenburg. A simple algebraic proof of the equational interpolation theorem. *Algebra Universalis*, 28:48–51, 1991.

[36] D. Sannella and A. Tarlecki. Specifications in an arbitrary institution. *Information and Computation*, 76:165–210, 1988.

[37] L. Schröder, T. Mossakowski, and C. Lüth. Type class polymorphism in an institutional framework. In José Fiadeiro, editor, *17th WADT*, Lecture Notes in Computer Science. Springer; Berlin; http://www.springer.de, 2004. To appear.

[38] A. Tarlecki. On the existence of free models in abstract algebraic institutions. *Theoretical Computer Science*, 37(3):269–304, 1985.

[39] A. Tarlecki. Bits and pieces of the theory of institutions. In D. Pitt, S. Abramsky, A. Poigné, and D. Rydeheard, editors, *Proc. CTCS*, volume 240 of *Lecture Notes in Computer Science*, pages 334–363. Springer-Verlag, 1986.

[40] A. Tarlecki. Quasi-varieties in abstract algebraic institutions. *Journal of Computer and System Sciences*, 33:333–360, 1986.

Till Mossakowski
DFKI Lab Bremen and Universität Bremen
Germany
e-mail: Till.Mossakowski@dfki.de

Joseph Goguen
University of California at San Diego
USA

Răzvan Diaconescu
Institute of Mathematics
Romanian Academy
Bucharest
Romania
e-mail: Razvan.Diaconescu@imar.ro

Andrzej Tarlecki
Institute of Informatics
Warsaw University

and

Institute of Computer Science
Polish Academy of Sciences
Warsaw
Poland
e-mail: tarlecki@mimuw.edu.pl

J.-Y. Beziau (Ed.), *Logica Universalis, 2nd edition*, 135–152
© 2007 Birkhäuser Verlag Basel/Switzerland

What is a Logic, and What is a Proof ?

Lutz Straßburger

Abstract. I will discuss the two problems of how to define identity between logics and how to define identity between proofs. For the identity of logics, I propose to simply use the notion of preorder equivalence. This might be considered to be folklore, but is exactly what is needed from the viewpoint of the problem of the identity of proofs: If the proofs are considered to be part of the logic, then preorder equivalence becomes equivalence of categories, whose arrows are the proofs. For identifying these, the concept of proof nets is discussed.

1. Introduction

When we study mathematical objects within a certain mathematical theory, we usually know when two of these objects are considered to be the same, i.e., are indistinguishable within the theory. For example in group theory two groups are indistinguishable if they are isomorphic, in topology two spaces are considered the same if they are homeomorphic, and in graph theory we have the notion of graph isomorphism. However, in proof theory the situation is different. Although we are able to manipulate and transform proofs in various ways, we have no satisfactory notion telling us when two proofs are the same, in the sense that they use the same argument. The reason is the lack of understanding of the essence of a proof, which in turn is caused by the bureaucracy involved in the syntactic presentation of proofs. It is therefore an important problem of research to find new ways of presenting proofs, that allow to grasp the essence of a proof by getting rid of bureaucratic syntax, and that identify proofs if and only if they use the same argument. As a matter of fact, the problem was already a concern of Hilbert, when he was preparing his famous lecture in 1900 [Thi03]. The history of mathematical logic and proof theory might have developed in a different way if he had included his "24th problem".

The text for the second edition has been updated by including some points that have been discussed at the UniLog meeting 2005 in Montreux.

Proofs are carried out within logical systems. We can, for example, have proofs in classical logic and proofs in linear logic. It should be obvious, that two proofs that are carried out in different logics must be distinguished (although every intuitionistic proof can also be seen as classical proof). Consequently, before expecting an answer to the question "When are two proofs the same?", we have first to give an answer to the question "When are two logics the same?".[1] The problem of identifying logics is not only of interest for proof theory, but for the whole area of logic, including mathematical logic as well as philosophical logic.

This means that we have to deal with two problems: the identity of proofs, and the identity of logics. Although the two problems are closely related, they are of a completely different nature.

For the identity of proofs, the actual problem is to find the right presentations of proofs that allow us to make the correct identifications. So far, proofs are presented as syntactical objects: we see Hilbert style proofs, natural deduction proofs, resolution proofs, sequent calculus proofs, proofs in the calculus of structures, tableau proofs, and many more — in particular, also proofs written up in natural language. Of course, the same proof can be written up in various different formalisms. And even in a single formalism, the same proof can take different shapes.

For the identity of logics, on the other hand, the actual problem is to find the "least common denominator" for a definition of logic. The reason is that there is no generally accepted consensus under logicians about the question what a logic actually is. Not only is the model theoreticians understanding of a logic ("a logic is something that has a syntax and semantics") different from the proof theoreticians understanding ("a logic is a deductive system that has the cut elimination property"), we also see in other areas of research various different notions of "logic", which are all tailored for a particular application.

But a clean definition of logic will immediately lead to a clean notion of equivalence of logics. In the next section, I will give (from the proof theoreticians viewpoint) such a definition together with its notion of equivalence. Although it could certainly be considered to be folklore knowledge — for long it has been used by logicians already — I discuss it here because it provides clear and firm grounds for investigating the problem of identifying proofs. This problem will be discussed in the last section of the paper.

2. What is a logic ?

Definition 2.1. A *logic* $\mathscr{L} = (\mathscr{A}_\mathscr{L}, \Rightarrow_\mathscr{L})$ is a set $\mathscr{A}_\mathscr{L}$ of *formulae*, together with a binary relation $\Rightarrow_\mathscr{L} \subseteq \mathscr{A}_\mathscr{L} \times \mathscr{A}_\mathscr{L}$, called the *consequence relation*, that is reflexive and transitive.

---

[1] Of course, this problem was of no concern for Hilbert, since at the time when he was thinking about the identity of proofs there was only one logic.

In other words, a logic is simply a preorder. The index $\mathscr{L}$ will be omitted for $\mathscr{A}$ and $\Rightarrow$, if no ambiguity is possible. The elements of $\mathscr{A}$ will be denoted by A, B, C, etc. Instead of $A \Rightarrow B$, we can also write $B \Leftarrow A$. Similarly, we write $A \Leftrightarrow B$ if $A \Rightarrow B$ and $A \Leftarrow B$. Observe that $\Leftrightarrow$ is always an equivalence relation. Let me make some comments about Definition 2.1:

- There are no cardinality restrictions on the set $\mathscr{A}$.
- The definition abstracts away from the structure of the set $\mathscr{A}$. For capturing the "purely logical part" of a logic, it is irrelevant, which and how many connectives, quantifiers, modalities, constants, variables, etc. are there. It is also not necessary (for the time being) what kind of objects the set $\mathscr{A}$ contains. This could be simply well-formed formulae, sets (finite or infinite) of well-formed formulae, mathematical structures like vector spaces, or even sentences of natural language, like "The book is green."[2]
- There are no computability, complexity, or compactness restrictions on the relation $\Rightarrow$, and there is no need to distinguish between syntax and semantics: It is of no relevance whether $\Rightarrow$ is defined in a model theoretic way ($A \Rightarrow B$ iff every model that makes A true does also make B true), by means of a deductive system ($A \Rightarrow B$ iff there is a proof of B from hypothesis A), or in some other way.
- The two properties of being reflexive and transitive are essential for our treatment of $\Rightarrow$. Reflexivity says that $A \Rightarrow A$ for every formula A. Transitivity says that whenever $A \Rightarrow B$ and $B \Rightarrow C$ then also $A \Rightarrow C$.[3]

Often the notion of a logic is presented such that the consequence relation is not defined as a subset of $\mathscr{A} \times \mathscr{A}$ but as a subset of $\mathfrak{P}_f(\mathscr{A}) \times \mathscr{A}$ or even of $\mathfrak{P}(\mathscr{A}) \times \mathscr{A}$, where $\mathfrak{P}(\mathscr{A})$ is the powerset of $\mathscr{A}$, i.e., the set of all subsets, and $\mathfrak{P}_f(\mathscr{A})$ is the set of all finite subsets of $\mathscr{A}$. Let us denote this new consequence relation by $\vdash$. It should be clear that such a definintion is perfectly equivalent to the one in 2.1, provided the structure of $\mathscr{A}$ has access to the concept of "conjunction", for example, via a connective $\wedge$. Then we have

$$\{A_1, \ldots, A_n\} \vdash B \qquad \text{iff} \qquad A_1 \wedge \cdots \wedge A_n \Rightarrow B \quad .$$

In the case of $\mathfrak{P}(\mathscr{A})$ we also need access to the concept of "infinite conjunction". Then we have

$$\Gamma \vdash B \qquad \text{iff} \qquad \bigwedge \Gamma \Rightarrow B \quad ,$$

where $\Gamma \subseteq \mathscr{A}$ is an arbitrary set of well-formed formulae and $\bigwedge \Gamma$ is their conjunction.

Alternatively, the notion of logic can be defined as a pair $(\mathscr{A}, \mathbb{T})$, where $\mathbb{T} \subseteq \mathscr{A}$ is the set of tautologies (or theorems). Again, this definition is perfectly

---

[2]With a sufficiently sophisticated definition of "well-formed fomula" the set $\mathscr{A}$ can in fact be restricted to that notion. For example we could allow something like "$\{\phi \mid \ldots\}$" to be a "well-formed fomula", and would by this also capture sets of formulae/propositions/sentences/whatever.

[3]Of course, these conditions can also be dropped, but then everything is possible.

equivalent to the one we have seen, provided the connectives that generate $\mathscr{A}$ have access to the concept of "implication" (either via a connective $\supset$, or via a disjunction together with a negation, or in any other way) and the concept of "truth" (for example via a constant $\top$). Then we have

$$A \Rightarrow B \quad \text{iff} \quad A \supset B \in \mathbb{T}$$

and

$$A \in \mathbb{T} \quad \text{iff} \quad \top \Rightarrow A \quad .$$

However, in both alternative definitions, we have to ensure the reflexivity and transitivity of the induced consequence relation. Because of the importance of these two conditions I prefer the definition as given in 2.1.

Let us now continue with some standard definitions for preorders.

Definition 2.2. The *skeleton* of a preorder is the partially ordered set $(\mathscr{A}/\Leftrightarrow, \leq)$, where $\mathscr{A}/\Leftrightarrow$ is the set of equivalence classes of $\mathscr{A}$ under $\Leftrightarrow$, and $\leq$ is defined by

$$[A]_\Leftrightarrow \leq [B]_\Leftrightarrow \quad \text{if and only if} \quad A \Rightarrow B \quad .$$

Observe that $\leq$ is anti-symmetric, and therefore a partial order.

Definition 2.3. A *homomorphism* F between two logics

$$\mathscr{L} = (\mathscr{A}, \Rightarrow_\mathscr{L}) \quad \text{and} \quad \mathscr{M} = (\mathscr{B}, \Rightarrow_\mathscr{M})$$

is a monotone function $F : \mathscr{A} \to \mathscr{B}$, i.e., if $A \Rightarrow_\mathscr{L} B$ then $F(A) \Rightarrow_\mathscr{M} F(B)$.

Definition 2.4. An *isomorphism* F between two logics $\mathscr{L} = (\mathscr{A}, \Rightarrow_\mathscr{L})$ and $\mathscr{M} = (\mathscr{B}, \Rightarrow_\mathscr{M})$ is a bijective function $F : \mathscr{A} \to \mathscr{B}$, where F as well as F^{-1} are both monotone, i.e., $A \Rightarrow_\mathscr{L} B$ if and only if $F(A) \Rightarrow_\mathscr{M} F(B)$. We say that two logics are *isomorphic* if there is an isomorphism between the two.

Definition 2.5. An *embedding* F of a logic $\mathscr{L} = (\mathscr{A}, \Rightarrow_\mathscr{L})$ into another logic $\mathscr{M} = (\mathscr{B}, \Rightarrow_\mathscr{M})$ is an injective function $F : \mathscr{A} \to \mathscr{B}$, such that $A \Rightarrow_\mathscr{L} B$ if and only if $F(A) \Rightarrow_\mathscr{M} F(B)$.

Definition 2.6. Two logics $\mathscr{L} = (\mathscr{A}, \Rightarrow_\mathscr{L})$ and $\mathscr{M} = (\mathscr{B}, \Rightarrow_\mathscr{M})$ are *equivalent* if there are two monotone functions $F : \mathscr{A} \to \mathscr{B}$ and $G : \mathscr{B} \to \mathscr{A}$ such that for all formulae $A \in \mathscr{A}$ we have $A \Leftrightarrow_\mathscr{L} G(F(A))$ and for all formulae $B \in \mathscr{B}$ we have $B \Leftrightarrow_\mathscr{M} F(G(B))$.

Although I am using here the standard order theoretic vocabulary, all these concepts have already been studied from the point of view of logic. In particular, note that skeleton of a logic (where $\mathscr{A}$ is the set of formulae) is simply the Lindenbaum-algebra.

To give another example, in [PU03], the terms *sound translation, exact translation,* and *translational equivalent* are used for the concepts of *homomorphism, embedding,* and *equivalent,* respectively.[4]

One could say that two logics are "the same" if and only if they are isomorphic, but there are also reasons to argue that the notion of isomorphism is too strong for identifying logics. In particular, under the notion of isomorphism we can only compare logics with the same cardinality of statements. Also from the point of view of proof theory, the the notion of equivalence (which is also used in category theory) seems more natural. In other words, I will follow the slogan:

> *Two logics are "the same"*
> *if and only if*
> *they are equivalent as preorders.*

We can make the following immediate observations:
- It is possible that two logics with different cardinality are equivalent.
- Under the notion of equivalence, we can say that the essence of a logic is captured by its skeleton, because we have that two logics are equivalent if and only if their skeletons are isomorphic.
- The proofs of the logic are not taken into account.

It is obvious that our notion of equivalence is able to identify all different formulations of classical propositional logic. For example we can generate the set $\mathscr{A}$ by using only conjunction and negation, and the set $\mathscr{B}$ by using only disjunction and negation. If $\Rightarrow_\mathscr{A}$ and $\Rightarrow_\mathscr{B}$ are the intended classical consequence relations, then $(\mathscr{A}, \Rightarrow_\mathscr{A})$ and $(\mathscr{B}, \Rightarrow_\mathscr{B})$ are equivalent, provided we start with the same number of propositional variables. We will not get an equivalence, if say $\mathscr{A}$ is generated from 5 propositional variables, and $\mathscr{B}$ from 2; and this certainly follows the intuition.

Furthermore, the notion of equivalence is able to successfully distinguish between classical logic and intuitionistic logic. Observe that both logics can use the same set $\mathscr{A}$ of statements, but the consequence relation is different for the two: for example, we have that $\neg\neg A \Rightarrow A$ (where $\neg A$ is the negation of A) in classical logic, but not in intuitionistic logic. In fact, in the case of classical logic, the skeleton is a Boolean algebra, and in the case of intuitionistic logic, it is a Heyting algebra.

Similarly, we can single out linear logic [Gir87] and its various fragments. For example the multiplicative fragment of linear logic (MLL) is not equivalent to the multiplicative additive fragment (MALL). In fact, all the known logics, that are considered to be different, can be distinguished by the notion of equivalence. For various modal logics, namely K, T, S4, and S5, this has been shown explicitly in [PU03]. But it can be shown straightforwardly also for other cases.

Notice that the notion of equivalence for logics that I use here is nothing but the category theoretical equivalence, restricted to preorders. The idea of using this well-known concept in the area of logic can be traced back at least to Lambek's

---

[4]But in [PU03] they are not defined in order-theoretic terms.

work [Lam68, Lam69]. However, since in [Uni05] the organizers write: *"Proposals such that one of [Pol98] or [PU03] apply only to some special situations."*, there might be an interest in some further comments:

In [Pol98], Pollard compares the notion of logic with the notion of function space, as it is studied in clone theory (see e.g. [PK79]). Although in the case of Boolean logic and the Boolean function spaces (as investigated by Post [Pos41]) this question has a certain interest, we should not be surprised by the "negative" result that the two notions do not coincide. The only disturbing fact in [Pol98] is, as pointed out by the author, that sometimes the projection function (in [Pol98] denoted by $=_1$) has a logical significance and sometimes not. The reason is that Pollard uses the notion of preorder isomorphism[5] for identifying logics, and this is too strong. Under the notion of preorder equivalence, the projection function is (as one would expect) irrelevant from the logical point of view, i.e., the logic does not change if $=_1$ is added as binary connective to the set of generators of $\mathscr{A}$.[6]

In [PU03], Pelletier and Urquhart make a convincing case that preorder equivalence[7] is the right notion of identifying logics. As mentioned already, they explicitly show how various modal logics are correctly distinguished under the notion of this equivalence. In the end of of the paper, the authors provide a concrete example illustrating the fact that two logics (i.e., preorders) which can be embedded into each other are not necessarily equivalent. Although this is not surprising from the order theoretic point of view, the example itself is instructive from the point of view of logic.

Also the notion of "equipollence" proposed in [CG05] coincides with preorder equivalence. The only difference is that in [CG05] a logic is defined to be a closure system (via Tarski's consequence operator) and not as a preorder. The disatvantage of that approach is that it does not scale when proofs enter the scene.

3. What is a proof ?

From now on we do no longer content ourselves with utterances like $A \Rightarrow B$, saying that "B is a logical consequence of A", but rather want to see a justification, or *proof*, of such a statement. In order not to end up in a triviality, we have to accept the fact that there can (and must) be different such justifications of the same statement. Instead of writing $A \Rightarrow B$, we will therefore write $f : A \to B$, in order to single out the proof f of the statement that B is a consequence of A.

More formally, this is the step that takes us from preorders to categories. This means that each pair (A, B) of statements is equipped with a (possibly empty) set

---

[5] [Pol98] does not use the order theoretic vocabulary but introduces the concept from the viewpoint of topology.

[6] This is so because we have $A =_1 B \Leftrightarrow A$, no matter what A and B are.

[7] As said before, in [PU03] the term "translational equivalence" is used and the relation to order theory is not mentioned. Their definition also relies on the existence of a connective $\leftrightarrow$ internalizing the logical equivalence. However, since classical implication is transitive and reflexive, the preorder structure is there, and the two definitions coincide.

of *proofs* (i.e., *morphisms* or *arrows* in the language of category theory) from A to B. The axioms of category theory demand that

1. for every formula A there is an identity proof $\mathrm{id}_A : A \to A$, and
2. for any two proofs $f : A \to B$ and $g : B \to C$ there is a uniquely defined proof $g \circ f : A \to C$, the *composite* of f and g.

Further, we demand that for every $f : A \to B$ we have that

$$\mathrm{id}_B \circ f = f = f \circ \mathrm{id}_A \quad ,$$

and for all $f : A \to B$ and $g : B \to C$ and $h : C \to D$, we have that

$$(h \circ g) \circ f = h \circ (g \circ f) \quad .$$

Under this refinement, a logic is no longer a preorder, but a category. This relation between category theory and proof theory has already been observed by Lambek in the early work [Lam68, Lam69]. What has been a homomorphism (or monotone function) between preorders, is now a functor F between categories $\mathscr{L}$ and $\mathscr{M}$. More precisely, F consists of a map from formulae of $\mathscr{L}$ into formulae of $\mathscr{M}$, and a map from proofs in $\mathscr{L}$ into proofs in $\mathscr{M}$ such that composition and identities are preserved.

Observe that from the point of view of proof theory, demanding the properties of a category is already quite a lot. For example in the sequent calculus the composition of proofs is given by cut elimination, and this is *per se* not necessarily an associative operation. Furthermore, in the sequent calculus for classical logic this operation is not even confluent, which means that the composition of proofs is not uniquely defined.

However, treating a logic as a category has several advantages. Not only do we get the right level of abstraction to investigate the question how to identify proofs, we also can still use our notion of equivalence of logics — two logics are equivalent iff they are equivalent as categories:

Definition 3.1. Two logics $\mathscr{L}$ and $\mathscr{M}$ are *equivalent* if there are functors $F : \mathscr{L} \to \mathscr{M}$ and $G : \mathscr{M} \to \mathscr{L}$, such that for all formulae A in $\mathscr{L}$ and all formulae B in $\mathscr{M}$ we have that $A \cong G(F(A))$ and $B \cong F(G(B))$.[8]

As useful as this might be for identifying logics, it does not tell us anything about the problem of identifying of proofs, which now becomes the problem of identifying arrows in a category. It should be clear, that the problem must be asked for every logic anew, and it has to be expected that it is of various difficulty in different logics.

There are essentially two different approaches towards this problem, which I will call here the *abstract approach* and the *concrete approach*. The abstract one is

---

[8]Here $A \cong G(F(A))$ means that the two are isomorphic in the category theoretical sense, i.e., there are proofs $f : A \to G(F(A))$ and $g : G(F(A)) \to A$, such that $f \circ g = \mathrm{id}_{G(F(A))}$ and $g \circ f = \mathrm{id}_A$; and similarly for $B \cong F(G(B))$.

purely algebraic. The idea is to find the right axioms covering enough properties of proofs such that the category theory meets exactly the proof theoretical intuition. Then the slogan is:

> **Two proofs are "the same"**
> **if and only if**
> **they are represented by the same morphism in a certain category.**

A successful example of this are the axioms of Cartesian closed categories which precisely capture proofs in propositional intuitionistic logic (see, e.g., [LS86] for an introduction). Furthermore, due to the Curry-Howard-correspondence [How80], we are able to name proofs in intuitionistic logic by λ-terms, which can be identified through the notion of normalization [Pra65, Pra71, ML75].

This leads us to the concrete approach towards the problem of the identity of proofs. Here, the basic idea is to find "concrete" mathematical objects capturing the essence of a proof by avoiding the syntactic bureaucracy that usually comes with a deductive system. In this sense, λ-terms can be seen as objects capturing the essence of intuitionistic proofs. Consider for example the following two proofs in the sequent calculus for intuitionistic logic:

$$
\cfrac{
  \cfrac{
    \cfrac{\mathsf{id}\ \cfrac{}{A \vdash A} \qquad \mathsf{id}\ \cfrac{}{A \vdash A}}
         {{\to}\mathsf{L}\ \cfrac{A \to A, A \vdash A}{} } \qquad \mathsf{id}\ \cfrac{}{A \vdash A}}
       {{\to}\mathsf{L}\ \cfrac{A \to A, A \to A, A \vdash A}{}}
  }{
  \mathsf{contL}\ \cfrac{A \to A, A \vdash A}{{\to}\mathsf{R}\ \cfrac{A \to A \vdash A \to A}{{\to}\mathsf{R}\ \cfrac{}{\vdash (A \to A) \to (A \to A)}}}
}
\tag{1}
$$

$$
\cfrac{}{
{\to}\mathsf{R}\ \cfrac{A \to A \vdash A \to A}{\vdash (A \to A) \to (A \to A)}}
\tag{2}
$$

Although they are different from each other in the sequent calculus, they both translate into the same λ-term

$$\lambda f_{A \to A}.\lambda x_A.ffx \quad .$$

On the other hand, the sequent calculus proof

$$
\text{→R} \cfrac{\text{→R} \cfrac{\text{weakL} \cfrac{\text{id} \cfrac{}{A \vdash A}}{A \to A, A \vdash A}}{A \to A \vdash A \to A}}{\vdash (A \to A) \to (A \to A)}
\tag{3}
$$

of the same formula $(A \to A) \to (A \to A)$ is translated into

$$
\lambda f_{A \to A}.\lambda x_A.x \quad .
$$

We can therefore say that the two proofs in (1) and (2) are the same, while the proof in (3) is different.[9]

Let us now turn to linear logic, or more precisely, the multiplicative fragment MLL. For this logic the essence of a proof is captured by *proof nets* [Gir87]. Very roughly speaking, proof nets are for linear logic, what λ-terms are for intuitionistic logic. The slogan here is:

> ***Two proofs are "the same"***
> ***if and only if***
> ***they are represented by the same proof net.***

These proof nets are geometric objects consisting of the formula tree (or sequent forest) extended by some additional graph structure, the so-called *axiom links*. This name is chosen because they represent the identity axioms appearing in the sequent proof:

$\leftarrow$ axiom links

$\leftarrow$ sequent forest

---

[9]We have here the two proofs representing the Church numerals 2 and 0.

The following example shows, how a sequent calculus proof in linear logic is translated into a proof net by using the *flow-graph* [Bus91] or *coherence graph* [EK66, KM71].

$$\downarrow \qquad\qquad (4)$$

$$\leftarrow$$

$$\vdash A^\perp \mathbin{⅋} (A \otimes A), A \mathbin{⅋} (A^\perp \otimes A^\perp)$$

The reader interested in the details is referred to [Str06]. The proof nets for MLL do not only allow us to make the right identifications on formal proofs presented in the sequent calculus (or any other formalism), but also allow us to construct the free *-autonomous category [Bar79, Blu93, SL04, LS06], and by this substantiate the connection between category theory and proof theory.

It should be a goal of the investigation in the proof theory for any logic to ensure that the abstract and the concrete approach yield the same notion of proof identity. And, in fact, for intuitionistic logic as well for multiplicative linear logic, the two approaches coincide:

morphisms in the free *Cartesian closed category* $=$	*proofs in* *intuitionistic logic* $=$	*typed λ-terms*
morphisms in the free **-autonomous category* $=$	*proofs in multipli-* *cative linear logic* $=$	*proof nets*

These notions of identifying proofs for linear logic and intuitionistic logic are particularly useful for computer science. However, for the logics which are most interesting for mathematics and philosophy, namely, classical logic and modal logics, no such notions exist (yet). This lack of ability of naming proofs in classical logic led Girard in [Gir91] to the statement: *"classical proof theory is inexistent"*.

In fact, any approach towards an identification of proofs in classical logic is facing problems from two sides:

1. From the category theoretical side: The obvious categorical axiomatization of classical logic leads to a collapse into a Boolean algebra; all proofs of a given formula B are identified.

2. From the proof theoretical side: as mentioned before, there is no clear notion of composing proofs in classical logic.

In a certain sense, both problems are incarnations of the same phenomenon, which can best be explained with the sequent calculus.[10] Suppose we have two proofs of the formula B in some sequent calculus system:

$$
\begin{array}{ccc}
\nabla_{\Pi_1} & & \nabla_{\Pi_2} \\
\vdash B & \text{and} & \vdash B
\end{array}
$$

Then we can with the help of the rules weakening, contraction, and cut

$$
\text{weak}\ \frac{\vdash \Gamma}{\vdash \Gamma, A} \qquad \text{cont}\ \frac{\vdash \Gamma, A, A}{\vdash \Gamma, A} \qquad \text{cut}\ \frac{\vdash \Gamma, A \quad \vdash \bar{A}, \Delta}{\vdash \Gamma, \Delta}
$$

form the following proof of B

$$
\text{cut}\ \frac{\text{weak}\ \dfrac{\overset{\Pi_1}{\vdash B}}{\vdash B, A} \qquad \text{weak}\ \dfrac{\overset{\Pi_2}{\vdash B}}{\vdash \bar{A}, B}}{\text{cont}\ \dfrac{\vdash B, B}{\vdash B}} \tag{5}
$$

If we eliminate the cut from this proof, we get either

$$
\text{cont}\ \frac{\text{weak}\ \dfrac{\overset{\Pi_1}{\vdash B}}{\vdash B, B}}{\vdash B} \qquad \text{or} \qquad \text{cont}\ \frac{\text{weak}\ \dfrac{\overset{\Pi_2}{\vdash B}}{\vdash B, B}}{\vdash B} \tag{6}
$$

depending on a nondeterministic choice. Now note that one can hardly find a reason why for any proof Π, the two proofs

$$
\text{cont}\ \frac{\text{weak}\ \dfrac{\overset{\Pi}{\vdash B}}{\vdash B, B}}{\vdash B} \qquad \text{and} \qquad \begin{array}{c}\nabla_{\Pi} \\ \vdash B\end{array} \tag{7}
$$

_____

[10]The argument is due to Yves Lafont [GLT89, Appendix B].

should be distinguished. After all, duplicating a formula and immediately after-
wards deleting one copy is not doing much. Also the laws of category theory tell
us to identify the two.

On the other hand, if we want the nice relationship between deductive system
and category theory, we need a confluent cut elimination, which means we have
to equate the two proofs in (6). Consequently, by (7), we have to equate Π_1 and
Π_2. Since there was no initial condition on Π_1 and Π_2, we conclude that any two
proofs of B must be equal.

Note that the problem with weakening can be solved by using the so called
mix rule

$$\text{mix}\ \frac{\vdash \Gamma \quad \vdash \Delta}{\vdash \Gamma, \Delta} \quad .$$

Then we can for the two proofs Π_1 and Π_2 give their sum $\Pi_1 + \Pi_2$:

$$\text{mix}\ \frac{\overset{\displaystyle \Pi_1}{\vdash B} \qquad \overset{\displaystyle \Pi_2}{\vdash B}}{\text{cont}\ \dfrac{\vdash B, B}{\vdash B}}$$

However, we run into similar problems with the contraction rule. If we try to
eliminate the cut from

$$\text{cut}\ \frac{\text{cont}\ \dfrac{\overset{\displaystyle \Pi_1}{\vdash \Gamma, A, A}}{\vdash \Gamma, A} \qquad \text{cont}\ \dfrac{\overset{\displaystyle \Pi_2}{\vdash \bar{A}, \bar{A}, \Delta}}{\vdash \bar{A}, \Delta}}{\vdash \Gamma, \Delta} \qquad (8)$$

we again have to make a nondeterministic choice. And here, mix is of no help.

Nonetheless, recently considerable progress has been made in the quest for
a decent proof theory for classical and modal logic. Through the development
of the calculus of structures [Gug02, GS01, BT01] it was possible to present new
formal systems for classical (propositional and predicate) logic [Brü03] and various
modal logics [Sto04]. The proof systems in the calculus of structures have a finer
granularity than in the sequent calculus, and by this allow new notions of proof
identifications. These led in [LS05b] to a novel kind of proof nets for classical logic.
The basic idea is again that the essence of a proof is captured by axiom links that
are put on top of the formula tree (or sequent forest). Consider for example the

following proof in the one-sided sequent calculus:

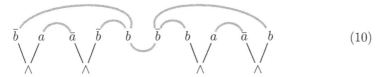

$$\text{(9)}$$

Following the idea used in (4), we can obtain a proof net by drawing the flow graph through the sequent proof. The result is

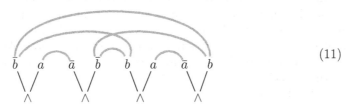

$$\text{(10)}$$

Now the reader is invited to do the following exercise: Take the proof in (9) and eliminate the cut via the usual procedure in the sequent calculus. Then translate the result into a proof net by the same method as above. The result will be either

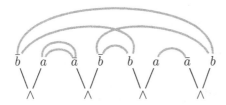

or

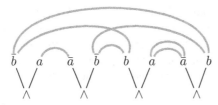

depending on a nondeterministic choice in the sequent calculus cut elimination. On the other hand, if we eliminate the cut directly from the proof net in (10), as described in [LS05b], then we obtain

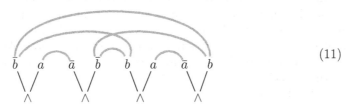

$$\text{(11)}$$

Unfortunately, there is no sequent calculus proof whose flow-graph translation into proof nets is (11). However, in the calculus of structures we can give such a proof [LS05b, Str06]:

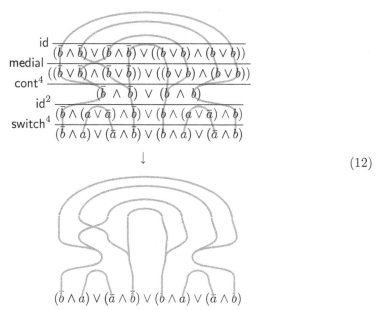

$$(12)$$

Here switch and medial are the inference rules

$$\text{switch}\ \frac{F\{(A \vee B) \wedge C\}}{F\{A \vee (B \wedge C)\}} \qquad \text{and} \qquad \text{medial}\ \frac{F\{(A \wedge B) \vee (C \wedge D)\}}{F\{(A \vee C) \wedge (B \vee D)\}}\quad,$$

formula context and A, B, C, and D are formula variables (see also [BT01, Brü03]. The important point here is that inference rules are applied deep inside formulae in the same way as we know it from term rewriting. This is reason why we can provide the proof (12) in the calculus of structures but not in the sequent calculus. For further details on the relation between deep inference and proof nets, the reader is referred to [Str05a].

Interestingly, the idea of capturing "the essence" of a proof with pairs of complementary atoms has already been used in Andrews' *matings* [And76] and in Bibel's *connection proofs* [Bib86]. But since they were only interested in proof search, they did not explore the possibility of *composing* proofs. This is done in [LS05b], where composition is defined via a strongly normalizing cut elimination. Therefore we indeed have a category, which could be called a "Boolean" category, since it is to a category what a Boolean algebra is to a poset. In [LS05a], a possible axiomatisation for these kind of categories is given. However, so far there is no axiomatisation that captures precisely the proof identificion induced by the proof nets sketched above (see also [Lam06, Str05b]).

From the point of view of category theory, an alternative approach is given in [FP04a, FP04b], where Führmann and Pym relax the equality on proofs defined by cut elimination to a partial order on proofs, and by this avoid the collapse into a Boolean algebra. Another work in that direction is [Hyl04], where Hyland exhibits concrete mathematical objects, e.g., Frobenius algebras, that can serve as denotations for classical proofs.

In [DP04] the authors use the concept of "coherence" (see, e.g., [Mac71, KM71]) to identify proofs: The category of proofs is defined together with a "graphical" category such that the canonical functor from the category of proofs into the "graphical" category is faithful. Two proofs are the same if they have the same "graphical" representation. In principle, this approach is in the same spirit as the approach based on proof nets: The category of proof nets plays the role of the "graphical" category.[11] However results based on proof nets are usually stronger than results based on "coherence": Not only do they tell when two proof are the same, but also whether a given object actually is a proof. This is usually done via a so-called correctness criterion.

4. Summary

Clearly, the question when two proofs are the same is mathematical more challenging than the question when two logics are the same, which simply reduces to the problem of finding a consensus on the definition. Nonetheless, if one is interested in defining identity between logics, one has to make up his mind whether one wants to ignore the proofs or whether one wants to take the proofs into account. For example, do we want to distinguish a cut-free system for intuitionistic propositional logic from the same system enriched with cut, or do we not? In the one case one can satisfy oneself with preorder equivalence, and in the other one has to take category equivalence.

To give another example, consider the conjuction-only fragment of classical and intuitionistic propositional logic. In both cases, the consequence relation is exactly the same. But the proofs are not: in intuitionistic logic there are two canonical proofs from $A \wedge A$ to A, namely, the two projections, which are also present in classical logic. But in classical logic, we can also form the sum of the two projections, which is not possible in intuitionistic logic.

Let me finish with mentioning some of the questions that are still waiting for an answer:

- Is there a philosophical justification for the identification of proofs made by proof nets?
- Are these identifications useful from the point of view of mathematics (i.e., can we use them for identifying real mathematical proofs)?

---

[11] In the case of unit-free multiplicative linear logic, proof nets coincide with Kelly-MacLane-graphs.

- Are there ways of extending the notion of proof net to the quantifiers (first order, second order, and higher order), for example by using Miller's *expansion trees* [Mil87]?
- Is it possible to include modalities into proof nets (e.g., by exploring the recent work by Stouppa [Sto04]), in order to get a way of identifying proofs in modal logics?

References

[And76]   Peter B. Andrews. Refutations by matings. *IEEE Transactions on Computers*, C-25:801–807, 1976.

[Bar79]   Michael Barr. **-Autonomous Categories*, volume 752 of *Lecture Notes in Mathematics*. Springer-Verlag, 1979.

[Bib86]   Wolfgang Bibel. A deductive solution for plan generation. *New Generation Computing*, 4:115–132, 1986.

[Blu93]   Richard Blute. Linear logic, coherence and dinaturality. *Theoretical Computer Science*, 115:3–41, 1993.

[Brü03]   Kai Brünnler. *Deep Inference and Symmetry for Classical Proofs*. PhD thesis, Technische Universität Dresden, 2003.

[BT01]   Kai Brünnler and Alwen Fernanto Tiu. A local system for classical logic. In R. Nieuwenhuis and A. Voronkov, editors, *LPAR 2001*, volume 2250 of *Lecture Notes in Artificial Intelligence*, pages 347–361. Springer-Verlag, 2001.

[Bus91]   Samuel R. Buss. The undecidability of k-provability. *Annals of Pure and Applied Logic*, 53:72–102, 1991.

[CG05]   Carlos Caleiro and Ricardo Gonçalves. Equipollent logical systems. In Jean-Yves Beziau, editor, *Logica Universalis*, pages 99–111. Birkhäuser, 2005.

[DP04]   Kosta Došen and Zoran Petrić. *Proof-Theoretical Coherence*. KCL Publications, London, 2004.

[EK66]   Samuel Eilenberg and Gregory Maxwell Kelly. A generalization of the functorial calculus. *Journal of Algebra*, 3(3):366–375, 1966.

[FP04a]   Carsten Führmann and David Pym. On the geometry of interaction for classical logic. preprint, 2004.

[FP04b]   Carsten Führmann and David Pym. On the geometry of interaction for classical logic (extended abstract). In *19th IEEE Symposium on Logic in Computer Science (LICS 2004)*, pages 211–220, 2004.

[Gir87]   Jean-Yves Girard. Linear logic. *Theoretical Computer Science*, 50:1–102, 1987.

[Gir91]   Jean-Yves Girard. A new constructive logic: Classical logic. *Mathematical Structures in Computer Science*, 1:255–296, 1991.

[GLT89]   Jean-Yves Girard, Yves Lafont, and Paul Taylor. *Proofs and Types*. Cambridge Tracts in Theoretical Computer Science. Cambridge University Press, 1989.

[GS01]   Alessio Guglielmi and Lutz Straßburger. Non-commutativity and MELL in the calculus of structures. In Laurent Fribourg, editor, *Computer Science Logic, CSL 2001*, volume 2142 of *LNCS*, pages 54–68. Springer-Verlag, 2001.

[Gug02] Alessio Guglielmi. A system of interaction and structure. To appear in *ACM Transactions on Computational Logic*, 2002.

[How80] W. A. Howard. The formulae-as-types notion of construction. In J. P. Seldin and J. R. Hindley, editors, *To H. B. Curry: Essays on Combinatory Logic, Lambda Calculus and Formalism*, pages 479–490. Academic Press, 1980.

[Hyl04] J. Martin E. Hyland. Abstract interpretation of proofs: Classical propositional calculus. In Jerzy Marcinkowski and Andrzej Tarlecki, editors, *Computer Science Logic, CSL 2004*, volume 3210 of *LNCS*, pages 6–21. Springer-Verlag, 2004.

[KM71] Gregory Maxwell Kelly and Saunders Mac Lane. Coherence in closed categories. *Journal of Pure and Applied Algebra*, 1:97–140, 1971.

[Lam68] Joachim Lambek. Deductive systems and categories. I: Syntactic calculus and residuated categories. *Math. Systems Theory*, 2:287–318, 1968.

[Lam69] Joachim Lambek. Deductive systems and categories. II. standard constructions and closed categories. In P. Hilton, editor, *Category Theory, Homology Theory and Applications*, volume 86 of *Lecture Notes in Mathematics*, pages 76–122. Springer, 1969.

[Lam06] François Lamarche. Exploring the gap between linear and classical logic, 2006. Submitted.

[LS86] Joachim Lambek and Phil J. Scott. *Introduction to higher order categorical logic*, volume 7 of *Cambridge studies in advanced mathematics*. Cambridge University Press, 1986.

[LS05a] François Lamarche and Lutz Straßburger. Constructing free Boolean categories. In *Proceedings of the Twentieth Annual IEEE Symposium on Logic in Computer Science (LICS'05)*, pages 209–218, 2005.

[LS05b] François Lamarche and Lutz Straßburger. Naming proofs in classical propositional logic. In Paweł Urzyczyn, editor, *Typed Lambda Calculi and Applications, TLCA 2005*, volume 3461 of *Lecture Notes in Computer Science*, pages 246–261. Springer-Verlag, 2005.

[LS06] François Lamarche and Lutz Straßburger. From proof nets to the free *-autonomous category. *Logical Methods in Computer Science*, 2(4:3):1–44, 2006.

[Mac71] Saunders Mac Lane. *Categories for the Working Mathematician*. Number 5 in Graduate Texts in Mathematics. Springer-Verlag, 1971.

[Mil87] Dale Miller. A compact representation of proofs. *Studia Logica*, 46(4):347–370, 1987.

[ML75] Per Martin-Löf. About models for intuitionistic type theories and the notion of definitional equality. In S. Kangar, editor, *Proceedings of the Third Scandinavian Logic Symposium*, pages 81–109. North-Holland Publishing Co., 1975.

[PK79] R. Pöschel and L. A. Kalužnin. *Funktionen- und Relationenalgebren, Ein Kapitel der Diskreten Mathematik*. Deutscher Verlag der Wissenschaften, Berlin, 1979.

[Pol98] Stephen Pollard. Homeomorphism and the equivalence of logical systems. *Notre Dame Journal of Formal Logic*, 39:422–435, 1998.

[Pos41] E. L. Post. *The Two-Valued Iterative Systems of Mathematical Logic*. Princeton University Press, Princeton, 1941.

[Pra65] Dag Prawitz. *Natural Deduction, A Proof-Theoretical Study*. Almquist and Wiksell, 1965.

[Pra71] Dag Prawitz. Ideas and results in proof theory. In J. E. Fenstad, editor, *Proceedings of the Second Scandinavian Logic Symposium*, pages 235–307. North-Holland Publishing Co., 1971.

[PU03]   Francis Jeffry Pelletier and Alasdair Urquhart. Synonymous logics. *Journal of Philosophical Logic*, 32:259–285, 2003.

[SL04]   Lutz Straßburger and François Lamarche. On proof nets for multiplicative linear logic with units. In Jerzy Marcinkowski and Andrzej Tarlecki, editors, *Computer Science Logic, CSL 2004*, volume 3210 of *LNCS*, pages 145–159. Springer-Verlag, 2004.

[Sto04] Finiki Stouppa. The design of modal proof theories: the case of S5. Master's thesis, Technische Universität Dresden, 2004.

[Str05a] Lutz Straßburger. From deep inference to proof nets. In *Structures and Deduction — The Quest for the Essence of Proofs (Satellite Workshop of ICALP 2005)*, 2005.

[Str05b] Lutz Straßburger. On the axiomatisation of Boolean categories with and without medial, 2005. Preprint, available at `http://arxiv.org/abs/cs.LO/0512086`.

[Str06] Lutz Straßburger. Proof nets and the identity of proofs, 2006. Lecture notes for ESSLLI'06. Available from `https://hal.inria.fr/inria-00107260/en/` and `http://arxiv.org/abs/cs.LO/0610123`.

[Thi03] Rüdiger Thiele. Hilbert's twenty-fourth problem. *American Mathematical Monthly*, 110:1–24, 2003.

[Uni05] Contest: How to define identity between logics? *1st World Congress and School on Universal Logic*, 2005. On the web at `http://www.uni-log.org/one2.html`.

Lutz Straßburger
INRIA Futurs, Projet Parsifal
École Polytechnique – LIX
Rue de Saclay
91128 Palaiseau Cedex
France
URL: `http://www.lix.polytechnique.fr/~lutz`

Part III
Tools and Concepts for Universal Logic

J.-Y. Beziau (Ed.), *Logica Universalis*, 2nd edition, 155–173

Non-deterministic Matrices and Modular Semantics of Rules

Arnon Avron

Abstract. We show by way of example how one can provide in a lot of cases simple modular semantics for rules of inference, so that the semantics of a system is obtained by joining the semantics of its rules in the most straightforward way. Our main tool for this task is the use of finite matrices, which are multi-valued structures in which the value assigned by a valuation to a complex formula can be chosen non-deterministically out of a certain nonempty set of options. The method is applied in the area of logics with a formal consistency operator (known as LFIs), allowing us to provide in a modular way effective, finite semantics for thousands of different LFIs.

Mathematics Subject Classification (2000). 03B22; 03B50; 03B53.

Keywords. Propositional logics, multiple-valued semantics, paraconsistency.

1. Introduction

It is well known that every propositional logic satisfying certain minimal conditions can be characterized semantically using a multi-valued matrix ([18]). However, there are many important decidable logics whose characteristic matrices necessarily consist of an infinite number of truth values. In such a case it might be quite difficult to find any of these matrices, or to use one when it is found. Even in case a logic does have a finite characteristic matrix it might be difficult to discover this fact, or to find such a matrix. The deep reason for these difficulties is that in an ordinary multi-valued semantics the rules and axioms of a system should be considered as a whole, and there is no method for separately determining the semantic effects of each rule or axiom alone.

In this paper we show how one can provide in a lot of cases simple modular semantics for rules of inference, so that the semantics of a system is obtained

---

This research was supported by THE ISRAEL SCIENCE FOUNDATION founded by The Israel Academy of Sciences and Humanities.

by joining the semantics for its rules in the most straightforward way. Our main tool for this task is the use of finite Nmatrices ([6, 4]. Nmatrices are multi-valued structures in which the value assigned by a valuation to a complex formula can be chosen non-deterministically out of a certain nonempty set of options. The use of finite structures of this sort has the benefit of preserving all the advantages of logics with ordinary finite-valued semantics (in particular: decidability and compactness), while it is applicable to a much larger family of logics. The central idea in using Nmatrices for providing semantics for rules is that the main effect of a "normal" rule is to reduce the degree of non-determinism of operations, by forbidding some options (in non-deterministic computations of truth values) which we could have had otherwise. This idea was first applied in [7, 6] for a very special (though extremely important) type of rules (which was called there "canonical rules"). For that type of rules 2-valued Nmatrices suffice. In this paper we show how by employing more than two values we can apply the method for a much larger class of rules. As a case study we have chosen the class of paraconsistent logics knows as LFIs, described in [11, 12] [1]. In what follows we use our method in order to modularly provide effective, finite semantics for thousands of different LFIs.

2. Preliminaries

2.1. Consequence Relations, Logics, and Pure Rules

Definition 2.1.

1. A *Scott consequence relation* (*scr* for short) for a language $\mathcal{L}$ is a binary relation $\vdash$ between sets of formulas of $\mathcal{L}$ that satisfies the following conditions:

   s-R   *strong reflexivity*:   if $\Gamma \cap \Delta \neq \emptyset$ then $\Gamma \vdash \Delta$.

   M   *monotonicity*:   if $\Gamma \vdash \Delta$ and $\Gamma \subseteq \Gamma'$, $\Delta \subseteq \Delta'$ then $\Gamma' \vdash \Delta'$.

   C   *Transitivity (cut)*:   if $\Gamma \vdash \psi, \Delta$ and $\Gamma', \psi \vdash \Delta'$ then $\Gamma, \Gamma' \vdash \Delta, \Delta'$.

2. An scr $\vdash$ for $\mathcal{L}$ is *structural* (or substitution-invariant) if for every uniform $\mathcal{L}$-substitution σ and every Γ and Δ, if $\Gamma \vdash \Delta$ then $\sigma(\Gamma) \vdash \sigma(\Delta)$. $\vdash$ is *finitary* if the following condition holds for all $\Gamma, \Delta \subseteq \mathcal{W}$: if $\Gamma \vdash \Delta$ then there exist finite $\Gamma' \subseteq \Gamma$ and $\Delta' \subseteq \Delta$ such that $\Gamma' \vdash \Delta'$. $\vdash$ is *consistent* (or *non-trivial*) if there exist non-empty Γ and Δ s.t. $\Gamma \not\vdash \Delta$. [2]

3. A propositional *logic* is a pair $\langle \mathcal{L}, \vdash \rangle$, where $\mathcal{L}$ is a propositional language and $\vdash$ is an scr for $\mathcal{L}$ which is structural and consistent. The logic $\langle \mathcal{L}, \vdash \rangle$ is finitary if $\vdash$ is finitary.

---

[1] The name "LFI" stands for "Logics of Formal Inconsistency". In our opinion it would make more sense to call them "logics of formal consistency", since they are obtained from classical logic by the addition of a new connective $\circ$, with the intended meaning of $\circ\varphi$ being: "φ is consistent".

[2] See [7, 6] for the importance of the consistency property.

Definition 2.2.

1. A *pure rule* in a propositional language $\mathcal{L}$ is any ordered pair $\langle \Gamma, \Delta \rangle$, where Γ and Δ are finite sets of formulas in $\mathcal{L}$ (We shall usually denote such a rule by $\Gamma \vdash \Delta$ rather than by $\langle \Gamma, \Delta \rangle$).
2. Let $\langle \mathcal{L}, \vdash_1 \rangle$ be a propositional logic, and let S be a set of rules in a propositional language $\mathcal{L}'$. By the extension of $\langle \mathcal{L}, \vdash_1 \rangle$ by S we mean the logic $\langle \mathcal{L}^*, \vdash^* \rangle$, where $\mathcal{L}^* = \mathcal{L} \cup \mathcal{L}'$, and $\vdash^*$ is the least *structural* scr $\vdash$ such that $\Gamma \vdash \Delta$ whenever $\Gamma \vdash_1 \Delta$ or $\langle \Gamma, \Delta \rangle \in S$.

Remark 2.3. Obviously, the extension of $\langle \mathcal{L}, \vdash_1 \rangle$ by S is well-defined (i.e., a logic) only if $\vdash^*$ is consistent. In all the cases we consider below this will easily be guaranteed by the semantics we provide (and so we shall not even mention it).

Remark 2.4. It is easy to see that $\vdash^*$ is the closure under cuts and weakenings of the set of all pairs $\langle \sigma(\Gamma), \sigma(\Delta) \rangle$, where σ is a uniform substitution in $\mathcal{L}^*$, and either $\Gamma \vdash_1 \Delta$ or $\langle \Gamma, \Delta \rangle \in S$. This in turn implies that an extension of a finitary logic by a set of pure rules is again finitary.

Remark 2.5. Most standard rules used in Gentzen-type systems are equivalent to finite sets of pure rules in the sense of Definition 2.2. For example, the usual $(\supset\Rightarrow)$ rule of classical logic is equivalent (using cuts, weakenings, and the reflexivity axioms $\varphi \vdash \varphi$) to the pure rule $\varphi, \varphi \supset \psi \vdash \psi$, while the classical $(\Rightarrow\supset)$ rule is equivalent to the set $\{\psi \vdash \varphi \supset \psi, \ \vdash \varphi, \varphi \supset \psi\}$.

2.2. Non-deterministic Matrices

Our main semantical tool in what follows will be the following generalization from [7, 6] of the concept of a matrix:[3]

Definition 2.6.

1. A *non-deterministic matrix* (*Nmatrix* for short) for a propositional language $\mathcal{L}$ is a tuple $\mathcal{M} = \langle \mathcal{V}, \mathcal{D}, \mathcal{O} \rangle$, where:
   (a) $\mathcal{V}$ is a non-empty set of *truth values*.
   (b) $\mathcal{D}$ is a non-empty proper subset of $\mathcal{V}$.
   (c) For every n-ary connective $\diamond$ of $\mathcal{L}$, $\mathcal{O}$ includes a corresponding n-ary function $\tilde{\diamond}$ from $\mathcal{V}^n$ to $2^{\mathcal{V}} - \{\emptyset\}$.
   We say that $\mathcal{M}$ is *(in)finite* if so is $\mathcal{V}$.
2. Let $\mathcal{W}$ be the set of formulas of $\mathcal{L}$. A *(legal) valuation* in an Nmatrix $\mathcal{M}$ is a function $v : \mathcal{W} \to \mathcal{V}$ that satisfies the following condition for every n-ary connective $\diamond$ of $\mathcal{L}$ and $\psi_1, \ldots, \psi_n \in \mathcal{W}$:
$$v(\diamond(\psi_1, \ldots, \psi_n)) \in \tilde{\diamond}(v(\psi_1), \ldots, v(\psi_n)).$$

---

[3] A special two-valued case of this definition was essentially introduced in [9]. Another particular case of the same idea, using a similar name, was used in [13]. It should also be noted that Carnielli's "possible-translations semantics" (see [10]) was originally called "non-deterministic semantics", but later the name was changed to the present one.

3. A valuation v in an Nmatrix $\mathcal{M}$ is a *model* of (or *satisfies*) a formula ψ in $\mathcal{M}$ (notation: $v \models^{\mathcal{M}} \psi$) if $v(\psi) \in \mathcal{D}$. v is a *model* in $\mathcal{M}$ of a set Γ of formulas (notation: $v \models^{\mathcal{M}} \Gamma$) if it satisfies every formula in Γ.

4. $\vdash_{\mathcal{M}}$, the consequence relation induced by the Nmatrix $\mathcal{M}$, is defined by: $\Gamma \vdash_{\mathcal{M}} \Delta$ if for every v such that $v \models^{\mathcal{M}} \Gamma$, there is $\varphi \in \Delta$ such that $v \models^{\mathcal{M}} \varphi$.

5. A logic $\mathbf{L} = \langle \mathcal{L}, \vdash_{\mathbf{L}} \rangle$ is *sound* for an Nmatrix $\mathcal{M}$ (where $\mathcal{L}$ is the language of $\mathcal{M}$) if $\vdash_{\mathbf{L}} \subseteq \vdash_{\mathcal{M}}$. $\mathbf{L}$ is *complete* for $\mathcal{M}$ if $\vdash_{\mathbf{L}} \supseteq \vdash_{\mathcal{M}}$. $\mathcal{M}$ is *characteristic* for $\mathbf{L}$ if $\mathbf{L}$ is both sound and complete for it (i.e., if $\vdash_{\mathbf{L}} = \vdash_{\mathcal{M}}$). $\mathcal{M}$ is *weakly-characteristic* for $\mathbf{L}$ if for every formula φ of $\mathcal{L}$, $\vdash_{\mathbf{L}} \varphi$ iff $\vdash_{\mathcal{M}} \varphi$.

Remark 2.7. We shall identify an ordinary (deterministic) matrix with an Nmatrix whose functions in $\mathcal{O}$ always return singletons.

Theorem 2.8 ([6]). *A logic which has a finite characteristic Nmatrix is finitary and decidable.*

Definition 2.9. Let $\mathcal{M}_1 = \langle \mathcal{V}_1, \mathcal{D}_1, \mathcal{O}_1 \rangle$ and $\mathcal{M}_2 = \langle \mathcal{V}_2, \mathcal{D}_2, \mathcal{O}_2 \rangle$ be Nmatrices for a language $\mathcal{L}$.

1. A *reduction* of $\mathcal{M}_1$ to $\mathcal{M}_2$ is a function $F : \mathcal{V}_1 \to \mathcal{V}_2$ such that:
   (a) For every $x \in \mathcal{V}_1$, $x \in \mathcal{D}_1$ iff $F(x) \in \mathcal{D}_2$ (i.e., $D_1 = F^{-1}[D_2]$).
   (b) $F(y) \in \widetilde{\diamond}_{\mathcal{M}_2}(F(x_1), \dots, F(x_n))$ for every n-ary connective $\diamond$ of $\mathcal{L}$ and every $x_1, \dots, x_n, y \in \mathcal{V}_1$ such that $y \in \widetilde{\diamond}_{\mathcal{M}_1}(x_1, \dots, x_n)$ (in other words: $\widetilde{\diamond}_{\mathcal{M}_1}(x_1, \dots, x_n) \subseteq F^{-1}[\widetilde{\diamond}_{\mathcal{M}_2}(F(x_1), \dots, F(x_n))]$).

2. A reduction of $\mathcal{M}_1$ to $\mathcal{M}_2$ is called *exact* if it has the following properties:
   (a) F is *onto* $\mathcal{V}_2$.
   (b) For every n-ary connective $\diamond$ of $\mathcal{L}$ and every $x_1, \dots, x_n, y \in \mathcal{V}_1$:

   $$F(y) \in \widetilde{\diamond}_{\mathcal{M}_2}(F(x_1), \dots, F(x_n)) \quad \text{iff} \quad y \in \widetilde{\diamond}_{\mathcal{M}_1}(x_1, \dots, x_n)$$

   (equivalently: if $\widetilde{\diamond}_{\mathcal{M}_1}(x_1, \dots, x_n) = F^{-1}[\widetilde{\diamond}_{\mathcal{M}_2}(F(x_1), \dots, F(x_n))]$).

3. $\mathcal{M}_1$ is a *refinement* of $\mathcal{M}_2$ if there exists a reduction of $\mathcal{M}_1$ to $\mathcal{M}_2$. It is an *exact refinement* of $\mathcal{M}_2$ if this reduction is exact.

Theorem 2.10.

1. *If $\mathcal{M}_1$ is a refinement of $\mathcal{M}_2$ then $\vdash_{\mathcal{M}_2} \subseteq \vdash_{\mathcal{M}_1}$.*
2. *If $\mathcal{M}_1$ is an exact refinement of $\mathcal{M}_2$ then $\vdash_{\mathcal{M}_2} = \vdash_{\mathcal{M}_1}$.*

Proof. For the first part, assume that F is a reduction of $\mathcal{M}_1$ to $\mathcal{M}_2$. We show that if v is a legal valuation in $\mathcal{M}_1$ then $v' = F \circ v$ (the composition of F and v) is a legal valuation in $\mathcal{M}_2$. Indeed, let $\diamond$ be an n-ary connective of $\mathcal{L}$, and let $\varphi_1, \dots, \varphi_n$ be n formulas of $\mathcal{L}$. We show that $v'(\diamond(\varphi_1, \dots, \varphi_n)) \in \widetilde{\diamond}_{\mathcal{M}_2}(v'(\varphi_1), \dots, v'(\varphi_n))$. Let $y = v(\diamond(\varphi_1, \dots, \varphi_n))$, and $x_i = v(\varphi_i)$ $(i = 1, \dots, n)$. Then $y \in \widetilde{\diamond}_{\mathcal{M}_1}(x_1, \dots, x_n)$, and so $F(y) \in \widetilde{\diamond}_{\mathcal{M}_2}(F(x_1), \dots, F(x_n))$. Since $v'(\diamond(\varphi_1, \dots, \varphi_n) = F(y)$ and $v'(\varphi_i) = F(x_i)$ $(i = 1, \dots, n)$, our claim follows.

Now assume that $\Gamma \vdash_{\mathcal{M}_2} \Delta$. We show that $\Gamma \vdash_{\mathcal{M}_1} \Delta$ as well. So let v be a model of Γ in $\mathcal{M}_1$. Then $v(\varphi) \in \mathcal{D}_1$ for every $\varphi \in \Gamma$. Hence $F(v(\varphi)) \in \mathcal{D}_2$ for every $\varphi \in \Gamma$. Since $F \circ v$ is a legal valuation in $\mathcal{M}_2$, this means that $F \circ v$ is a model of Γ

in $\mathcal{M}_2$, and so $F(v(\psi)) = (F \circ v)(\psi) \in \mathcal{D}_2$ for some $\psi \in \Delta$. Since F is a reduction function, this implies that $v(\psi) \in \mathcal{D}_1$ for some $\psi \in \Delta$, as required.

For the second part note that if F is an exact reduction of $\mathcal{M}_1$ to $\mathcal{M}_2$, then every right inverse G of F [4] can easily be shown to be a reduction of $\mathcal{M}_2$ to $\mathcal{M}_1$. Thus by the first part $\vdash_{\mathcal{M}_1} \subseteq \vdash_{\mathcal{M}_2}$ too, and so $\vdash_{\mathcal{M}_2} = \vdash_{\mathcal{M}_1}$.   □

Remark 2.11. An important case in which $\mathcal{M}_1 = \langle \mathcal{V}_1, \mathcal{D}_1, \mathcal{O}_1 \rangle$ is a refinement of $\mathcal{M}_2 = \langle \mathcal{V}_2, \mathcal{D}_2, \mathcal{O}_2 \rangle$ is when $\mathcal{V}_1 \subseteq \mathcal{V}_2$, $\mathcal{D}_1 = \mathcal{D}_2 \cap \mathcal{V}_1$, and $\tilde{\diamond}_{\mathcal{M}_1}(\vec{x}) \subseteq \tilde{\diamond}_{\mathcal{M}_2}(\vec{x})$ for every n-ary connective $\diamond$ of $\mathcal{L}$ and every $\vec{x} \in \mathcal{V}_1^n$. It is easy to see that the identity function on $\mathcal{V}_1$ is in this case a reduction of $\mathcal{M}_1$ to $\mathcal{M}_2$. A refinement of this sort will be called *simple*.[5]

2.3. Positive Classical Logic

Definition 2.12. Let $\mathbf{CL}^+ = \langle \mathcal{L}_{cl}^+, \vdash_{cl}^+ \rangle$, where $\mathcal{L}_{cl}^+ = \{\wedge, \vee, \supset\}$, and $\vdash_{cl}^+$ is the classical consequence relation in the language $\mathcal{L}_{cl}^+$ (i.e., $\Gamma \vdash_{cl}^+ \Delta$ iff every classical two-valued model of Γ is a model of at least one formula in Δ).

Remark 2.13. For any pure rule in a propositional language containing $\mathcal{L}_{cl}^+$ it is possible to find an equivalent rule of the form $\vdash \varphi$ (by translating the condition $\varphi_1, \ldots, \varphi_n \vdash \psi_1, \ldots, \psi_k$ into $\vdash \varphi_1 \wedge \ldots \wedge \varphi_n \supset \psi_1 \vee \ldots \vee \psi_k$ in case $k > 0$, and to $\vdash \varphi_1 \wedge \ldots \wedge \varphi_n \supset q$, where q is an atomic formula not occurring in $\varphi_1, \ldots, \varphi_n$, in case k=0. Hence it is possible to construct a sound and complete Hilbert-type system (with MP as the sole rule of inference) for any extension of $\mathbf{CL}^+$ by a finite set of pure rules. On the other hand any pure rule is equivalent above $\mathbf{CL}^+$ to a finite set of rules in which none of the formulas has either $\vee, \wedge$ or $\supset$ as its principal connective. For example, a condition of the form $\varphi \wedge \psi, \Gamma \vdash \Delta$ can be replaced by $\varphi, \psi, \Gamma \vdash \Delta$, while $\Gamma \vdash \Delta, \varphi \wedge \psi$ can be replaced by $\{\Gamma \vdash \Delta, \varphi \ , \ \Gamma \vdash \Delta, \psi\}$.

Definition 2.14. Let $\mathcal{M} = \langle \mathcal{V}, \mathcal{D}, \mathcal{O} \rangle$ be an Nmatrix for a language which includes $\mathcal{L}_{cl}^+$. We say that $\mathcal{M}$ is *suitable* for $\mathbf{CL}^+$ if the following conditions are satisfied:

- If $a \in \mathcal{D}$ and $b \in \mathcal{D}$ then $a \tilde{\wedge} b \subseteq \mathcal{D}$.
- If $a \notin \mathcal{D}$ then $a \tilde{\wedge} b \subseteq \mathcal{V} - D$.
- If $b \notin \mathcal{D}$ then $a \tilde{\wedge} b \subseteq \mathcal{V} - D$.

- If $a \in \mathcal{D}$ then $a \tilde{\vee} b \subseteq \mathcal{D}$.
- If $b \in \mathcal{D}$ then $a \tilde{\vee} b \subseteq \mathcal{D}$.
- If $a \notin \mathcal{D}$ and $b \notin \mathcal{D}$ then $a \tilde{\vee} b \subseteq \mathcal{V} - D$.

- If $a \notin \mathcal{D}$ then $a \tilde{\supset} b \subseteq \mathcal{D}$.
- If $b \in \mathcal{D}$ then $a \tilde{\supset} b \subseteq \mathcal{D}$.
- If $a \in \mathcal{D}$ and $b \notin \mathcal{D}$ then $a \tilde{\supset} b \subseteq \mathcal{V} - D$.

---

[4]By this one means a function $G : \mathcal{V}_2 \to \mathcal{V}_1$ such that $F(G(x)) = x$ for every $x \in \mathcal{V}_2$. Such a function G exists here, since F is onto $\mathcal{V}_2$.

[5]What we call here "a simple refinement" is what was called "a refinement" in [2]. The present definition of "a refinement" is a refinement of the definition given to that concept there.

Theorem 2.15. *Suppose $\mathcal{M} = \langle \mathcal{V}, \mathcal{D}, \mathcal{O} \rangle$ is suitable for $\mathbf{CL}^+$. Let $\mathcal{M}' = \langle \mathcal{V}, \mathcal{D}, \mathcal{O}' \rangle$, where $\mathcal{O}'$ is the subset of $\mathcal{O}$ which corresponds to the connectives of $\mathcal{L}_{\mathrm{cl}}^+$. Then $\vdash_{\mathrm{cl}}^+ = \vdash_{\mathcal{M}'}$. Hence $\sigma(\Gamma) \vdash_{\mathcal{M}} \sigma(\Delta)$ whenever $\Gamma \vdash_{\mathrm{cl}}^+ \Delta$ and σ is a substitution in the language of $\mathcal{M}$.*

Proof. Let $\mathcal{M}_{\mathbf{CL}^+}$ be the classical two-valued matrix, where the two truth values are $\mathbf{t}$ and $\mathbf{f}$. Since $\mathcal{M}$ is suitable for $\mathbf{CL}^+$, the function

$$\lambda x \in \mathcal{V}. \begin{cases} \mathbf{t} & x \in \mathcal{D} \\ \mathbf{f} & x \notin \mathcal{D} \end{cases}$$

is a reduction of $\mathcal{M}'$ to $\mathcal{M}_{\mathbf{CL}^+}$. Hence $\vdash_{\mathrm{cl}}^+ \subseteq \vdash_{\mathcal{M}'}$. That $\vdash_{\mathrm{cl}}^+ = \vdash_{\mathcal{M}'}$ follows from the well-known fact that $\mathbf{CL}^+$ is a maximal nontrivial logic in its language. $\qquad \square$

2.4. Formal Systems with a Formal Consistency Operation

2.4.1. The Basic Logic. Let $\mathcal{L}_{\mathrm{cl}} = \{\wedge, \vee, \supset, \neg\}$. $\mathcal{L}_{\mathrm{cl}}$ is the standard language of the classical propositional logic $\mathbf{CL}$. The latter may be characterized as the extension of $\mathbf{CL}^+$ by the rules $\neg\varphi, \varphi \vdash$ and $\vdash \neg\varphi, \varphi$. The two main ideas of da-Costa's school of paraconsistent logics ([14, 11, 12]) are to limit the applicability of the first of these two rules (which amounts to "a single contradiction entails everything") to the case where φ is "consistent", and to express the assumption of this consistency of φ within the language. The easiest way to implement these ideas is to add to the language of $\mathbf{CL}$ a new connective $\circ$, with the intended meaning of $\circ\varphi$ being "φ is consistent". Then one can explicitly add the assumption of the consistency of φ to the problematic (from a paraconsistent point of view) classical rule concerning $\neg$. This leads to the basic system $\mathbf{B}$ described below.[6]

Definition 2.16. Let $\mathcal{L}_{\mathrm{C}} = \{\wedge, \vee, \supset, \neg, \circ\}$.

Definition 2.17. The logic $\mathbf{B}$ is the minimal logic in $\mathcal{L}_{\mathrm{C}}$ which extends $\mathbf{CL}^+$ and satisfies the following two conditions:

    (t): $\vdash \neg\varphi, \varphi$
    (b): $\circ\varphi, \neg\varphi, \varphi \vdash$

Lemma 2.18. *Let LK be the standard Gentzen calculus for classical propositional logic, and let GB be obtained from LK by replacing the $(\neg \Rightarrow)$ rule by:*

$$(\circ, \neg \Rightarrow) \qquad \frac{\Gamma \Rightarrow \Delta, \varphi}{\circ\varphi, \neg\varphi, \Gamma \Rightarrow \Delta}$$

Then for every finite Γ and Δ, $\Gamma \vdash_B \Delta$ iff $\Gamma \Rightarrow \Delta$ has a cut-free proof in GB.

Proof. Using cuts, it is straightforward to show that for every finite Γ and Δ, $\Gamma \vdash_B \Delta$ iff $\Gamma \Rightarrow \Delta$ has a proof in GB. The cut-elimination theorem can then be proved for GB by the usual syntactic method of Gentzen (i.e., by using double induction on the complexity of the cut formula and on the height of the cut). $\quad \square$

---

[6]The logic $\mathbf{B}$ is called mbC in [12]. We prefer to use here a shorter name.

Remark 2.19. By Remark 2.13, a Hilbert-type system which is sound and complete for **B** can be obtained by adding the following two axioms to some standard Hilbert-type system for $\mathbf{CL}^+$ having MP as the sole rule of inference:

(t): $\neg\varphi \vee \varphi$
(b): $(\circ\varphi \wedge \neg\varphi \wedge \varphi) \supset \psi$

The main property of **B** is given in the next theorem from [11, 12]:

Theorem 2.20. *Let $\vdash_{cl}$ be the classical consequence relation (in the language $\mathcal{L}_{cl}$). Then $\Gamma \vdash_{cl} \Delta$ iff there exists a subset Σ of the set of subformulas of $\Gamma \cup \Delta$ such that $\circ\Sigma, \Gamma \vdash_B \Delta$ (where $\circ\Sigma = \{\circ\psi \mid \psi \in \Sigma\}$).*

Proof. Suppose $\Gamma \vdash_{cl} \Delta$. Then there are finite subsets Γ' and Δ' of Γ and Δ (respectively) such that the sequent $\Gamma' \Rightarrow \Delta'$ has a cut-free proof in LK. Replace in this proof any application of the classical $(\neg \Rightarrow)$ rule by an application of $(\circ, \neg \Rightarrow)$. The result will be a cut-free proof in GB of a sequent of the form $\circ\Sigma, \Gamma' \Rightarrow \Delta'$, where Σ is a subset of the set of subformulas of $\Gamma' \Rightarrow \Delta'$. Hence $\circ\Sigma, \Gamma \vdash_B \Delta$.

For the converse, assume that $\circ\Sigma, \Gamma \vdash_B \Delta$. Then there are finite subsets Γ' of Γ, Δ' of Δ, and Σ' of Σ, such that $\circ\Sigma', \Gamma' \Rightarrow \Delta'$ has a cut free-proof in GB. Replace in that proof every application of $(\circ, \neg \Rightarrow)$ by an application of the classical $(\neg \Rightarrow)$. Since $\circ$ does not occur in $\Gamma' \Rightarrow \Delta'$, the result is a proof in LK of this sequent. It follows that $\Gamma \vdash_{cl} \Delta$. $\qquad\qquad\square$

2.4.2. Other Logics with a Formal Consistency Operation. Rule (b) provides the most basic property expected of $\circ$. There are of course many others which might seem plausible to assume. The next two definitions provide a list of rules and systems (not all!) that have been considered in the literature on LFIs.[7]

Definition 2.21. Let $RULES$ be the set consisting of the following 10 rules:

(c): $\neg\neg\varphi \vdash \varphi$
(e): $\varphi \vdash \neg\neg\varphi$
(k1): $\vdash \circ\varphi, \varphi$
(k2): $\vdash \circ\varphi, \neg\varphi$
(i1): $\neg\circ\varphi \vdash \varphi$
(i2): $\neg\circ\varphi \vdash \neg\varphi$
(a$_\neg$): $\circ\varphi \vdash \circ(\neg\varphi)$
(a$_\sharp$): $\circ\varphi, \circ\psi \vdash \circ(\varphi\sharp\psi)$ $(\sharp \in \{\wedge, \vee, \supset\})$

Definition 2.22. For $S \subseteq RULES$, $\mathbf{B}[S]$ is the extension of **B** by the rules in S.

---

[7]In [11, 12] what is considered instead of (**i1**) and (**i2**) is actually their combination, the rule (**i**): $\neg\circ\varphi \vdash \varphi \wedge \neg\varphi$. This rule has been split here into two rules as described in Remark 2.13. Conditions (**k1**) and (**k2**) were not considered in [11, 12], but they are natural weaker versions of (**i1**) and (**i2**), respectively.

3. Semantics for the Basic System

The system **B** treats the positive classical connectives exactly as classical logic does. Hence an Nmatrix for B should most naturally be sought among the Nmatrices which are suitable for $\mathbf{CL}^+$. In such Nmatrices the answer to the question whether a sentence of the form $\varphi \sharp \psi$ ($\sharp \in \{\vee, \wedge, \supset\}$) is true or not relative to a given valuation v (i.e., whether $v(\varphi \sharp \psi) \in \mathcal{D}$ or not) is completely determined by the answers to the same question for φ and ψ. The situation with respect to the unary connectives $\neg$ and $\circ$ is different. The truth/falsity of $\neg\varphi$ or $\circ\varphi$ is not completely determined by the truth/falsity of φ. More data is needed for this. Now the central idea of the semantics we are about to present is to include all the relevant data concerning a sentence φ in the truth value from $\mathcal{V}$ which is assigned to φ. In our case the relevant data beyond the truth/falsity of φ is the truth/falsity of $\neg\varphi$ and $\circ\varphi$. This leads to the use of elements from $\{0,1\}^3$ as our truth values, where the intended intuitive meaning of $v(\varphi) = \langle x, y, z\rangle$ is the following:

- $x = 1$ iff φ is "true" (i.e., $v(\varphi) \in \mathcal{D}$).
- $y = 1$ iff $\neg\varphi$ is "true" (i.e., $v(\neg\varphi) \in \mathcal{D}$).
- $z = 1$ iff $\circ\varphi$ is "true" (i.e., $v(\circ\varphi) \in \mathcal{D}$).

However, because of the special principles of **B** not all triples can be used. Thus rule (**t**) means that at least one element of the pair $\{\varphi, \neg\varphi\}$ should be true. Hence the truth-values $\langle 0,0,1\rangle$ and $\langle 0,0,0\rangle$ should be rejected. Similarly, rule (**b**) means that φ, $\neg\varphi$, and $\circ\varphi$ cannot all be true. Hence $\langle 1,1,1\rangle$ should be rejected. We are left with 5 truth-values. Among them those which are designated are those which can be assigned to true formulas, i.e., those whose first component is 1. Then we define the operations in the most liberal way which is coherent with the intended meaning of the truth-values, and with the need to use an Nmatrix suitable for $\mathbf{CL}^+$. The resulting Nmatrix is described in the next definition.

Definition 3.1. The Nmatrix $\mathcal{M}_5^B = \langle \mathcal{V}_5, \mathcal{D}_5, \mathcal{O}_5^B\rangle$ is defined as follows:

- $\mathcal{V}_5 = \{t, t_I, I, f_I, f\}$ where:

$$
\begin{aligned}
t &= \langle 1,0,1\rangle \\
t_I &= \langle 1,0,0\rangle \\
I &= \langle 1,1,0\rangle \\
f &= \langle 0,1,1\rangle \\
f_I &= \langle 0,1,0\rangle
\end{aligned}
$$

- $\mathcal{D}_5 = \{t, I, t_I\}$ $(= \{\langle x,y,z\rangle \in \mathcal{V}_5 \mid x = 1\})$.
- Let $\mathcal{D} = \mathcal{D}_5$, $\mathcal{F} = \mathcal{V}_5 - \mathcal{D}$. The operations in $\mathcal{O}_5^B$ are defined by:

$$
a \,\widetilde{\vee}\, b = \begin{cases} \mathcal{D} & \text{if either } a \in \mathcal{D} \text{ or } b \in \mathcal{D}, \\ \mathcal{F} & \text{if } a, b \in \mathcal{F} \end{cases}
$$

$$
a \,\widetilde{\supset}\, b = \begin{cases} \mathcal{D} & \text{if either } a \in \mathcal{F} \text{ or } b \in \mathcal{D} \\ \mathcal{F} & \text{if } a \in \mathcal{D} \text{ and } b \in \mathcal{F} \end{cases}
$$

$$a \tilde{\wedge} b = \begin{cases} \mathcal{D} & \text{if } a \in \mathcal{D} \text{ and } b \in \mathcal{D} \\ \mathcal{F} & \text{if either } a \in \mathcal{F} \text{ or } b \in \mathcal{F} \end{cases}$$

$$\tilde{\neg} a = \begin{cases} \mathcal{D} & \text{if } a \in \{I, f, f_I\} \\ \mathcal{F} & \text{if } a \in \{t, t_I\} \end{cases}$$

$$\tilde{\circ} a = \begin{cases} \mathcal{D} & \text{if } a \in \{t, f\} \\ \mathcal{F} & \text{if } a \in \{I, t_I, f_I\} \end{cases}$$

An Explanation. The rules in **B** related to the positive classical connectives impose no constraints on the truth/falsity of $\neg\varphi$ or $\circ\varphi$. Hence they affect only the first component of truth-values. Thus if the first component of $v(\varphi)$ is 1 (i.e., if $v(\varphi)$ is in $\mathcal{D}_5$) then also the first component of $v(\psi \supset \varphi)$ should be 1, but there are no limitations in this case on the other two components of $v(\psi \supset \varphi)$. Hence $v(\psi \supset \varphi)$ may in this case be any element of $\mathcal{D}_5$. This implies that $a \tilde{\supset} b$ should be $\mathcal{D}_5$ in case $b \in \mathcal{D}_5 = \{t, t_I, I\}$. The other parts of the definitions of $\tilde{\supset}, \tilde{\vee},$ and $\tilde{\wedge}$ are derived similarly. The truth-value of $\neg\varphi$, on the other hand, is dictated by the second component of $v(\varphi)$. If it is 1 then $\neg\varphi$ should be true, implying that $v(\neg\varphi)$ should be an element of $\mathcal{D}_5$. Since **B** imposes no further constraints on $v(\neg\varphi)$ in this case, we get the condition that $\tilde{\neg} a$ should be $\mathcal{D}_5$ in case $a \in \{I, f, f_I\}$. The other parts of the definitions of $\tilde{\neg}$ and $\tilde{\circ}$ are derived similarly (note that in the case of $\circ$ the relevant component is the third).

The five-valued $\mathcal{M}_5^B$ is our basic Nmatrix. In the next section we shall obtain semantics for a lot of extensions of **B** by refining this Nmatrix (where our refinements will be of the special type described in Remark 2.11). However, in many of the systems we discuss (including **B** itself), one needs to include in the truth-value assigned to a formula φ only information concerning the truth/falsity of φ and $\neg\varphi$. Hence a 3-valued Nmatrix consisting of pairs from $\{0,1\}^2$ (where the pair $\langle 0,0 \rangle$ is rejected because of rule (**t**)) would suffice. The basic 3-valued Nmatrix corresponding to **B** is given in the next Definition.

Definition 3.2. The Nmatrix $\mathcal{M}_3^B = \langle \mathcal{V}_3, \mathcal{D}_3, \mathcal{O}_3^B \rangle$ is defined as follows:

- $\mathcal{V}_3 = \{\mathbf{t}, \mathbf{I}, \mathbf{f}\}$ where:
  - $\mathbf{t} = \langle 1, 0 \rangle$
  - $\mathbf{I} = \langle 1, 1 \rangle$
  - $\mathbf{f} = \langle 0, 1 \rangle$
- $\mathcal{D}_3 = \{\mathbf{t}, \mathbf{I}\}$ ($= \{\langle x, y \rangle \in \mathcal{V}_3 \mid x = 1\}$).
- Let this time $\mathcal{D} = \mathcal{D}_3$, $\mathcal{F} = \mathcal{V}_3 - \mathcal{D} = \{\mathbf{f}\}$. The operations in $\mathcal{O}_3^B$ corresponding to $\wedge, \vee$ and $\supset$ are defined like in $\mathcal{M}_5^B$. The other two operations are defined as follows:

$$\tilde{\neg} a = \begin{cases} \mathcal{F} & \text{if } a \in \{\mathbf{t}\} \\ \mathcal{D} & \text{if } a \in \{\mathbf{I}, \mathbf{f}\} \end{cases}$$

$$\tilde{\circ} a = \begin{cases} \mathcal{V}_3 & \text{if } a \in \{\mathbf{t}, \mathbf{f}\} \\ \mathcal{F} & \text{if } a = \mathbf{I} \end{cases}$$

Lemma 3.3. $\vdash_B \subseteq \vdash_{\mathcal{M}_3^B}$

Proof. $\mathcal{M}_3^B$ is suitable for $\mathbf{CL}^+$. Hence by Theorem 2.15 it suffices to check that rules (**t**) and (**b**) are satisfied by $\vdash_{\mathcal{M}_3^B}$. This is easy. $\qquad\square$

Lemma 3.4. $\vdash_{\mathcal{M}_3^B} \subseteq \vdash_{\mathcal{M}_5^B}$.

Proof. The function f defined by $f(\langle x, y, z \rangle) = \langle x, y \rangle$ is easily seen to be a reduction of $\mathcal{M}_5^B$ to $\mathcal{M}_3^B$. Hence the lemma follows from Theorem 2.10. $\qquad\square$

Lemma 3.5. $\vdash_{\mathcal{M}_5^B} \subseteq \vdash_B$.

Proof. Suppose $\Gamma \nvdash_B \Delta$. We construct a model of Γ in $\mathcal{M}_5^B$ which is not a model of any formula in Δ. For this extend Γ to a maximal set Γ^* of formulas such that $\Gamma^* \nvdash_B \Delta$. Γ^* has the following properties:

1. $\varphi \notin \Gamma^*$ iff $\Gamma^*, \varphi \vdash_B \Delta$.
2. $\varphi \vee \psi \in \Gamma^*$ iff either $\varphi \in \Gamma^*$ or $\psi \in \Gamma^*$.
3. $\varphi \wedge \psi \in \Gamma^*$ iff both $\varphi \in \Gamma^*$ and $\psi \in \Gamma^*$.
4. $\varphi \supset \psi \in \Gamma^*$ iff either $\varphi \notin \Gamma^*$ or $\psi \in \Gamma^*$.
5. For every sentence φ of $\mathcal{L}_C$ either $\varphi \in \Gamma^*$ or $\neg\varphi \in \Gamma^*$.
6. If $\neg\varphi$ and φ are both in Γ^* then $\circ\varphi \notin \Gamma^*$.

The first property in this list follows from the maximality property of Γ^*. The last from rule (**b**). To show the second property, assume first that $\varphi \vee \psi \notin \Gamma^*$. Then $\Gamma^*, \varphi \vee \psi \vdash_B \Delta$. Since also $\varphi \vdash_B \varphi \vee \psi$, we get that $\Gamma^*, \varphi \vdash_B \Delta$, and so $\varphi \notin \Gamma^*$. Similarly, also $\psi \notin \Gamma^*$ in this case. Now assume that neither $\varphi \in \Gamma^*$ nor $\psi \in \Gamma^*$. Then $\Gamma^*, \varphi \vdash_B \Delta$, and $\Gamma^*, \psi \vdash_B \Delta$. Since also $\varphi \vee \psi \vdash_B \varphi, \psi$ (since $\vdash_B$ is an extension of $\vdash_{cl}^+$), we get that $\Gamma^*, \varphi \vee \psi \vdash_B \Delta$, and so $\varphi \vee \psi \notin \Gamma^*$.

The proofs of the other parts are similar (for the fifth property we use the fact that $\vdash_B$ satisfies rule (**t**)).

Define now a valuation v by $v(\varphi) = \langle x(\varphi), y(\varphi), z(\varphi) \rangle$, where:

$$x(\varphi) = \begin{cases} 1 & \varphi \in \Gamma^* \\ 0 & \varphi \notin \Gamma^* \end{cases} \qquad y(\varphi) = \begin{cases} 1 & \neg\varphi \in \Gamma^* \\ 0 & \neg\varphi \notin \Gamma^* \end{cases} \qquad z(\varphi) = \begin{cases} 1 & \circ\varphi \in \Gamma^* \\ 0 & \circ\varphi \notin \Gamma^* \end{cases}$$

It is easy to check that the above properties of Γ^* imply that v is a legal valuation in $\mathcal{M}_5^B$. Obviously, v is a model of Γ which is not a model of any formula in Δ. $\quad\square$

Theorem 3.6. *Both $\mathcal{M}_5^B$ and $\mathcal{M}_3^B$ are characteristic Nmatrices for* **B**.

Proof. This is immediate from the last three lemmas. $\qquad\square$

Corollary 3.7. **B** *is decidable.*

Proof. This follows from Theorems 3.6 and 2.8. $\qquad\square$

4. Semantics for the Extensions of B Induced by RULES

One of the main virtues of our semantics is that for many syntactic conditions concerning $\neg$ and $\circ$, it is easy to compute corresponding semantic conditions on *simple* refinements (Remark 2.11) of $\mathcal{M}_5^B$. In the next definition we list the semantic conditions induced by the rules in $RULES$ (Definition 2.21).

Definition 4.1.

1. The refining conditions induced by the conditions in $RULES$ are:

   C(**c**): If $x \in \{f, f_I\}$ then $\tilde{\neg} x \subseteq \{t, t_I\}$.

   C(**e**): $\tilde{\neg} I = \{I\}$

   C(**k1**): f_I should be deleted.

   C(**k2**): t_I should be deleted.

   C(**a$_\neg$**): $\{t, f\}$ is closed under $\tilde{\neg}$ (implying $\tilde{\neg} t = \{f\}$, $\tilde{\neg} f = \{t\}$).

   C(**a$_\sharp$**): $\{t, f\}$ is closed under $\tilde{\sharp}$.

   C(**i1**): f_I should be deleted, and $\tilde{\circ}(f) \subseteq \{t, t_I\}$

   C(**i2**): t_I should be deleted, and $\tilde{\circ}(t) = \{t\}$

2. For $S \subseteq RULES$, let $C(S) = \{Cr \mid r \in S\}$

Here are some examples of how these conditions have been derived:

Notation. Let $P_1(\langle a, b, c \rangle) = a$, $P_2(\langle a, b, c \rangle) = b$, $P_3(\langle a, b, c \rangle) = c$.

C(**c**): A refutation of this rule is a valuation v in $\mathcal{M}_5^B$ such that $v(\varphi) \notin \mathcal{D}_5$ (i.e., $P_1(v(\varphi)) = 0$), but $v(\neg\neg\varphi) \in \mathcal{D}_5$ (i.e., $P_2(v(\neg\varphi)) = 1$). This will be impossible iff for every $x \in \mathcal{V}_5$ such that $P_1(x) = 0$ (i.e., for every $x \in \{f, f_I\}$), it is the case that if $y \in \tilde{\neg} x$ then $P_2(y) = 0$ (i.e., $y \in \{t, t_I\}$).

C(**e**): A refutation of this rule is a valuation v in $\mathcal{M}_5^B$ such that $v(\varphi) \in \mathcal{D}_5$ (i.e., $P_1(v(\varphi)) = 1$), but $v(\neg\neg\varphi) \notin \mathcal{D}_5$ (i.e., $P_2(v(\neg\varphi)) = 0$). This will be impossible iff for every $x \in \mathcal{V}_5$ such that $P_1(x) = 1$ (i.e., for every x in $\{t, t_I, I\}$), if $y \in \tilde{\neg} x$ then also $P_2(y) = 1$. For $x \in \{t, t_I\}$ this is already true in $\mathcal{M}_5^B$. For $x = I$ the only element y in $\tilde{\neg}_B x$ which satisfies this condition is $y = I$ (where $\tilde{\neg}_B$ is the interpretation of $\neg$ in $\mathcal{M}_5^B$).

C(**k1**): A refutation of this rule is a valuation v in $\mathcal{M}_5^B$ for which both $v(\varphi)$ and $v(\circ\varphi)$ are not in $\mathcal{D}_5$. This will be impossible iff f_I is not available.

C(**a$_\sharp$**): A refutation of this rule is a valuation v in $\mathcal{M}_5^B$ s.t. $P_3(v(\varphi)) = 1$, $P_3(v(\psi)) = 1$, and $P_3(v(\varphi \sharp \psi)) = 0$ (i.e., $v(\varphi) \in \{t, f\}$, $v(\psi) \in \{t, f\}$, but $v(\varphi \sharp \psi) \notin \{t, f\}$). This will be impossible iff $\{t, f\}$ is closed under $\sharp$.

C(**i2**): A refutation of this rule is a valuation v in $\mathcal{M}_5^B$ s.t. $v(\neg\varphi) \notin \mathcal{D}_5$ (i.e., $P_2(v(\varphi)) = 0$), but $v(\neg\circ\varphi) \in \mathcal{D}_5$ (i.e., $P_2(v(\circ\varphi)) = 1$). This will be impossible iff for every $x \in \mathcal{V}_5$ such that $P_2(x) = 0$, also $P_2(\tilde{\circ} x) = 0$. In other words: iff for every $x \in \{t, t_I\}$, $\tilde{\circ} x \subseteq \{t, t_I\}$. For $x = t_I$ this is incoherent with the value of $\tilde{\circ}(t_I)$ in $\mathcal{M}_5^B$. Hence t_I should be deleted, and so necessarily $\tilde{\circ}(t) = \{t\}$.

Definition 4.2. For $S \subseteq RULES$, let $\mathcal{M}_S$ be the weakest simple refinement (Remark 2.11) of $\mathcal{M}_5^B$ in which the conditions in $C(S)$ are all satisfied. In other words: $\mathcal{M}_S = \langle \mathcal{V}_S, \mathcal{D}_S, \mathcal{O}_S \rangle$, where $\mathcal{V}_S$ is the set of values from $\mathcal{V}_5$ which are not deleted by any condition in S, $\mathcal{D}_S = \mathcal{D}_5 \cap \mathcal{V}_S$, and for any connective $\diamond \in \mathcal{O}$ and any $\vec{x} \in \mathcal{V}_S^n$ (where n is the arity of $\diamond$), the interpretation in $\mathcal{O}_S$ of $\diamond$ assigns to $\vec{x}$ the set $\widetilde{\diamond}_{\mathcal{M}_S}(\vec{x})$ of all the values in $\widetilde{\diamond}_B(\vec{x})$ which are not forbidden by any condition in $C(S)$ (where $\widetilde{\diamond}_B$ is the interpretation of $\diamond$ in $\mathcal{M}_5^B$).

An Example. Let $S = \{(\mathbf{i1}), (\mathbf{a}_\neg)\}$. Then $\mathcal{M}_S = \langle \mathcal{V}_S, \mathcal{D}_S, \mathcal{O}_S \rangle$, where:

- $\mathcal{V}_S = \{t, t_I, I, f\}$
- $\mathcal{D}_S = \{t, I, t_I\}$
- $\widetilde{\vee}, \widetilde{\wedge}$ and $\widetilde{\supset}$ are defined like in the case of $\mathcal{M}_5^B$ (but now $\mathcal{F} = \{f\}$).
- $\widetilde{\neg}t = \widetilde{\neg}t_I = \{f\} \quad \widetilde{\neg}I = \mathcal{D}_S \quad \widetilde{\neg}f = \{t\}$
- $\widetilde{\circ}t = \mathcal{D}_S \quad \widetilde{\circ}t_I = \widetilde{\circ}I = \{f\} \quad \widetilde{\circ}f = \{t, t_I\}$

Remark 4.3. It is not difficult to see that for all $S \subseteq RULES$, $\{t, f, I\} \subseteq \mathcal{V}_S$, $\{t, I\} \subseteq \mathcal{D}_S$, and $\widetilde{\diamond}_{\mathcal{M}_S}(\vec{x})$ is never empty (in fact, $\widetilde{\diamond}_{\mathcal{M}_S}(\vec{x}) \cap \{t, f, I\}$ is never empty). Hence $\mathcal{M}_S$ is a well-defined Nmatrix.

Remark 4.4. It can easily be checked that in any simple refinement of $\mathcal{M}_5^B$ which satisfies $C(\mathbf{a}_\neg)$, $\widetilde{\neg}$ behaves on $\{t, f\}$ like classical negation (i.e., $\widetilde{\neg}t = \{f\}, \widetilde{\neg}f = \{t\}$). Similarly, if $\sharp \in \{\vee, \wedge, \supset\}$ then in simple refinements of $\mathcal{M}_5^B$ which satisfy $C(\mathbf{a}_\sharp)$, $\widetilde{\sharp}$ behaves on $\{t, f\}$ like the classical $\sharp$.

Theorem 4.5. *For all $S \subseteq RULES$, $\mathcal{M}_S$ is a characteristic Nmatrix for $\mathbf{B}[S]$.*

Proof. It is easy to verify, that for any $r \in RULES$, the satisfaction of $C(r)$ in some simple refinement of $\mathcal{M}_5^B$ guarantees the validity of r in that refinement. This entails the soundness of $\mathbf{B}[S]$ with respect to $\mathcal{M}_S$. For completeness we repeat the construction done in the proof of Lemma 3.5. It is not difficult to show that the rules of S force the resulting valuation to be a legal valuation in $\mathcal{M}_S$. We do here the case where $S = \{(\mathbf{i1}), (\mathbf{a}_\neg)\}$ as an example. So suppose that $\Gamma \nvdash_{\mathbf{B}[S]} \Delta$. Construct the set Γ^* and the valuation v like in the proof of Lemma 3.5, using $\mathbf{B}[S]$ instead of $\mathbf{B}$. This v is legal for $\mathcal{M}_5^B$, and it is a model of Γ which is not a model of any formula in Δ. Now the presence of $(\mathbf{i1})$ implies that $v(\varphi) \neq f_I$ for every φ (because there can be no formula φ such that both $\varphi \notin \Gamma^*$ and $\circ\varphi \notin \Gamma^*$. Indeed: if $\varphi \notin \Gamma^*$ then because of $(\mathbf{i1})$ $\neg\circ\varphi \notin \Gamma^*$, and so $\circ\varphi \in \Gamma^*$). Hence v is actually a valuation in $\mathcal{V}_S$. It remains to show that it is legal for $\mathcal{M}_S$. Since v is legal for $\mathcal{M}_5^B$, it suffices to show that it respects the conditions imposed by $(\mathbf{i1})$ and $(\mathbf{a}_\neg)$:

C$(\mathbf{i1})$: Since f_I is not used by v, respecting $C(\mathbf{i1})$ amounts in the present case to $v(\circ\varphi)$ being in $\{t, t_I\}$ in case $v(\varphi) = f$. But here $v(\varphi) = f$ iff $\varphi \notin \Gamma^*$, and the latter implies (because of $(\mathbf{i1})$) that $\neg\circ\varphi \notin \Gamma^*$, which means (by definition of v) that indeed $v(\circ\varphi) \in \{t, t_I\}$.

C$(\mathbf{a}_\neg)$: Again since f_I is not used by v, respecting $C(\mathbf{a}_\neg)$ amounts in the present case to $v(\neg\varphi) = t$ in case $v(\varphi) = f$. But $v(\varphi) = f$ iff $\varphi \notin \Gamma^*$, $\neg\varphi \in \Gamma^*$, and $\circ\varphi \in \Gamma^*$. Because of $(\mathbf{a}_\neg)$ the latter implies that $\circ\neg\varphi \in \Gamma^*$. Since also $\neg\varphi \in \Gamma^*$, necessarily $v(\neg\varphi) = t$. $\qquad\square$

5. Applications

5.1. Decidability

A first important corollary of our semantics is the following:

Corollary 5.1. $\mathbf{B}[S]$ *is decidable for any* $S \subseteq RULES$.

Proof. Immediate from Theorems 4.5 and 2.8. □

5.2. Dependencies between the Conditions

Not all the 1024 systems of the form $\mathbf{B}[S]$ (where $S \subseteq RULES$) are different from each other. Using Theorem 4.5 it is a mechanical matter to check the relations among them, finding what rules in $RULES$ follow from what subsets of the other rules in $RULES$. The next theorem sums up all existing dependencies:

Theorem 5.2. *The following is an exhaustive list of all the dependencies among the rules in RULES:*

- (**k1**) *follows from* (**i1**).
- (**k2**) *follows from* (**i2**).
- (**c**) *follows from* $\{(\mathbf{a}_\neg), (\mathbf{k1})\}$ *(and from* $\{(\mathbf{a}_\neg), (\mathbf{i1})\}$*)*.
- ($\mathbf{a}_\neg$) *follows from* $\{(\mathbf{c}), (\mathbf{k1}), (\mathbf{k2})\}$ *(and of course also from* $\{(\mathbf{c}), (\mathbf{k1}), (\mathbf{i2})\}$, $\{(\mathbf{c}), (\mathbf{i1}), (\mathbf{k2})\}$, *and* $\{(\mathbf{c}), (\mathbf{i1}), (\mathbf{i2})\}$*)*.

Proof. The first two items on this list trivially follow from the corresponding semantic conditions. For the third, note that without f_I (i.e., in the presence of (**k1**) or (**i1**)), condition C(**c**) reduces to $\tilde{\neg}f \subseteq \{t, t_I\}$ and this condition immediately follows from C($\mathbf{a}_\neg$). Finally, the forth clause is immediate from the fact that if t, f, and I are the only available truth-values, then $\mathcal{F} = \{f\}$, and both conditions C(**c**) and C($\mathbf{a}_\neg$) reduce to $\tilde{\neg}f = \{t\}$.

A not too difficult examination of the corresponding 10 conditions given in Definition 4.1 (together with the Definition of $\mathcal{M}_5^B$) reveals that the above list is indeed exhaustive. □

Corollary 5.3. *Rules* (**c**) *and* ($\mathbf{a}_\neg$) *are equivalent in the presence of rules* (**k1**) *and* (**k2**). *In particular, they are equivalent in the system* $\mathbf{Bi}$, *obtained from* $\mathbf{B}$ *by adding the following schema from* [11, 12]:

(**i**): $\neg \circ \varphi \vdash \varphi \wedge \neg \varphi$

5.3. Cases Where 3-valued Nmatrices Suffice

In Section 3 We have seen that for the basic system $\mathbf{B}$ a three-valued reduction of $\mathcal{M}_5^B$ (in which the truth-values include information only on the truth/falsity of a sentence and its negation) suffices. The argument remains almost the same if either (**c**), (**e**) or both are added to $\mathbf{B}$, since these rules do not involve $\circ$. In the corresponding refinements of $\mathcal{M}_3^B$ we should have $\tilde{\neg}f = \{t\}$ in case (**c**) is added, and $\tilde{\neg}I = \{I\}$ in case (**e**) is added.

Another obvious case in which a logic $\mathbf{B}[S]$ ($S \subseteq RULES$) has a characteristic 3-valued Nmatrix is when both (**k1**) and (**k2**) are derivable in it (i.e., if either (**k1**) or (**i1**) is in S, and also either (**k2**) or (**i2**) is in S). In this case Theorem 4.5 directly provides such an Nmatrix.

Conjecture. Except for the cases we have just described, no other system $\mathbf{B}[S]$ ($S \subseteq RULES$) has any characteristic 3-valued Nmatrix.

What we can prove for *all* the systems considered here is the following:

Theorem 5.4. $\mathbf{B}[S]$ *($S \subseteq RULES$) does not have a characteristic 2-valued Nmatrix.*

Proof. Suppose $\mathcal{M}$ is a 2-valued Nmatrix for which $\mathbf{B}[S]$ is sound and complete. We may assume that the two truth-values of $\mathcal{M}$ are 1 and 0, where 1 is designated and 0 is not. Since condition (**t**) is valid in $\mathcal{M}$, necessarily $\tilde{\neg}0 = \{1\}$. Hence it suffices to consider the following 3 cases:

- Suppose $\tilde{\neg}1 = \{0\}$. Then $\neg\varphi, \varphi \vdash_{\mathcal{M}}$ for all φ. Since $\mathbf{B}[S]$ is paraconsistent (because by assigning $v(p) = v(\neg p) = I$ we get a model of $\{p, \neg p\}$ in $\mathcal{M}_S$), we get a contradiction.
- Suppose $\tilde{\circ}1 = \{0\}$. Then $\circ\varphi, \varphi \vdash_{\mathcal{M}}$ for all φ. However, assigning t to both p and $\circ p$ is legal in $\mathcal{M}_S$ (for every $S \subseteq RULES$). Hence $\circ\varphi, \varphi \nvdash_{\mathbf{B}[S]}$, and we get a contradiction.
- Suppose that 1 is in both $\tilde{\neg}1$ and $\tilde{\circ}1$. Then assigning 1 to all the sentences in $\{p, \neg p, \circ p\}$ is legal in $\mathcal{M}$, contradicting the validity of (**b**) in $\mathcal{M}$.

We got a contradiction in all possible cases. Hence no such Nmatrix exists. □

6. Other Plausible Extensions of the Basic System

In addition to the rules considered so far (that were basically taken from [11, 12]), it is of course possible to consider many other rules that might seem plausible. We consider now two natural groups of rules that may also be added to the basic system $\mathbf{B}$, and are very easy to handle in our framework.

6.1. Rules for Combinations of Negation with the Classical Connectives

(**c**) and (**e**) are just two of the standard classically valid rules concerning negation which are derived from the classical equivalences of $\neg\neg\varphi$ with φ, $\neg(\varphi \wedge \psi)$ with $\neg\varphi \vee \neg\psi$, $\neg(\varphi \vee \psi)$ with $\neg\varphi \wedge \neg\psi$, and $\neg(\varphi \supset \psi)$ with $\varphi \wedge \neg\psi$. By splitting the last 3 equivalences into simple rules (see Remark 2.13) we get the following list:

Definition 6.1. Let DM be the set consisting of the following 9 rules:

$(\neg \wedge 1)$: $\neg\varphi \vdash \neg(\varphi \wedge \psi)$

$(\neg \wedge 2)$: $\neg\psi \vdash \neg(\varphi \wedge \psi)$

$(\neg \wedge 3)$: $\neg(\varphi \wedge \psi) \vdash \neg\varphi, \neg\psi$

$(\neg \vee 1)$: $\neg(\varphi \vee \psi) \vdash \neg\varphi$

$(\neg \vee 2)$: $\neg(\varphi \vee \psi) \vdash \neg\psi$

$(\neg \vee 3)$: $\neg\varphi, \neg\psi \vdash \neg(\varphi \vee \psi)$

$(\neg \supset 1)$: $\neg(\varphi \supset \psi) \vdash \varphi$

$(\neg \supset 2)$: $\neg(\varphi \supset \psi) \vdash \neg\psi$
$(\neg \supset 3)$: $\varphi, \neg\psi \vdash \neg(\varphi \supset \psi)$

In [1] we have shown how to modularly provide 3-valued non-deterministic semantics for these rules, where the basic logic is $CLuN$ (which is the logic in $\mathcal{L}_{cl}$ obtained from **B** by deleting the schema (**b**)). It is straightforward to adapt those results to the present context, with **B** as the basic logic. All we need to do is to find for the rules in DM equivalent semantic conditions on refinements of $\mathcal{M}_5^B$, using considerations of the type applied for the rules in $RULES$. For this, one should only note that for the rules in DM only the first two components of our truth-values are relevant. We present here as an example the derived semantic conditions which are equivalent to the rules corresponding to the equivalence between $\neg(\varphi \wedge \psi)$ and $\neg\varphi \vee \neg\psi$:

C($\neg \wedge 1$): If $x \in \mathcal{D}$ then $I \widetilde{\wedge} x = \{I\}$.
C($\neg \wedge 2$): If $x \in \mathcal{D}$ then $x \widetilde{\wedge} I = \{I\}$.
C($\neg \wedge 3$): If $x \in \{t, t_I\}$ and $y \in \{t, t_I\}$ then $x \wedge y \subseteq \{t, t_I\}$.

Like in Definition 4.2, We can now define $\mathcal{M}_S$ for every $S \subseteq RULES \cup DM$. It is then easy to prove the following generalization of Theorem 4.5:

Theorem 6.2. *For all $S \subseteq RULES \cup DM$, $\mathcal{M}_S$ is a characteristic Nmatrix for* **B**$[S]$.

Corollary 6.3.

1. $(\neg \wedge 3)$ *is derivable in* **B**$[\{(\mathbf{k2}), (\mathbf{a}_\wedge)\}]$.
2. $(\neg \wedge 3)$ *and* $(\mathbf{a}_\wedge)$ *are equivalent in any extension of* **B**$[\{(\mathbf{k1}), (\mathbf{k2})\}]$

Proof. In the presence of (**k2**) the truth-value t_I is not available. Hence in this case C($\neg \wedge 3$) reduces to $t \widetilde{\wedge} t = \{t\}$. This last condition follows from C($\mathbf{a}_\wedge$), implying the first part of the corollary. Now in the presence of (**k1**) only f is not in the set $\mathcal{D}$ of designated values, and so in this case C($\mathbf{a}_\wedge$) too reduces to $t \widetilde{\wedge} t = \{t\}$. Hence the equivalence in the second part. □

Remark 6.4. One of the principles behind the construction of da Costa's C-systems ([14, 11]) has been that the consistent formulas should be closed under the classical connectives. This has been the reason for including the schemes of the form $(\mathbf{a}_\sharp)$ in the systems. From Corollaries 5.3 and 6.3 it follows that under weak conditions (which are satisfied, e.g., in the presence of axiom (**i**)), the axioms expressing the applications of this principle to $\neg$ and $\wedge$ can be replaced by well-known classical tautologies in which $\circ$ is not mentioned.[8]

Remark 6.5. It is interesting to note that the semantics we get for the system **B**$[RULES \cup DM]$ itself (or just for the system **B**$[DM \cup \{(\mathbf{c}), (\mathbf{e}), (\mathbf{i1}), (\mathbf{i2})\}]$) is

---

[8]This fact might give some justification why also (**c**) (and not only (**t**)) has been included in the original basic system C_1 of da Costa ([14]).

a characteristic 3-valued (ordinary, deterministic) *matrix*. This is the famous 3-valued matrix characteristic for the paraconsistent logic called $LFI1$ in [11, 12], to which $\mathbf{B}[RULES \cup DM]$ is equivalent.[9]

6.2. Rules Concerning ∘

Finally we turn to rules involving the connective ∘ but not ¬. We briefly consider two types of rules of this sort.

Strengthening the closure rules: The rules of the form $(\mathbf{a}_\sharp)$ express the assumption that if φ and ψ are consistent, then so is $\varphi\sharp\psi$. Now it is plausible to consider stronger assumptions. One alternative that is investigated in [11, 12] is that $\varphi\sharp\psi$ should be consistent if either φ or ψ is consistent. There is no problem to handle this stronger assumption within our framework by finding corresponding semantic conditions. First, the assumption for $\sharp$ split into the following two rules:

- $(\mathbf{o}_\sharp^1)$ $\circ\varphi \vdash \circ(\varphi\sharp\psi)$
- $(\mathbf{o}_\sharp^2)$ $\circ\psi \vdash \circ(\varphi\sharp\psi)$

Now the first of them, for example, translates to the condition: if $P_3(x) = 1$ then $P_3(x\widetilde{\sharp}y) = 1$. In other words: If $x \in \{t, f\}$ then $x\widetilde{\sharp}y \in \{t, f\}$. What this implies in specific cases depends on the semantics of $\sharp$. Thus for $\wedge$ we get:

$C(\mathbf{o}_\wedge^1)$: $f\widetilde{\wedge}y = \{f\}$ for every y, while $t\widetilde{\wedge}y = \{t\}$ for $y \neq t$.

Note that in the presence of $\mathbf{k1+k2}$, $C(\mathbf{o}_\wedge^1)$ reduces to $t\widetilde{\wedge}y = \{t\}$ for $y \in \{t, I\}$.

It is important to note that by using $C(\mathbf{o}_\wedge^1)$, the rule $(\mathbf{o}_\wedge^1)$ can be added to $RULES$ without essentially affecting the validity of Theorem 4.5. However a new situation arises if we consider $(\mathbf{o}_\wedge^1)$ together with $(\neg\wedge 2)$. $C(\neg\wedge 2)$ implies that $t\widetilde{\wedge}I = \{I\}$, while $C(\mathbf{o}_\wedge^1)$ implies that $t\widetilde{\wedge}I = \{t\}$. This means that we cannot use both t and I in constructing models for theories based on the logic $\mathbf{B}' = \mathbf{B}[\{(\mathbf{o}_\wedge^1), (\neg\wedge 2)\}]$. However, the combination of the corresponding conditions does not decisively rule out any of these two truth-values. Hence the framework we have developed here does not provide a unique characteristic Nmatrix for $\mathbf{B}'$. However, it can be shown that it does provide *two* finite Nmatrices $\mathcal{M}_1$ and $\mathcal{M}_2$ such that $\vdash_{\mathbf{B}'}=\vdash_{\mathcal{M}_1} \cap \vdash_{\mathcal{M}_2}$.

Some common modal rules: We end with considering the effects of the counterparts for ∘ of the three modal axioms of the modal logic $S4$:

(K): $\circ\varphi, \circ(\varphi \supset \psi) \vdash \circ\psi$

(4): $\circ\varphi \vdash \circ\circ\varphi$

(T): $\circ\varphi \vdash \varphi$

In the context of extensions of $\mathbf{B}$ (i.e., refinements of $\mathcal{M}_5^B$) the corresponding semantic conditions can easily be found to be:

$C(\mathbf{K})$: If $x \in \{t, f\}$ and $y \in \{I, t_I, f_I\}$ then $x \supset y \subseteq \{I, t_I, f_I\}$.

[9] This logic was originally introduced in [19]. Later it was reintroduced (together with its 3-valued deterministic semantics) in [15, 16], and was called there J_3 (see also [17]).

C(**4**): If $x \in \{t, f\}$ then $\tilde{\circ}x = \{t\}$

C(**T**): The truth-value f should be deleted.

Note that C(**T**) is in direct conflict with C(**k1**), since together they leave no nondesignated element, implying that any formulas is a theorem of the resulting logic (this can of course be verified directly, using a cut). A more interesting observation is that the combined effect of C(**k1**), C(**k2**), and C(**4**) is identical to the combined effect of C(**i1**) and C(**i2**). Hence the axioms (**k1**), (**k2**), and (**4**) are together equivalent to the axiom (**i**) (which is standard in C-systems — see [11]).

7. Conclusions and Further Research

We have presented an extensive study of the use of Nmatrices in deriving useful semantic for thousands of extensions of one particular basic system: **B**. It should be clear from this case-study that the method has a very large range of applications (far beyond the framework of **B**). However, it is still necessary to formulate it in exact, general terms, and to determine its scope. Another important task is to develop extensions of the framework for cases in which the method used in this paper is too weak. Two such cases (and related questions and tasks) are:

1. The primary constraint on rules to which our method applies seems to be *purity*. A good example of a context in which this constraint is violated, is provided by normal modal logics. As we have seen in the previous section, the usual *axioms* used in these logics pose no real problem. However, the necessitation rule, as it is used in modal logics, is *impure*: if $\vdash$ is supposed to be an extension of the classical consequence relation, then the necessitation rule cannot be translated into $\varphi \vdash \Box\varphi$. Indeed, in classical logic we have that $\Box\varphi \vdash \varphi \supset \Box\varphi$, and that $\vdash \varphi, \varphi \supset \Box\varphi$. Together with $\varphi \vdash \Box\varphi$ these facts entail $\vdash \varphi \supset \Box\varphi$ (using cuts). However, $\varphi \supset \Box\varphi$ is not valid in any interesting modal logic. It seems therefore that extra machinery, like the use of non-deterministic Kripke structures, should be added in order to handle rules of this sort. Steps in this direction have been taken in [1, 2], where hybrid semantics, employing both Nmatrices and Kripke structures, has been provided for many extensions of positive intuitionistic logic[10] (which is another logic which employs impure rules).

2. Two common features of all the rules considered in this paper are that each of them is concerned with at most two different connectives, and also the nesting depth of each formula used in their schematic description is at most two. An example of a rule which lacks both features is rule (**l**) from [11, 12]:

   (**l**): $\neg(\varphi \wedge \neg\varphi) \vdash \circ\varphi$

   Now in [3] it is shown that **B**[{(**c**), (**l**)}] (which is called there Cl) has no finite characteristic Nmatrix. Hence at least one of the two features we have

---

[10] One of those extensions is da Costa's basic paraconsistent system C_ω from [14].

mentioned should be essential. Which one? And what can be done in its absence? Concerning the last question, it should be noted that *Cl* has also been shown in [3] to have an *infinite* characteristic Nmatrix, which is simple enough to yield a decision procedure. Can this fact be generalized?

Another natural (and important) line for further research is to use the semantic ideas presented here for systematically producing tableaux-style proof-systems for the various logics dealt with in this paper. Now general systems of this type have in fact been developed in [5] for every logic which has finite characteristic Nmatrix. However, the central idea of those systems is to use signed formulas, where the signs are (essentially) the truth-values of the characteristic Nmatrix (and so the number of signs equals the number of the truth-values of that Nmatrix). Here it might be more effective to use 6 signs rather than 5, according to the two possible values of the three *components* of each of the five truth-values (or four signs in the cases where 3-valued versions suffice).

Finally, an obvious crucial line of further research is to extend the results and methods of this paper to first-order languages. First steps in this direction have been made in [8].

References

[1] A. Avron, *Non-deterministic Semantics for Families of Paraconsistent Logics*, To appear in "Paraconsistency with no Frontiers" (J.-Y. Beziau and W. Carnielli, eds.).

[2] A. Avron, *A Non-deterministic View on Nonclassical Negations*, Studia Logica 80, 159-194 (2005).

[3] A. Avron, *Non-deterministic Semantics for Logics with a Consistency Operators*, Forthcoming in the International Journal of Approximate Reasoning.

[4] A. Avron, *Logical Non-determinism as a Tool for Logical Modularity: An Introduction*, In "We Will Show Them: Essays in Honor of Dov Gabbay", Vol 1 (S. Artemov, H. Barringer, A. S. d'Avila Garcez, L. C. Lamb, and J. Woods, eds.), 105-124, College Publications, 2005.

[5] A. Avron, and B. Konikowska, *Multi-valued Calculi for Logics Based on Non-determinism*, Logic Journal of the IGPL 13, 365-387, (2005).

[6] A. Avron, and I. Lev, *Non-deterministic Multiple-valued Structures*, Journal of Logic and Computation 15, 241-261 (2005).

[7] A. Avron and I. Lev, *Canonical Propositional Gentzen-Type Systems*, in "Proceedings of the 1st International Joint Conference on Automated Reasoning (IJCAR 2001)" (R. Goré, A Leitsch, T. Nipkow, Eds), LNAI 2083, 529-544, Springer Verlag, 2001.

[8] A. Avron, and A. Zamanski, *Quantification in Non-deterministic Multi-valued Structures*, In "Proceedings of the 35th IEEE International Symposium on Multiple-Valued Logic (ISMVL 2005)", 296-301, IEEE Computer Society Press, 2005.

[9] D. Batens, K. De Clercq, and N. Kurtonina, *Embedding and Interpolation for Some Paralogics. The Propositional Case*, Reports on Mathematical Logic 33 (1999), 29-44.

[10] W. A. Carnielli, *Possible-translations Semantics for Paraconsistent Logics*, in "Frontiers of Paraconsistent Logic" (D. Batens, C. Mortensen, G. Priest, and J. P. Van Bendegem, eds.), 149-163. King's College Publications, Research Studies Press, Baldock, UK, 2000.

[11] W. A. Carnielli and J. Marcos, *A Taxonomy of C-systems*, in "Paraconsistency — the logical way to the inconsistent" (W. A. Carnielli, M. E. Coniglio, I. L. M. D'ottaviano, eds.), Lecture Notes in Pure and Applied Mathematics, 1-94, Marcel Dekker, 2002.

[12] W. A. Carnielli, M. E. Coniglio, and J. Marcos, *Logics of Formal Inconsistency*, to appear in "Handbook of Philosophical Logic" (D. Gabbay and F. Guenthner, eds).

[13] J. M. Crawford and D. W. Etherington, *A Non-deterministic Semantics for Tractable Inference*, in "Proc. of the 15th International Conference on Artificial Intelligence and the 10th Conference on Innovative Applications of Artificial Intelligence", 286-291, MIT Press, Cambridge, 1998.

[14] N. C. A. da Costa, *On the theory of inconsistent formal systems*, Notre Dame Journal of Formal Logic 15 (1974), 497–510.

[15] I. L. M. D'Ottaviano and N. C. A. da Costa, *Sur un problème de Jaśkowski*, Comptes Rendus de l'Academie de Sciences de Paris (A-B) 270 (1970), 1349–1353.

[16] I. L. M. D'Ottaviano, *The completeness and compactness of a three-valued first-order logic*, Revista Colombiana de Matematicas, vol. XIX (1985), 31–42.

[17] R. L. Epstein, "The semantic foundation of logic", vol. I: Propositional Logics, ch. IX, Kluwer Academic Publisher, 1990.

[18] J.Łoś and R. Suszko, *Remarks on Sentential Logics*, Indagationes Mathematicae 20 (1958), 177–183.

[19] K. Schütte, "Beweistheorie", Springer, Berlin, 1960.

Arnon Avron
School of Computer Science
Tel-Aviv University
Ramat Aviv 69978
Israel
e-mail: aa@math.tau.ac.il

J.-Y. Beziau (Ed.), *Logica Universalis*, 2nd edition, 175–194
© 2007 Birkhäuser Verlag Basel/Switzerland

Two's Company:
"The Humbug of Many Logical Values"

Carlos Caleiro, Walter Carnielli,
Marcelo E. Coniglio and João Marcos

> *How was it possible that*
> *the humbug of many logical values*
> *persisted over the last fifty years?*
> — Roman Suszko, 1976.

Abstract. The Polish logician Roman Suszko has extensively pleaded in the
1970s for a restatement of the notion of many-valuedness. According to him,
as he would often repeat, "there are but two logical values, true and false." As
a matter of fact, a result by Wójcicki-Lindenbaum shows that any tarskian
logic has a many-valued semantics, and results by Suszko-da Costa-Scott show
that any many-valued semantics can be reduced to a two-valued one. So, why
should one even consider using logics with more than two values? Because, we
argue, one has to decide how to deal with bivalence and settle down the trade-
off between logical 2-valuedness and truth-functionality, from a pragmatical
standpoint.

This paper will illustrate the ups and downs of a two-valued reduction of
logic. Suszko's reductive result is quite non-constructive. We will exhibit here a
way of effectively constructing the two-valued semantics of any logic that has a
truth-functional finite-valued semantics and a sufficiently expressive language.
From there, as we will indicate, one can easily go on to provide those logics
with adequate canonical systems of sequents or tableaux. The algorithmic
methods developed here can be generalized so as to apply to many non-finitely
valued logics as well — or at least to those that admit of computable quasi
tabular two-valued semantics, the so-called dyadic semantics.

Mathematics Subject Classification (2000). Primary 03B22; Secondary 03B50.

Keywords. Suszko's Thesis, bivalence, truth-functionality.

The work of the first and the fourth authors was partially supported by FEDER (European
Union) and FCT (Portugal), namely via the Projects POCTI / MAT / 37239 / 2001 FibLog
and POCTI / MAT / 55796 / 2004 QuantLog of the Security and Quantum Information Group
(SQIG / IST, Portugal), and the grant SFRH / BD / 8825 / 2002. The second author was
partially supported by CNPq (Brazil) and by a senior scientist research grant from the SQIG /
IST.

1. Suszko's Thesis

"After 50 years we still face an illogical paradise of many truths and falsehoods". Thus spoke Suszko in 1976, at the 22nd Conference on the History of Logic, in Cracow (cf. [25]). He knew all too well who was the first to blame for that state of affairs: "Łukasiewicz is the chief perpetrator of a magnificent conceptual deceipt lasting out in mathematical logic to the present day." Suszko was perfectly aware, of course, that there are logics that can only be characterized truth-functionally with the help of n-valued matrices, for $n > 2$. He also knew that there were logics, such as most logics proposed by Lukasiewicz or by Post, that were characterizable by finite-valued matrices, and he knew that there were logics, such as Łukasiewicz's L_ω, intuitionistic logic, or all the usual modal systems, that could only be characterized by infinite-valued matrices. Suszko was even ready to concede, in his reconstruction of the Fregean distinction between 'sense' and 'reference' of sentences, that the talk about many truth-values, in a sense, could not be avoided, "unless one agrees that thought is about nothing, or, rather, stops talking with sentences" (cf. [23]).

Still, Suszko insisted that "obviously any multiplication of logical values is a mad idea" (cf. [25]). How come? The point at issue is, according to Suszko, a distinction between the *algebraic truth-values* of many-valued logics, that were supposed to play a merely referential role, while only two *logical truth-values* would really exist. The philosophical standpoint according to which "there are but two logical values, true and false" receives nowadays the label of *Suszko's Thesis* (cf. [16, 18, 26]).

Suszko illustrated his proposition by showing how Łukasiewicz's 3-valued logic L_3 could be given a 2-valued (obviously non-truth-functional) semantics (cf. [24]). He did not explain though how he obtained the latter semantics, or how the procedure could be effectively applied to other logics. The present paper shows how that can be done for a large class of finite-valued logics. It also illustrates some uses for that 2-valued reduction in the mechanization of proof procedures. Our initial related explorations in the field appeared in our reports [8, 6]. A detailed appraisal and an extended investigation of both the technical and the philosophical issues involved in Suszko's Thesis can be found in our [7].

The plan of the present paper is as follows. Section 2 explains the general reductive results that make tarskian logics n-valued and 2-valued. Section 3 presents the technology for separating truth-values, which is the cornerstone of our reductive procedure. Section 4 introduces gentzenian semantics as an appropriate format for presenting bivaluation axioms, and proposes dyadic semantics so as to define the class of computable 2-valued semantics. Section 5 obtains, in an effective way, 2-valued semantics for many-valued logics, applying the algorithm from the main Theorem 5.2. Several detailed examples are given in Section 6. The question of obtaining 'bivalent' tableaux for such logics is treated in Section 7. Finally, Section 8 briefly summarizes the obtained results and prompts for a continuation of the present investigations.

2. Reductive results

Let $\mathcal{S}$ denote a non-empty set of *formulas* and let $\mathcal{V}$ denote a non-empty set of *truth-values*. Any $\Gamma \subseteq \mathcal{S}$ will be called a *theory*. Assume $\mathcal{V} = \mathcal{D} \cup \mathcal{U}$ for suitable disjoint sets $\mathcal{D}$ and $\mathcal{U}$ of *designated* and *undesignated* values. Any mapping $\S_k^{\mathcal{V}} :$ $\mathcal{S} \to \mathcal{V}_k$ is called a (*n-valued*) *valuation*, where n is $|\mathcal{V}_k|$ (the cardinality of $\mathcal{V}_k = \mathcal{D}_k \cup \mathcal{U}_k$); if both $\mathcal{D}_k$ and $\mathcal{U}_k$ are singletons, $\S_k^{\mathcal{V}}$ is called a *bivaluation*. Any collection sem of valuations is called a (*n-valued*) *semantics*, where n is the cardinality of the largest $\mathcal{V}_k$ such that $\S_k^{\mathcal{V}} \in$ sem. A *model* of a formula φ is any valuation $\S_k^{\mathcal{V}}$ such that $\S_k^{\mathcal{V}}(\varphi) \in \mathcal{D}_k$. A canonical notion of *entailment* given by a *consequence relation* $\vDash_{\mathsf{sem}} \subseteq \mathrm{Pow}(\mathcal{S}) \times \mathcal{S}$ associated to the semantics sem can be defined by saying that a formula $\varphi \in \mathcal{S}$ follows from a set of formulas $\Gamma \subseteq \mathcal{S}$ whenever all models of all formulas of Γ are also models of φ, that is,

$$\Gamma \vDash_{\mathsf{sem}} \varphi \quad \text{iff} \quad \S_k^{\mathcal{V}}(\varphi) \in \mathcal{D}_k \text{ whenever } \S_k^{\mathcal{V}}(\Gamma) \subseteq \mathcal{D}_k, \text{ for every } \S_k^{\mathcal{V}} \in \mathsf{sem}. \quad \text{(DER)}$$

That much for a semantic (many-valued) account of consequence. Now, for an abstract account of consequence, consider the following set of properties:

(CR1) $\Gamma, \varphi, \Delta \Vdash \varphi$;   (inclusion)
(CR2) If $\Delta \Vdash \varphi$, then $\Gamma, \Delta \Vdash \varphi$;   (dilution)
(CR3) $(\forall \beta \in \Delta)(\Gamma \Vdash \beta$ and $\Delta \Vdash \alpha)$ implies $\Gamma \Vdash \alpha$.   (cut for sets)

A *logic* $\mathcal{L}$ will in this section be defined simply as a set of formulas together with a consequence relation defined over it. Logics respecting axioms (CR1)–(CR3) are called *tarskian*. Notice, in particular, that when sem is a singleton, one also defines a tarskian logic. Furthermore, an arbitrary intersection of tarskian logics also defines a tarskian logic. Given some logic $\mathcal{L} = \langle \mathcal{S}, \Vdash \rangle$, a theory $\Gamma \subseteq \mathcal{S}$ will be called *closed* in case it contains all of its consequences; the closure $\overline{\Gamma}$ of a theory Γ may be obtained by setting $\varphi \in \overline{\Gamma}$ iff $\Gamma \Vdash \varphi$. A *Lindenbaum matrix* for a theory Γ is defined by taking $\mathcal{V} = \mathcal{S}$, $\mathcal{D} = \overline{\Gamma}$ and $\mathsf{sem}[\Gamma] = \{\mathbf{id}_{\mathcal{S}}\}$ (the identity mapping on the set of formulas).

It is easy to check that every n-valued logic is tarskian. It can be shown that the converse is also true:

Theorem 2.1. (Wójcicki's Reduction)

Every tarskian logic $\mathcal{L} = \langle \mathcal{S}, \Vdash \rangle$ is n-valued, for some $n \leq |\mathcal{S}|$.

Proof. For each theory Γ of $\mathcal{L}$, notice that the corresponding Lindenbaum matrix defines a sound semantics for that logic, that is, $\Gamma \Vdash \varphi$ implies $\Gamma \vDash_{\mathsf{sem}[\Gamma]} \varphi$. To obtain completeness, one can now consider the intersection of all Lindenbaum matrices and check that $\Vdash = \cap_{\Gamma \subseteq \mathcal{S}} \vDash_{\mathsf{sem}[\Gamma]}$. $\qquad \qquad \square$

This result shows that the above semantic and the abstract accounts of consequence define exactly the same class of logics. While we know that classical propositional logic can be characterized in fact by a collection of 2-valued matrices, and several other tarskian logics can be similarly characterized by other collections of finite-valued matrices, Wójcicki's Reduction shows that any tarskian logic has, in general, an infinite-valued characteristic matrix. Apart from the 'many truths and falsehoods' allowed by many-valued semantics, it should be observed that such semantics retain, in a sense, a shadow of bivalence, as reflected in the distinction between designated and undesignated values. Capitalizing on that distinction, one can show in fact that every tarskian logic also has an adequate 2-valued semantics:

Theorem 2.2. (Suszko's Reduction)

Every tarskian n-valued logic can also be characterized as 2-valued.

Proof. For any n-valuation $\S$ of a given semantics $\mathsf{sem}(n)$, and every consequence relation based on $\mathcal{V}_n$ and $\mathcal{D}_n$, define $\mathcal{V}_2 = \{T, F\}$ and $\mathcal{D}_2 = \{T\}$ and set the characteristic total function $b_\S : \mathcal{S} \to \mathcal{V}_2$ to be such that $b_\S(\varphi) = T$ iff $\S(\varphi) \in \mathcal{D}$. Now, collect all such bivaluations $b_\S$'s into a new semantics $\mathsf{sem}(2)$, and notice that $\Gamma \vDash_{\mathsf{sem}(2)} \varphi$ iff $\Gamma \vDash_{\mathsf{sem}(n)} \varphi$. $\square$

The above results deserve a few brief comments. First of all, the standard formulations of Wójcicki's Reduction (cf. [27]) and of Suszko's Reduction (cf. [18]) usually presuppose more about the set of formulas (more specifically, they assume that it is a free algebra) and about the consequence relation (among other things, they assume that it is structural, i.e., that it allows for uniform substitutions). As we have seen, however, such assumptions are unnecessary for the more general formulation of the reductive results. From the next section on, however, we will incorporate those assumptions in our logics. Second, reductive theorems similar in spirit to Suszko's Reduction have in fact been independently proposed in the 70s by other authors, such as Newton da Costa and Dana Scott (a summary of important results from the theory of bivaluations that sprang from those approaches can be found in [4]). Third, it might seem paradoxical that the same logic is characterized by an n-valued semantics, for some sufficiently large n, and also by a 2-valued semantics. As we will see, though, the tension is resolved when we notice that the whole issue involves a trade-off between 'algebraic' truth-functionality and 'logical' bivalence. From the point of view of Suszko's Thesis, explained in the last section, these results can only lend some plausibility to the idea that "there are but two logical values, true and false": At the very least, we now know that the assertion makes perfect sense once we are talking about tarskian logics. Last, but not least, it should be noticed that the above reductive results are quite non-constructive. In case the logic *has* a finite-valued truth-functional semantics, Wójcicki's Reduction tells you nothing, in general, about how it can be obtained. Furthermore, Suszko's Reduction does not give you any hint, in general, on how a 2-valued semantics could be determined by anything like a finite recursive set of clauses, even for the case of logics with finite-valued truth-functional semantics.

In the present paper we obtain an effective method that assigns a useful 2-valued semantics to every finite-valued truth-functional logic provided that the 'algebraic values' of the semantics can be individualized by means of the linguistic resources of the logic.

3. Separating truth-values

Let's begin by adding some standard structure to the sets of formulas of our logics. Let $ats = \{p_1, p_2, \ldots\}$ be a denumerable set of *atomic sentences*, and let $\Sigma = \{\Sigma_n\}_{n \in \mathbb{N}}$ be a propositional signature, where each Σ_n is a set of *connectives* of arity n. Let $cct = \bigcup_{n \in \mathbb{N}} \Sigma_n$ be the whole set of connectives. The set of formulas $\mathcal{S}$ is then defined as the algebra freely generated by ats over Σ. Let's also add here some structure to the set of truth-values of our logics. Unless explicitly stated otherwise, from now on $\mathcal{L}$ will stand for a propositional finite-valued logic. Additionally, $\mathbb{V}$ will be a fixed Σ-algebra defining a truth-functional semantics for $\mathcal{L}$ over a finite non-empty set of truth-values $\mathcal{V} = \mathcal{D} \cup \mathcal{U}$. Assume that $\mathcal{D} = \{d_1, \ldots, d_i\}$ and $\mathcal{U} = \{u_1, \ldots, u_j\}$ are the sets of designated and undesignated truth-values, respectively, with $\mathcal{D} \cap \mathcal{U} = \emptyset$. Assume also that the valuations composing the semantics of *genuinely n-valued* logics (logics having n-valued characterizing matrices, but no m-valued such matrices, for $m < n$) are given by the homomorphisms $\S : \mathcal{S} \to \mathbb{V}$. A *uniform substitution* is an endomorphism $\varepsilon : \mathcal{S} \to \mathcal{S}$. Let us denote by $\varphi(p_1, \ldots, p_n)$ a formula φ whose set of atomic sentences appear among $p_1, \ldots, p_n$. From now on, we write $\varphi(p_1/\alpha_1, \ldots, p_n/\alpha_n)$ instead of $\varepsilon(\varphi(p_1, \ldots, p_n))$ whenever $\varepsilon(p_k) = \alpha_k$. Given a genuinely n-valued logic $\mathcal{L}$ whose semantics is determined by $\langle \mathcal{V}, cct, \mathcal{D} \rangle$, we shall denote by $\mathcal{L}^{\mathbf{c}}$ any functionally complete genuinely n-valued (conservative) extension of it (extending, if necessary, the signature Σ), that is, a logic $\mathcal{L}^{\mathbf{c}}$ with the same number of (un)designated values as $\mathcal{L}$, but which can define all n-valued matrices — had they not been already definable from the start.

Def. 3.1. A set of *interpretation maps* $[.] : \mathcal{V}^n \to \mathcal{V}$ over $\mathcal{S}$, for each $n \in \mathbb{N}^+$, is defined as follows, given $\vec{v} = (v_1, \ldots, v_n) \in \mathcal{V}^n$:

 (i) $[p_k](\vec{v}) = v_k$, if $1 \le k \le n$;
 (ii) $[\otimes(\varphi_1, \ldots, \varphi_m)](\vec{v}) = \otimes([\varphi_1](\vec{v}), \ldots, [\varphi_m](\vec{v}))$, where $\otimes$ is an m-ary connective and $\otimes$ is identified with the corresponding operator in the algebra $\mathbb{V}$.

Remark 3.2. Given formulas $\varphi(p)$ and α of $\mathcal{L}$, and a homomorphism $\S : \mathcal{S} \to \mathbb{V}$, then we have:

$$[\varphi](\S(\alpha)) = \S(\varphi(p/\alpha)). \qquad (*)$$

Def. 3.3. Let $v_1, v_2 \in \mathcal{V}$. We say that v_1 and v_2 are *separated*, and we write $v_1 \sharp v_2$, in case v_1 and v_2 belong to different classes of truth-values, that is, in case either $v_1 \in \mathcal{D}$ and $v_2 \in \mathcal{U}$, or $v_1 \in \mathcal{U}$ and $v_2 \in \mathcal{D}$. Given some genuinely n-valued logic $\mathcal{L}$, there is always some formula $\varphi(p)$ of $\mathcal{L}^{\mathbf{c}}$ which *separates* v_1 and v_2, that is, such that $[\varphi](v_1)\sharp[\varphi](v_2)$ (or else one of these two values would be redundant, and the logic would thus not be genuinely n-valued). Equivalently, one can say that $\varphi(p)$ separates v_1 and v_2 if the truth-values obtained in the truth-table for φ when p takes the values v_1 and v_2 are separated. We say that v_1 and v_2 are *effectively separated* by a logic $\mathcal{L}$ in case there is some separating formula $\varphi(p)$ to be found among the original set of formulas of $\mathcal{L}$. In that case we will also say that the values v_1 and v_2 of $\mathcal{L}$ are *effectively separable*.

Example 3.4. Clearly, if $v_1 \sharp v_2$ then p separates v_1 and v_2. Therefore, every pair of separated truth-values is always effectively separable. As another example, note that $\varphi(p) = \neg p$ separates 0 and $\frac{1}{2}$ in Łukasiewicz's logic L$_3$ (see the formulation of its matrices at Example 3.9), given that $[\neg p](0) = \neg 0 = 1$, $[\neg p](\frac{1}{2}) = \neg\frac{1}{2} = \frac{1}{2}$, and $1 \sharp \frac{1}{2}$. The separability of the truth-values of a logic $\mathcal{L}$ clearly depends on the original expressibility of this logic, i.e., the range of matrices that it can define by way of interpretations of its formulas. The truth-values of a functionally complete logic, for instance, are all obviously separable. Consider, in contrast, a logic whose semantics is given by $\langle\{0, \frac{1}{2}, 1\}, \{\otimes\}, \{1\}\rangle$, where $v_1 \otimes v_2 = v_1$ if $v_1 = v_2$, otherwise $v_1 \otimes v_2 = 1$. The values 0 and $\frac{1}{2}$ of this logic are obviously not separable.

Assumption 3.5. (Separability)
From this point on we will assume that, for any finite-valued logic we consider, every pair $\langle v_1, v_2 \rangle \in \mathcal{D}^2 \cup \mathcal{U}^2$ such that $v_1 \neq v_2$ is effectively separable.

It follows from the last assumption that it is possible to individualize every truth-value in terms of its membership to $\mathcal{D}$ (to be represented here by the 'logical' value T) or to $\mathcal{U}$ (to be represented by the 'logical' value F). As it will be shown, together with this assumption about the expressibility of our logics, the residual bivalence embodied in the distinction between designated and undesignated values will permit us to effectively reformulate our original n-valued semantics using at most two truth-values.

Remark 3.6. Consider the mapping $t : \mathcal{V} \to \{T, F\}$ such that $t(v) = T$ iff $v \in \mathcal{D}$, for some logic $\mathcal{L}$. Note that:
$$\varphi \text{ separates } v_1 \text{ and } v_2 \text{ iff } t([\varphi](v_1)) \neq t([\varphi](v_2)). \qquad (**)$$
Now, suppose that φ_{mn} separates d_m and d_n (for $1 \leq m < n \leq i$), and ψ_{mn} separates u_m and u_n (for $1 \leq m < n \leq j$). Given a variable x and $d \in \mathcal{D}$, consider the equation:
$$t([\varphi_{mn}](x)) = q_{mn}^d$$
where $q_{mn}^d = t([\varphi_{mn}](d))$. Observe that $q_{mn}^d \in \{T, F\}$ and $q_{mn}^{d_m} \neq q_{mn}^{d_n}$, using $(**)$. Thus, if $\vec{\varphi}_d(x)$ is the sequence $(t([\varphi_{mn}](x)) = q_{mn}^d)_{1 \leq m < n \leq i}$, the distinguished truth-value d can then be characterized through the sequence of equations $Q_d(x)$: $(t(x) = T, \vec{\varphi}_d(x))$, where commas represent conjunctions. That is,
$$x = d \text{ iff } t(x) = T \wedge \bigwedge_{1 \leq m < n \leq i} t([\varphi_{mn}](x)) = q_{mn}^d$$

characterizes d in terms of membership to $\mathcal{D}$ or to $\mathcal{U}$ (or, equivalently, in terms of T/F), as desired. Analogously, if r_{mn}^u is $t([\psi_{mn}](u))$ for $1 \leq m < n \leq j$ and $u \in \mathcal{U}$, then the sequence of equations $R_u(x)$: $(t(x) = F, \vec{\psi}_u(x))$ characterizes u in terms of T/F, where $\vec{\psi}_u(x) = (t([\psi_{mn}](x)) = r_{mn}^u)_{1 \leq m < n \leq j}$. That is,
$$x = u \text{ iff } t(x) = F \wedge \bigwedge_{1 \leq m < n \leq j} t([\psi_{mn}](x)) = r_{mn}^u$$

characterizes u in terms of T/F using t.

Remark 3.7. If $\mathcal{D} = \{d\}$ then we simply write $x = d$ iff $t(x) = T$. Analogously, if $\mathcal{U} = \{u\}$ then we simply write $x = u$ iff $t(x) = F$.

Remark 3.8. For any given logic $\mathcal{L}$, the composition $b = t \circ \S$ gives us exactly Suszko's 2-valued reduction, viz. a 2-valued (usually non-truth-functional) semantic presentation of $\mathcal{L}$. Given a logic that respects our Separability Assumption 3.5, we will see in the next section how this 2-valued semantics can be mechanically written down in terms of 'dyadic semantics'. A later section will show how such semantics can provide us with classic-like tableaux for those same logics.

Example 3.9. Consider the n-valued logics of Łukasiewicz, $n > 2$, which can be formulated by way of:

$$\mathrm{L}_n = \langle \{0, \tfrac{1}{n-1}, \ldots, \tfrac{n-2}{n-1}, 1\}, \{\neg, \Rightarrow, \vee, \wedge\}, \{1\} \rangle.$$

The above operations over the truth-values can be defined as follows:

$$\neg v_1 := 1 - v_1; \qquad (v_1 \Rightarrow v_2) := \mathsf{Min}(1, 1 - v_1 + v_2);$$
$$(v_1 \vee v_2) := \mathsf{Max}(v_1, v_2); \quad (v_1 \wedge v_2) := \mathsf{Min}(v_1, v_2).$$

Consider now the particular case of L_5. Then we can take, for instance:

$$\psi_{0\frac{1}{4}} = \psi_{0\frac{2}{4}} = \psi_{0\frac{3}{4}} = \neg p; \quad \psi_{\frac{1}{4}\frac{2}{4}} = \psi_{\frac{1}{4}\frac{3}{4}} = (\neg p \Rightarrow p); \quad \psi_{\frac{2}{4}\frac{3}{4}} = (p \Rightarrow \neg p).$$

To save on notation, take $\triangle(p) = \psi_{\frac{1}{4}\frac{2}{4}}$ and $\triangledown(p) = \psi_{\frac{2}{4}\frac{3}{4}}$, and consider next the table:

v	$\neg v$	$\triangle(v)$	$\triangledown(v)$
0	1	0	1
$\frac{1}{4}$	$\frac{3}{4}$	$\frac{2}{4}$	1
$\frac{2}{4}$	$\frac{2}{4}$	1	1
$\frac{3}{4}$	$\frac{1}{4}$	1	$\frac{2}{4}$

Note that (the reduced version of) each $\vec{\psi}_k(x)$ is as follows:

$$\vec{\psi}_0(x) = \langle t(\neg x) = T, t(\triangle(x)) = F, t(\triangledown(x)) = T \rangle,$$
$$\vec{\psi}_{\frac{1}{4}}(x) = \langle t(\neg x) = F, t(\triangle(x)) = F, t(\triangledown(x)) = T \rangle,$$
$$\vec{\psi}_{\frac{2}{4}}(x) = \langle t(\neg x) = F, t(\triangle(x)) = T, t(\triangledown(x)) = T \rangle,$$
$$\vec{\psi}_{\frac{3}{4}}(x) = \langle t(\neg x) = F, t(\triangle(x)) = T, t(\triangledown(x)) = F \rangle.$$

We obtain thus the following characterizations of the truth-values:

$$x = 0 \text{ iff } t(x) = F \wedge t(\neg x) = T \wedge t(\triangle(x)) = F \wedge t(\triangledown(x)) = T,$$
$$x = \tfrac{1}{4} \text{ iff } t(x) = F \wedge t(\neg x) = F \wedge t(\triangle(x)) = F \wedge t(\triangledown(x)) = T,$$
$$x = \tfrac{2}{4} \text{ iff } t(x) = F \wedge t(\neg x) = F \wedge t(\triangle(x)) = T \wedge t(\triangledown(x)) = T,$$
$$x = \tfrac{3}{4} \text{ iff } t(x) = F \wedge t(\neg x) = F \wedge t(\triangle(x)) = T \wedge t(\triangledown(x)) = F.$$

Obviously, the sole distinguished truth-value 1 is characterized simply by:

$$x = 1 \quad \text{iff} \quad t(x) = T.$$

A similar procedure can be applied to all the remaining finite-valued logics of Łu-kasiewicz, making use for instance of the well-known Rosser-Turquette (definable) functions so as to produce the appropriate effective separations of truth-values.

4. Dyadic semantics

Suszko's Reduction is quite general, and it applies to any tarskian logic, be it truth-functional or not. In the next section we will exhibit our algorithmic reductive method for automatically obtaining 2-valued formulations of any sufficiently expressive finite-valued logic. To that purpose, it will be convenient to make use of an appropriate equational language, made explicit in the following.

Def. 4.1. A *gentzenian semantics* for a logic $\mathcal{L}$ is an adequate (sound and complete) set of 2-valued valuations $b : \mathcal{S} \to \{T, F\}$ given by conditional clauses $(\Phi \to \Psi)$ where both Φ and Ψ are (meta)formulas of the form $\top$ (top), $\bot$ (bottom) or:

$$b(\varphi_1^1) = w_1^1, \ldots, b(\varphi_1^{n_1}) = w_1^{n_1} \mid \ldots \mid b(\varphi_m^1) = w_m^1, \ldots, b(\varphi_m^{n_m}) = w_m^{n_m}. \quad (G)$$

Here, $w_i^j \in \{T, F\}$, each φ_i^j is a formula of $\mathcal{L}$, commas "," represent conjunctions, and bars "|" represent disjunctions. The (meta)logic governing these clauses is FOL, First-Order Classical Logic (further on, $\to$ will be used to represent the implication connective from this metalogic). We may alternatively write a clause of the form (G) as $\bigvee_{1 \le k \le m} \bigwedge_{1 \le s \le n_m} b(\varphi_k^s) = w_k^s$.

A dyadic semantics will consist in a specialization of a gentzenian semantics, in a deliberate intent to capture the computable class of such semantics, as follows. It should be noticed, at any rate, that not all decidable 2-valued semantics will come with a built-in gentzenian presentation. Moreover, as shown in Example 4.6, there are many logics that are characterizable by gentzenian or even by dyadic semantics, yet not by any genuinely finite-valued semantics.

Remark 4.2. (i) Given an algebra of formulas $\mathcal{S}$, an appropriate *measure of complexity* of these formulas may be defined as the output of some schematic mapping $\ell : \mathcal{S} \to \mathbb{N}$, with the restriction that $\ell(p_k) = 0$, for each $p_k \in ats$. As a particular case, the *canonical* measure of complexity of $\varphi = \otimes(\varphi_1, \ldots, \varphi_m)$ has the additional restriction that $\ell(\varphi) = 1 + \ell(\varphi_1) + \ldots + \ell(\varphi_m)$, for each $\otimes \in cct$.

(ii) Let $var : \mathcal{S} \to \mathsf{Pow}(ats)$ be a mapping that associates to each formula its set of atomic subformulas. Given an algebra of formulas $\mathcal{S}$, denote by $\mathcal{S}[n]$, for $n \ge 1$, the set $\mathcal{S}[n] = \{\varphi \in \mathcal{S} : var(\varphi) = \{p_1, \ldots, p_n\}\}$. There are surely non-empty (and possibly finite) families of formulas $(\psi_i)_{i \in I}$, for some $I = \{1, 2, \ldots\} \subseteq \mathbb{N}^+$, and there are $1 \le n_i \le \aleph_0$, for each $i \in I$, with $\psi_i \in \mathcal{S}[n_i]$, which cover the whole set of formulas up to some substitution, that is, such that $\mathcal{S} = \bigcup_{i \in I}\{\varepsilon(\psi_i) : \varepsilon \text{ is a substitution}\}$. A minimal example of such a covering family is given by $\{\otimes(p_1, \ldots, p_n) : \otimes \in \Sigma_n \text{ and } n \in \mathbb{N}\}$.

Def. 4.3. A logic $\mathcal{L}$ is said to be *quasi tabular* in case:

(i) There is some measure of complexity ℓ and there is some covering family of formulas $\{\psi_i\}_{i \in I}$, with $\psi_i \in \mathcal{S}[n_i]$, for some (possibly finite) set $I = \{1, 2, \ldots\} \subseteq \mathbb{N}^+$ such that for each ψ_i there is a finite sequence $\langle \phi_s^i \rangle_{s=1,\ldots,k_i}$ of formulas such that $var(\phi_s^i) \subseteq \{p_1, \ldots, p_{n_i}\}$, and $\ell(\phi_s^i) < \ell(\psi_i)$, for $1 \leq s \leq k_i$.

(ii) There is an adequate $|\mathcal{V}|$-valued set of valuations $\S : \mathcal{S} \to \mathcal{V}$ for $\mathcal{L}$, for some finite set of truth-values $\mathcal{V}$, such that for each $i \in I$ there is some recursive function $[.]_i : \mathcal{V}^{k_i} \to \mathcal{V}$ according to which, if $\phi = \varepsilon(\psi_i)$ for some substitution ε, then $\S(\phi) \bowtie_i \lceil \S(\varepsilon(\phi_1^i)), \ldots, \S(\varepsilon(\phi_{k_i}^i)) \rceil_i$ for every $\S$, where $\bowtie_i$ is one of the following partial ordering relations defined on $\mathcal{V}$: $=$, $\leq$, or $\geq$.

The reader will have remarked that the above definition of quasi-tabularity extends, in a sense, the usual Fregean notion of semantic compositionality.

Def. 4.4. A quasi tabular logic is called *tabular* in case ℓ can be taken to be the canonical measure of complexity and, accordingly, for each $i \in I$, one can take $\langle \phi_s^i \rangle_{s=1,\ldots,k_i}$ as the immediate subformulas of ψ_i. In that case, also, the covering set $\{\psi_i\}_{i \in I}$ can be taken to be the minimal one (check Remark 4.2(ii)), and each $\bowtie_i$ can be limited to the equality symbol $=$.

Tabular logics define exactly the class of *truth-functional logics*, given that the former logics are always genuinely n-valued, for some $1 \leq n \leq |\mathcal{V}|$.

Def. 4.5. A quasi tabular logic $\mathcal{L}$ is said to have a *dyadic semantics* in case the set $\mathcal{V}$ of Def. 4.3(ii) is $\{T, F\}$, and additionally $\mathcal{L}$ can be endowed with an adequate gentzenian semantics.

The class of quasi tabular logics is quite wide: Genuinely finite-valued logics are but a very special case of them, and the former class in fact coincides with the class of logics which can be given a so-called 'society semantics with complex base' (cf. [17]). It even includes logics that cannot be characterized as genuinely finite-valued, as the following example shows:

Example 4.6. Consider the paraconsistent logic $\mathcal{C}_1$ (cf. [14]). It is well known that this logic has no genuinely finite-valued characterizing semantics, though it *can* be decided by way of 'quasi matrices' (cf. [15]). In fact, a dyadic semantics for $\mathcal{C}_1$ is promptly available (cf. [9]). To that effect, recall that α° abbreviates $\neg(\alpha \wedge \neg\alpha)$ in $\mathcal{C}_1$, and consider the following bivaluational axioms (where $\sqcap, \sqcup, -$ are the usual lattice operators):

(4.6.1) $b(\neg\alpha) \geq -b(\alpha)$;
(4.6.2) $b(\neg\neg\alpha) \leq b(\alpha)$;
(4.6.3) $b(\alpha \wedge \beta) = b(\alpha) \sqcap b(\beta)$;
(4.6.4) $b(\alpha \vee \beta) = b(\alpha) \sqcup b(\beta)$;
(4.6.5) $b(\alpha \Rightarrow \beta) = -b(\alpha) \sqcup b(\beta)$;
(4.6.6) $b(\alpha^\circ) = -b(\alpha) \sqcup -b(\neg\alpha)$;
(4.6.7) $b((\alpha \otimes \beta)^\circ) \geq (-b(\alpha) \sqcup -b(\neg\alpha)) \sqcap (-b(\beta) \sqcup -b(\neg\beta))$, for $\otimes \in \{\wedge, \vee, \Rightarrow\}$.

184   C. Caleiro, W. Carnielli, M.E. Coniglio and J. Marcos

As it will be clear further on, in case it is possible to obtain a tableau decision procedure from a gentzenian semantics $\mathcal{B}$ for a logic $\mathcal{L}$ then $\mathcal{B}$ is a dyadic semantics for $\mathcal{L}$.

5. From finite matrices to dyadic valuations

Let $\otimes$ be some connective of $\mathcal{L}$; for the sake of simplicity, suppose that $\otimes$ is binary. If an entry of the truth-table for $\otimes$ states that $\otimes(v_1, v_2) = v$ then we can express this situation as follows:

$$\text{If } x = v_1 \text{ and } y = v_2, \text{ then } \otimes(x, y) = v.$$

Now, recall from Remark 3.6 the mapping $t : \mathcal{V} \to \{T, F\}$ such that $t(v) = T$ iff $v \in \mathcal{D}$. If the previous situation is expressed in terms of T/F using this mapping, we will get, respectively, systems of equations $E_{v_1}(x)$, $E_{v_2}(y)$ and $E_v(\otimes(x,y))$, and consequently the following statement in terms of T/F:

$$\text{if } E_{v_1}(x) \text{ and } E_{v_2}(y) \text{ then } E_v(\otimes(x,y)).$$

In the formal metalanguage of a gentzenian semantics (Def. 4.1), this statement is of the form:

$$\begin{aligned}
t([\beta_1](x)) = w_1, \ldots, t([\beta_m](x)) = w_m, \\
t([\gamma_1](y)) = w_1', \ldots, t([\gamma_{m'}](y)) = w_{m'}' \\
\to t([\delta_1](\otimes(x,y))) = w_1'', \ldots, t([\delta_{m''}](\otimes(x,y))) = w_{m''}'',
\end{aligned} \qquad (\ast\ast\ast)$$

where $w_n, w_{k'}', w_{s''}'' \in \{T, F\}$ for $1 \le n \le m$, $1 \le k' \le m'$ and $1 \le s'' \le m''$.

Now, suppose that v is $\S(\alpha)$ for some formula α. Then, using $(\ast)$ (check Remarks 3.2 and 3.8) we obtain:

$$t([\varphi](v)) = t([\varphi](\S(\alpha))) = t(\S(\varphi(p/\alpha))) = b(\varphi(p/\alpha))$$

for every formula $\varphi(p)$. Using this in $(\ast\ast\ast)$ we obtain an axiom for $\mathcal{B}$ of the form:

$$\begin{aligned}
b(\beta_1(p/\alpha)) = w_1, \ \ldots, \ b(\beta_m(p/\alpha)) = w_m, \\
b(\gamma_1(p/\beta)) = w_1', \ \ldots, \ b(\gamma_{m'}(p/\beta)) = w_{m'}' \\
\to b(\delta_1(p/\otimes(\alpha,\beta))) = w_1'', \ \ldots, \ b(\delta_{m''}(p/\otimes(\alpha,\beta))) = w_{m''}'',
\end{aligned}$$

for $w_n, w_{k'}', w_{s''}'' \in \{T, F\}$ etc. Obviously, we can repeat this process for each entry of each connective $\otimes$ of $\mathcal{L}$. For 0-ary connectives there is no input at the left-hand side; in such case, you should write conditional clauses of the form $(\top \to \Psi)$.

Example 5.1. In L_5 we have, for instance, the following entry in the truth-table for $\wedge$: If $v_1 = \frac{2}{4}$ and $v_2 = 1$ then $v_1 \wedge v_2 = \frac{2}{4}$. Or, in other words: If $\S(\alpha) = \frac{2}{4}$ and $\S(\beta) = 1$ then $\S(\alpha \wedge \beta) = \frac{2}{4}$, for any formulas α and β, and any homomorphism $\S$. From Example 3.9 we obtain, using t and $b = t \circ \S$:

$$b(\alpha) = F, \ b(\neg\alpha) = F, \ b(\triangle(\alpha)) = T, \ b(\triangledown(\alpha)) = T, \ b(\beta) = T$$
$$\to b(\alpha \wedge \beta) = F, \ b(\neg(\alpha \wedge \beta)) = F, \ b(\triangle(\alpha \wedge \beta)) = T, b(\triangledown(\alpha \wedge \beta)) = T.$$

As explained and illustrated above, each entry of the truth-table of each connective $\otimes$ of $\mathcal{L}$ determines an axiom for a gentzenian valuation $b : \mathcal{S} \rightarrow \{T, F\}$. We obtain thus, through the above method, a kind of unique (partial) 'dyadic print' of the original truth-functional logic.

Theorem 5.2. *Given a logic $\mathcal{L}$, let $\mathcal{B}$ be the set of gentzenian valuations $b : \mathcal{S} \rightarrow \{T, F\}$ satisfying the axioms obtained from the truth-tables of $\mathcal{L}$ using the above method, plus the following axioms:*

(C1): $\top \;\rightarrow\; b(\alpha) = T \mid b(\alpha) = F$;
(C2): $b(\alpha) = T, \; b(\alpha) = F \;\rightarrow\; \bot$;
(C3): $b(\alpha) = T \;\rightarrow\; \bigvee_{d \in \mathcal{D}} \bigwedge_{1 \leq m < n \leq i} b(\varphi_{mn}(p/\alpha)) = q_{mn}^{d}$;
(C4): $b(\alpha) = F \;\rightarrow\; \bigvee_{u \in \mathcal{U}} \bigwedge_{1 \leq m < n \leq j} b(\psi_{mn}(p/\alpha)) = r_{mn}^{u}$,

for every $\alpha \in \mathcal{S}$ (here, q_{mn}^{d} and r_{mn}^{u} are as in Remark 3.6). Then $b \in \mathcal{B}$ iff $b = t \circ \S$ for some homomorphism $\S : \mathcal{S} \rightarrow \mathbb{V}$.

Proof. Given $b \in \mathcal{B}$, define a homomorphism $\S : \mathcal{S} \rightarrow \mathbb{V}$ such that:

(i) $\S(\alpha) = d$ iff $b(\alpha) = T$ and $b(\varphi_{mn}(p/\alpha)) = q_{mn}^{d}$ for every $1 \leq m < n \leq i$;
(ii) $\S(\alpha) = u$ iff $b(\alpha) = F$ and $b(\psi_{mn}(p/\alpha)) = r_{mn}^{u}$ for every $1 \leq m < n \leq j$,

where α ranges over the atomic sentences $ats \in \mathcal{S}$. Note that $\mathcal{S}$ is well-defined as a total functional assignment because $b \in \mathcal{B}$ satisfies conditions (**C1**)–(**C2**) above. Since b satisfies all the axioms obtained from all the entries of the truth-tables of $\mathcal{L}$, it is straightforward to prove, by induction on the complexity of the formula $\alpha \in \mathcal{S}$, that (i) and (ii) hold when α ranges over all the formulas in $\mathcal{S}$. (Indeed, note that, in the light of conditions (**C2**)–(**C4**), given $b \in \mathcal{B}$ and $b(\alpha) = T$ we can conclude that there exists a unique $d \in \mathcal{D}$ such that $\bigwedge_{1 \leq m < n \leq i} b(\varphi_{mn}(p/\alpha)) = q_{mn}^{d}$; similarly, given $b(\alpha) = F$ we can conclude that there exists a unique $u \in \mathcal{U}$ such that $\bigwedge_{1 \leq m < n \leq j} b(\psi_{mn}(p/\alpha)) = r_{mn}^{u}$.) From this we obtain that $\S(\varphi) \in \mathcal{D}$ iff $b(\varphi) = T$, therefore $b = t \circ \S$ as desired. The converse (if $b = t \circ \S$ for some homomorphism $\S$, then $b \in \mathcal{B}$) is immediate. $\square$

Thus, a new 2-valued adequate semantics based on but two 'logical values' can now be seen to realize Suszko's Thesis, through the above constructive method.

Corollary 5.3. (i) *For every bivaluation $b : \mathcal{S} \rightarrow \{T, F\}$ in $\mathcal{B}$ there exists a homomorphism $\S_b : \mathcal{S} \rightarrow \mathbb{V}$ such that:*

$$\S_b(\alpha) \in \mathcal{D} \quad iff \quad b(\alpha) = T, \text{ for any } \alpha \in \mathcal{S}. \tag{1}$$

(ii) *For every $\S : \mathcal{S} \rightarrow \mathbb{V}$ there exists a $b_\S \in \mathcal{B}$ such that:*

$$b_\S(\alpha) = T \quad iff \quad \S(\alpha) \in \mathcal{D}, \text{ for any } \alpha \in \mathcal{S}. \tag{2}$$

We now have two notions of semantic entailment for $\mathcal{L}$. The first one, $\models$, uses the truth-tables given by $\mathbb{V}$ and its corresponding homomorphic valuations, whereas the second one, $\models_\mathcal{B}$, uses the related gentzenian semantics $\mathcal{B}$. But both notions are in a sense 'talking about the same thing':

Theorem 5.4. *The set $\mathcal{B}$ of gentzenian valuations for $\mathcal{L}$ is adequate, that is, for any $\Gamma \cup \{\varphi\} \subseteq \mathcal{S}$:*

$$\Gamma \models \varphi \quad \textit{iff} \quad \Gamma \models_{\mathcal{B}} \varphi.$$

Proof. Suppose that $\Gamma \models \varphi$, and let $b \in \mathcal{B}$ be such that $b(\Gamma) \subseteq \{T\}$, if possible. By Corollary 5.3(i) there exists a homomorphism $\S_b$ such that $\S_b(\Gamma) \subseteq \mathcal{D}$. By hypothesis we get $\S_b(\varphi) \in \mathcal{D}$, whence $b(\varphi) = T$ by (1). This shows that $\Gamma \models_{\mathcal{B}} \varphi$. The converse is proven in an analogous way, using Corollary 5.3(ii). $\qquad\square$

6. Some Illustrations

In this section we will give examples of gentzenian semantics for several genuinely finite-valued paraconsistent logics, obtained through applications of the reductive algorithm proposed in the last section. Instead of writing extensive lists of bivaluational axioms, one for each entry of each truth-table, plus some complementing axioms, we shall be using First-Order Classical Logic, FOL, in what follows, in order to manipulate and simplify the clauses written in our equational metalanguage. Moreover, we will often seek to reformulate things so as to make them more convenient for a tableaux-oriented approach, as in the next section.

Example 6.1. The paraconsistent logic $\mathbf{P}_3^1 = \langle \{0, \frac{1}{2}, 1\}, \{\neg, \Rightarrow\}, \{\frac{1}{2}, 1\} \rangle$, was introduced by Sette in [22] (where it was called P^1), having as truth-tables:

	0	$\frac{1}{2}$	1
$\neg$	1	1	0

$\Rightarrow$	0	$\frac{1}{2}$	1
0	1	1	1
$\frac{1}{2}$	0	1	1
1	0	1	1

Note that $\neg p$ separates $\frac{1}{2}$ and 1. Indeed:

$$[\neg p](1) = 0, \ [\neg p](\tfrac{1}{2}) = 1,$$

and $0 \sharp 1$. Thus:

$$
\begin{aligned}
x = 0 \quad &\text{iff} \quad t(x) = F; \\
x = \tfrac{1}{2} \quad &\text{iff} \quad t(x) = T, t(\neg x) = T; \\
x = 1 \quad &\text{iff} \quad t(x) = T, t(\neg x) = F.
\end{aligned}
$$

Applying our reductive algorithm to the truth-tables of $\neg$ and $\Rightarrow$ we may, after some simplification, obtain the following axioms for b:

(i) $b(\alpha) = F \rightarrow b(\neg \alpha) = T, b(\neg\neg\alpha) = F$;

(ii) $b(\alpha) = T, b(\neg\alpha) = T \rightarrow b(\neg\alpha) = T, b(\neg\neg\alpha) = F$;

(iii) $b(\alpha) = T, b(\neg\alpha) = F \rightarrow b(\neg\alpha) = F$;

(iv) $b(\alpha) = F \mid b(\beta) = T \rightarrow b(\alpha \Rightarrow \beta) = T, b(\neg(\alpha \Rightarrow \beta)) = F$;

(v) $b(\alpha) = T, b(\beta) = F \rightarrow b(\alpha \Rightarrow \beta) = F$.

In this case, axiom **(C3)** corresponds to $b(\alpha) = T \;\to\; b(\neg\alpha) = T \mid b(\neg\alpha) = F$, which can be derived from **(C1)**. Axiom **(C4)** corresponds to $b(\alpha) = F \to b(\neg\alpha) = T$, which is derivable from the above clause (i). Using FOL we may rewrite clauses (i)–(v) equivalently as:

(6.1.1) $b(\neg\alpha) = F \;\to\; b(\alpha) = T$;

(6.1.2) $b(\neg\neg\alpha) = T \;\to\; b(\neg\alpha) = F$;

(6.1.3) $b(\alpha \Rightarrow \beta) = T \;\to\; b(\alpha) = F \mid b(\beta) = T$;

(6.1.4) $b(\alpha \Rightarrow \beta) = F \;\to\; b(\alpha) = T,\ b(\beta) = F$;

(6.1.5) $b(\neg(\alpha \Rightarrow \beta)) = T \;\to\; b(\alpha) = T,\ b(\beta) = F$.

Note that (6.1.3)–(6.1.5) axiomatize a sort of 'classic-like' implication. Axioms (6.1.1)–(6.1.5) plus **(C1)**–**(C2)** characterize a dyadic semantics for $\mathbf{P}_3^1$.

Example 6.2. The paraconsistent logic $\mathbf{P}_4^1 = \langle\{0, \frac{1}{3}, \frac{2}{3}, 1\}, \{\neg, \Rightarrow\}, \{\frac{1}{3}, \frac{2}{3}, 1\}\rangle$, was introduced in [11] and [19], and studied under the name P^2 in [17]. The truth-tables of its connectives are as follows:

	0	$\frac{1}{3}$	$\frac{2}{3}$	1
$\neg$	1	$\frac{2}{3}$	1	0

$\Rightarrow$	0	$\frac{1}{3}$	$\frac{2}{3}$	1
0	1	1	1	1
$\frac{1}{3}$	0	1	1	1
$\frac{2}{3}$	0	1	1	1
1	0	1	1	1

It is easy to see that $\neg p$ separates 1 and $\frac{1}{3}$, as well as 1 and $\frac{2}{3}$. On the other hand, $\neg\neg p$ separates $\frac{1}{3}$ and $\frac{2}{3}$. From this we get:

$x = 0$ iff $t(x) = F$;

$x = \frac{1}{3}$ iff $t(x) = T, t(\neg x) = T, t(\neg\neg x) = T$;

$x = \frac{2}{3}$ iff $t(x) = T, t(\neg x) = T, t(\neg\neg x) = F$;

$x = 1$ iff $t(x) = T, t(\neg x) = F, t(\neg\neg x) = T$.

From the truth-table for $\neg$ we obtain, after applying FOL:

(6.2.1) $b(\neg\alpha) = F \;\to\; b(\alpha) = T$;

(6.2.2) $b(\neg\neg\alpha) = T \;\to\; b(\alpha) = T$;

(6.2.3) $b(\neg\neg\neg\alpha) = T \;\to\; b(\neg\neg\alpha) = F$.

Once again, axiom **(C3)** is derivable from **(C1)**, and axiom **(C4)** is derivable from the clauses above. The implication $\Rightarrow$ is again 'classic-like', in the same sense as in the last example. Therefore, axioms (6.2.1)–(6.2.3), (6.1.3)–(6.1.5) and **(C1)**–**(C2)** characterize together a dyadic semantics for $\mathbf{P}_4^1$. Similar procedures can be applied to each paraconsistent logic of the hierarchy $\mathbf{P}_{n+2}^1(= P^n$, from [17]), for $n \in \mathbb{N}^+$.

Example 6.3. Having already used negation in the two above examples in order to separate truth-values, let us now make it differently. Consider the paraconsistent propositional logic $\mathbf{LFI1} = \langle\{0, \frac{1}{2}, 1\}, \{\neg, \bullet, \Rightarrow, \wedge, \vee\}, \{\frac{1}{2}, 1\}\rangle$, studied in detail in [13], whose matrices are:

	0	$\frac{1}{2}$	1
$\neg$	1	$\frac{1}{2}$	0
$\bullet$	0	1	0

$\Rightarrow$	0	$\frac{1}{2}$	1
0	1	1	1
$\frac{1}{2}$	0	$\frac{1}{2}$	1
1	0	$\frac{1}{2}$	1

plus conjunction $\wedge$ and disjunction $\vee$ defined as in Łukasiewicz's logics (see Example 3.9). Clearly, $\bullet p$ separates 1 and $\frac{1}{2}$. Thus:

$x = 0$ iff $t(x) = F$;

$x = \frac{1}{2}$ iff $t(x) = T$, $t(\bullet x) = T$;

$x = 1$ iff $t(x) = T$, $t(\bullet x) = F$.

From the truth-table for $\neg$, and using FOL, we obtain:

(6.3.1) $b(\neg\alpha) = T \;\rightarrow\; b(\alpha) = F \mid b(\bullet\alpha) = T$;

(6.3.2) $b(\neg\alpha) = F \;\rightarrow\; b(\alpha) = T,\; b(\bullet\alpha) = F$.

Axiom **(C3)** is again derivable from **(C1)**; axiom **(C4)** is derivable from (6.3.2). Now, these are the axioms for $\bullet$:

(6.3.3) $b(\bullet\alpha) = T \;\rightarrow\; b(\alpha) = T$;

(6.3.4) $b(\bullet\bullet\alpha) = T \;\rightarrow\; b(\bullet\alpha) = F$;

(6.3.5) $b(\bullet\neg\alpha) = T \;\rightarrow\; b(\bullet\alpha) = T$;

(6.3.6) $b(\bullet\neg\alpha) = F \;\rightarrow\; b(\neg\alpha) = F \mid b(\alpha) = F$.

From the truth-tables for the binary connectives, and using FOL, we obtain:

(6.3.7) $b(\alpha \wedge \beta) = T \;\rightarrow\; b(\alpha) = T,\; b(\beta) = T$;

(6.3.8) $b(\alpha \wedge \beta) = F \;\rightarrow\; b(\alpha) = F \mid b(\beta) = F$;

(6.3.9) $b(\alpha \vee \beta) = T \;\rightarrow\; b(\alpha) = T \mid b(\beta) = T$;

(6.3.10) $b(\alpha \vee \beta) = F \;\rightarrow\; b(\alpha) = F,\; b(\beta) = F$;

(6.3.11) $b(\alpha \Rightarrow \beta) = T \;\rightarrow\; b(\alpha) = F \mid b(\beta) = T$;

(6.3.12) $b(\alpha \Rightarrow \beta) = F \;\rightarrow\; b(\alpha) = T,\, b(\beta) = F$.

To those we may add, furthermore:

(6.3.13) $b(\bullet(\alpha \wedge \beta)) = T$
    $\rightarrow\; b(\alpha) = T,\; b(\bullet\beta) = T \mid b(\beta) = T,\; b(\bullet\alpha) = T$;

(6.3.14) $b(\bullet(\alpha \wedge \beta)) = F$
    $\rightarrow\; b(\alpha) = F \mid b(\beta) = F \mid b(\alpha) = T,\; b(\bullet\alpha) = F,\; b(\beta) = T,\; b(\bullet\beta) = F$;

(6.3.15) $b(\bullet(\alpha \vee \beta)) = T$
    $\rightarrow\; b(\alpha) = F,\; b(\bullet\beta) = T \mid b(\beta) = F,\; b(\bullet\alpha) = T \mid b(\bullet\alpha) = T,\; b(\bullet\beta) = T$;

(6.3.16) $b(\bullet(\alpha \vee \beta)) = F$
    $\rightarrow\; b(\alpha) = F,\; b(\beta) = F \mid b(\alpha) = T,\; b(\bullet\alpha) = F \mid b(\beta) = T,\; b(\bullet\beta) = F$;

(6.3.17) $b(\bullet(\alpha \Rightarrow \beta)) = T \;\rightarrow\; b(\alpha) = T,\; b(\bullet\beta) = T$;

(6.3.18) $b(\bullet(\alpha \Rightarrow \beta)) = F \;\rightarrow\; b(\alpha) = F \mid b(\bullet\beta) = F$.

So, if the above axioms are taken together with **(C1)**–**(C2)**, then we obtain a natural dyadic semantics for **LFI1**. Two slightly different (non-gentzenian) bivaluation semantics for **LFI1** were explored in [13].

Example 6.4. Belnap's paraconsistent and paracomplete 4-valued logic (cf. [2]), $B_4 = \langle \{0, \frac{1}{3}, \frac{2}{3}, 1\}, \{\neg, \wedge, \vee\}, \{\frac{2}{3}, 1\}\rangle$, can be presented by way of the following matrices:

¬		$\frac{1}{3}$	$\frac{2}{3}$	
	0			1

| ¬ | 0 | $\frac{2}{3}$ | $\frac{1}{3}$ | 1 |

$\wedge$	0	$\frac{1}{3}$	$\frac{2}{3}$	1
0	0	0	0	0
$\frac{1}{3}$	0	$\frac{1}{3}$	0	$\frac{1}{3}$
$\frac{2}{3}$	0	0	$\frac{2}{3}$	$\frac{2}{3}$
1	0	$\frac{1}{3}$	$\frac{2}{3}$	1

$\vee$	0	$\frac{1}{3}$	$\frac{2}{3}$	1
0	0	$\frac{1}{3}$	$\frac{2}{3}$	1
$\frac{1}{3}$	$\frac{1}{3}$	$\frac{1}{3}$	1	1
$\frac{2}{3}$	$\frac{2}{3}$	1	$\frac{2}{3}$	1
1	1	1	1	1

Clearly, $\neg p$ separates 1 and $\frac{2}{3}$ and also separates $\frac{1}{3}$ and 1. Thus:

$x = 0$ iff $t(x) = F, t(\neg x) = F$;
$x = \frac{1}{3}$ iff $t(x) = F, t(\neg x) = T$;
$x = \frac{2}{3}$ iff $t(x) = T, t(\neg x) = F$;
$x = 1$ iff $t(x) = T, t(\neg x) = T$.

Now, from the truth-table for $\neg$, and using FOL, we obtain:

(6.4.1) $b(\neg\neg\alpha) = T \;\to\; b(\alpha) = T$;
(6.4.2) $b(\neg\neg\alpha) = F \;\to\; b(\alpha) = F$.

Both axioms **(C3)** and **(C4)** are now derivable from **(C1)**. From the truth-tables of conjunction and disjunction, using FOL, we obtain the positive clauses (6.3.7)–(6.3.10) again, but also:

(6.4.3) $b(\neg(\alpha \wedge \beta)) = T \;\to\;$ $b(\alpha) = F, \; b(\neg\alpha) = T, \; b(\beta) = F, \; b(\neg\beta) = T \mid$
$b(\alpha) = F, \; b(\neg\alpha) = T, \; b(\beta) = T, \; b(\neg\beta) = T \mid$
$b(\alpha) = T, \; b(\neg\alpha) = T, \; b(\beta) = F, \; b(\neg\beta) = T \mid$
$b(\alpha) = T, \; b(\neg\alpha) = T \; b(\beta) = T, \; b(\neg\beta) = T$;

(6.4.4) $b(\neg(\alpha \wedge \beta)) = F \;\to\;$ $b(\alpha) = F, \; b(\neg\alpha) = F \mid b(\alpha) = T, \; b(\neg\alpha) = F \mid$
$b(\beta) = F, \; b(\neg\beta) = F \mid b(\beta) = T, \; b(\neg\beta) = F$;

(6.4.5) $b(\neg(\alpha \vee \beta)) = T \;\to\;$ $b(\alpha) = F, \; b(\neg\alpha) = T \mid b(\alpha) = T, \; b(\neg\alpha) = T \mid$
$b(\beta) = F, \; b(\neg\beta) = T \mid b(\beta) = T, \; b(\neg\beta) = T$;

(6.4.6) $b(\neg(\alpha \vee \beta)) = F \;\to\;$ $b(\alpha) = F, \; b(\neg\alpha) = F, \; b(\beta) = F, \; b(\neg\beta) = F \mid$
$b(\alpha) = F, \; b(\neg\alpha) = F, \; b(\beta) = T, \; b(\neg\beta) = F \mid$
$b(\alpha) = T, \; b(\neg\alpha) = F, \; b(\beta) = F, \; b(\neg\beta) = F \mid$
$b(\alpha) = T, \; b(\neg\alpha) = F \; b(\beta) = T, \; b(\neg\beta) = F$.

A dyadic semantics for B_4 is given by the above axioms, plus **(C1)**–**(C2)**.

7. Application: tableaux for logics with dyadic semantics

In the examples from the last section we found axioms for the set $\mathcal{B}$ of bivaluation mappings b (defining a gentzenian semantics for a genuinely finite-valued logic $\mathcal{L}$) expressed as conditional clauses of the form:

$$b(\alpha) = w$$
$$\rightarrow b(\alpha_1^1) = w_1^1, \ldots, b(\alpha_1^{n_1}) = w_1^{n_1} \mid \ldots \mid b(\alpha_m^1) = w_m^1, \ldots, b(\alpha_m^{n_m}) = w_m^{n_m},$$

where $w, w_k^s \in \{T, F\}$ and α_k^s has smaller complexity, under some appropriate measure (recall Remark 4.2 and Def. 4.3), than α. Each clause as above generates a tableau rule for $\mathcal{L}$ as follows: Translate $b(\beta) = T$ as the signed formula $T(\beta)$, and $b(\beta) = F$ as the signed formula $F(\beta)$. Then, a conditional clause such as the one above induces the following tableau-rule:

$$w(\alpha)$$

$$\begin{array}{ccc} \diagup & \cdots & \diagdown \\ w_1^1(\alpha_1^1) & & w_m^1(\alpha_m^1) \\ \vdots & & \vdots \\ w_1^{n_1}(\alpha_1^{n_1}) & & w_m^{n_m}(\alpha_m^{n_m}) \end{array}$$

where $w, w_k^s \in \{T, F\}$. In that case, it is routine to prove that the set of tableau rules for $\mathcal{L}$ obtained from the clauses for $\mathcal{B}$ characterizes a sound and complete tableau system for $\mathcal{L}$ (check [7] for details). We are supposing that there exists a basic common rule known as *branching rule*, as follows:

$$\frac{\cdots}{T(\varphi) \mid F(\varphi)}$$

This rule is generated by clause (C1) of Theorem 5.2. In certain cases it may be possible to dispense with such rule, but taking into consideration that tableau rules are not mandatory but permissive there is little loss of generality in keeping such rule. The branching rule is not analytic, but can be bounded in certain cases so as to guarantee the termination of the decidable tableau procedure. Moreover, the variables occurring in the formula φ must in general be contained in the finite collection of variables occurring in the tableau branch.

The structural similarity between the tableau rules so obtained and the classical ones is not fortuitous. Applying the above idea to the gentzenian semantics obtained in the last section for a large class of many-valued logics, one can devise two-signed tableau systems for them. Many-signed tableau systems for many-valued logics, constructed with the help of their many truth-values used as labels may be obtained as in [10]. Here, though, we learn that we can forget about those 'algebraic truth-values' and work only with the 'logical values' T and F, just like in the classical case. While the former many-signed tableaux enjoy the so-called *subformula property*, according to which each formula α_k^s obtained from the application to α of a tableau rule as the one above is a subformula of the initial formula α, the latter related two-signed tableaux obtained through our method will often fail this property, reflecting the loss of the truth-functionality of the many-valued homomorphisms in transforming them into bivaluations. We will still have, though, a *shortening property* which is as advantageous for efficiency as the subformula property: Each formula α_k^s will be *less complex* (under some appropriate measure, recall Def. 4.3) than the initial formula α being analyzed by the tableau rules, the only exception being the above mentioned branching rule.

Example 7.1. The following set of rules characterizes a tableau system for the paraconsistent logic $\mathbf{P}_3^1$, according to clauses (6.1.1)–(6.1.5) of Example 6.1:

$$(7.1.1) \quad \frac{F(\neg\alpha)}{T(\alpha)} \qquad (7.1.2) \quad \frac{T(\neg\neg\alpha)}{F(\neg\alpha)}$$

$$(7.1.3) \quad \frac{T(\alpha \Rightarrow \beta)}{F(\alpha) \mid T(\beta)} \qquad (7.1.4) \quad \frac{F(\alpha \Rightarrow \beta)}{T(\alpha),\, F(\beta)} \qquad (7.1.5) \quad \frac{T(\neg(\alpha \Rightarrow \beta))}{T(\alpha),\, F(\beta)}$$

Example 7.2. Following Example 6.2, an adequate set of tableau rules for the paraconsistent logic $\mathbf{P}_4^1$ is given by (7.1.3)–(7.1.5) plus:

$$(7.2.1) \quad \frac{F(\neg\alpha)}{T(\alpha)} \qquad (7.2.2) \quad \frac{T(\neg\neg\alpha)}{T(\alpha)} \qquad (7.2.3) \quad \frac{T(\neg\neg\neg\alpha)}{F(\neg\neg\alpha)}$$

Example 7.3. Here is a tableau system for the paraconsistent logic **LFI1** (see Example 6.3), based on its dyadic semantics:

$$(7.3.1) \quad \frac{T(\neg\alpha)}{F(\alpha) \mid T(\bullet\alpha)} \qquad (7.3.2) \quad \frac{F(\neg\alpha)}{T(\alpha), F(\bullet\alpha)} \qquad (7.3.3) \quad \frac{T(\bullet\alpha)}{T(\alpha)}$$

$$(7.3.4) \quad \frac{T(\bullet\bullet\alpha)}{F(\bullet\alpha)} \qquad (7.3.5) \quad \frac{T(\bullet\neg\alpha)}{T(\bullet\alpha)} \qquad (7.3.6) \quad \frac{F(\bullet\neg\alpha)}{F(\neg\alpha) \mid F(\alpha)}$$

$$(7.3.7) \quad \frac{T(\alpha \wedge \beta)}{T(\alpha),\, T(\beta)} \qquad (7.3.8) \quad \frac{F(\alpha \wedge \beta)}{F(\alpha) \mid F(\beta)}$$

$$(7.3.9) \quad \frac{T(\alpha \vee \beta)}{T(\alpha) \mid T(\beta)} \qquad (7.3.10) \quad \frac{F(\alpha \vee \beta)}{F(\alpha),\, F(\beta)}$$

$$(7.3.11) \quad \frac{T(\alpha \Rightarrow \beta)}{F(\alpha) \mid T(\beta)} \qquad (7.3.12) \quad \frac{F(\alpha \Rightarrow \beta)}{T(\alpha), F(\beta)}$$

$$(7.3.13) \quad \frac{T(\bullet(\alpha \wedge \beta))}{\begin{array}{c} T(\alpha), \mid T(\beta), \\ T(\bullet\beta) \mid T(\bullet\alpha) \end{array}} \qquad (7.3.14) \quad \frac{F(\bullet(\alpha \wedge \beta))}{\begin{array}{c} F(\alpha) \mid F(\beta) \mid T(\alpha), T(\beta), \\ \mid F(\bullet\alpha), F(\bullet\beta) \end{array}}$$

$$(7.3.15) \quad \frac{T(\bullet(\alpha \vee \beta))}{\begin{array}{c} F(\alpha), \mid F(\beta), \mid T(\bullet\alpha), \\ T(\bullet\beta) \mid T(\bullet\alpha) \mid T(\bullet\beta) \end{array}} \qquad (7.3.16) \quad \frac{F(\bullet(\alpha \vee \beta))}{\begin{array}{c} F(\alpha), \mid T(\alpha), \mid T(\beta), \\ F(\beta) \mid F(\bullet\alpha) \mid F(\bullet\beta) \end{array}}$$

$$(7.3.17) \quad \frac{T(\bullet(\alpha \Rightarrow \beta))}{T(\alpha), T(\bullet\beta)} \qquad (7.3.18) \quad \frac{F(\bullet(\alpha \Rightarrow \beta))}{F(\alpha) \mid F(\bullet\beta)}$$

192   C. Caleiro, W. Carnielli, M.E. Coniglio and J. Marcos

Compare this tableau system for **LFI1** with the tableau system for this same logic presented in [12]. The latter is based on a non-gentzenian semantics. As a result, (decidable) tableaux without the shortening property (in fact, tableaux allowing for loops) were thereby obtained.

Example 7.4. A tableau system for Belnap's 4-valued logic (see Example 6.4), B_4, can be obtained by adding to (7.3.7)–(7.3.10) the following rules:

$$(7.4.1) \ \frac{T(\neg\neg\alpha)}{T(\alpha)} \qquad (7.4.2) \ \frac{F(\neg\neg\alpha)}{F(\alpha)}$$

$$(7.4.3) \ \frac{T(\neg(\alpha \wedge \beta))}{F(\alpha),T(\neg\alpha), \ | \ F(\alpha),T(\neg\alpha), \ | \ T(\alpha),T(\neg\alpha), \ | \ T(\alpha),T(\neg\alpha), \\ F(\beta),T(\neg\beta) \ | \ T(\beta),T(\neg\beta) \ | \ F(\beta),T(\neg\beta) \ | \ T(\beta),T(\neg\beta)}$$

$$(7.4.4) \ \frac{F(\neg(\alpha \wedge \beta))}{F(\alpha),F(\neg\alpha) \ | \ T(\alpha),F(\neg\alpha) \ | \ F(\beta),F(\neg\beta) \ | \ T(\beta),F(\neg\beta)}$$

$$(7.4.5) \ \frac{T(\neg(\alpha \vee \beta))}{F(\alpha),T(\neg\alpha) \ | \ T(\alpha),T(\neg\alpha) \ | \ F(\beta),T(\neg\beta) \ | \ T(\beta),T(\neg\beta)}$$

$$(7.4.6) \ \frac{F(\neg(\alpha \vee \beta))}{F(\alpha),F(\neg\alpha), \ | \ F(\alpha),F(\neg\alpha), \ | \ T(\alpha),F(\neg\alpha), \ | \ T(\alpha),F(\neg\alpha), \\ F(\beta),F(\neg\beta) \ | \ T(\beta),F(\neg\beta) \ | \ F(\beta),F(\neg\beta) \ | \ T(\beta),F(\neg\beta)}$$

As done in [5], similar algorithmic procedures can be devised so as to provide adequate sequent systems to all the 2-valued semantics hereby constructed.

8. Conclusions

While Suszko's *Thesis* is a philosophical stance concerning the scope of Universal Logic as a general theory of logical structures (cf. [3]), Suszko's *Reduction* is presented in this paper as a general non-constructive result about the comprehensive class of tarskian logics.

We have exhibited here a method for the effective implementation of Suszko's Reduction by transforming any finite-valued truth-functional semantics whose truth-values can be individualized in the sense of Assumption 3.5 into homologous 2-valued semantics. The specific form of the gentzenian axioms we obtain permits us then to automatically define a (decidable) tableau system for each logic subjected to that 2-valued reduction. The same methods can be applied to many other well-known logics such as Łukasiewicz's Ł$_n$, Kleene's K_3, Gödel's G_3 etc. Our reductive method builds bulk in the reductive results from [20, 21] and [1].

It is an open problem to extend our 2-valued reductive procedure so as to cover other classes of logics such as modal or infinite-valued logics.

References

[1] D. Batens, *A bridge between two-valued and many-valued semantic systems: n-tuple semantics*, Proceedings of the XII International Symposium on Multiple-Valued Logic, IEEE Computer Science Press, 1982, pp. 318–322.

[2] N. D. Belnap, *A useful four-valued logic*, Modern Uses of Multiple-Valued Logic (J. M. Dunn, ed.), D. Reidel Publishing, Boston, 1977, pp. 8–37.

[3] J.-Y. Béziau, *Universal Logic*, Logica'94, Proceedings of the VIII International Symposium (T. Childers and O. Majers, eds.), Czech Academy of Science, Prague, CZ, 1994, pp. 73–93.

[4] J.-Y. Béziau, *Recherches sur la logique abstraite: les logiques normales*, Acta Universitatis Wratislaviensis no. 2023, Logika **18** (1998), 105–114.

[5] J.-Y. Béziau, *Sequents and bivaluations*, Logique et Analyse (N.S.) **44** (2001), no. 176, 373–394.

[6] C. Caleiro, W. A. Carnielli, M. E. Coniglio, and J. Marcos, *Dyadic semantics for many-valued logics*, Preprint available at: http://wslc.math.ist.utl.pt/ftp/pub/CaleiroC/03-CCCM-dyadic2.pdf.

[7] C. Caleiro, W. A. Carnielli, M. E. Coniglio, and J. Marcos, *How many logical values are there? Dyadic semantics for many-valued logics*, Preprint.

[8] C. Caleiro, W. A. Carnielli, M. E. Coniglio, and J. Marcos, *Suszko's Thesis and dyadic semantics*, Preprint available at: http://wslc.math.ist.utl.pt/ftp/pub/CaleiroC/03-CCCM-dyadic1.pdf.

[9] C. Caleiro and J. Marcos, *Non-truth-functional fibred semantics*, Proceedings of the International Conference on Artificial Intelligence (IC-AI'2001), held in Las Vegas, USA, June 2001 (H. R. Arabnia, ed.), vol. II, CSREA Press, Athens GA, USA, 2001, pp. 841–847. http://wslc.math.ist.utl.pt/ftp/pub/CaleiroC/01-CM-fiblog10.ps.

[10] W. A. Carnielli, *Systematization of the finite many-valued logics through the method of tableaux*, The Journal of Symbolic Logic **52** (1987), 473–493.

[11] W. A. Carnielli and M. Lima-Marques, *Society semantics for multiple-valued logics*, Advances in Contemporary Logic and Computer Science (W. A. Carnielli and I. M. L. D'Ottaviano, eds.), Contemporary Mathematics Series, vol. 235, American Mathematical Society, 1999, pp. 33–52.

[12] W. A. Carnielli and J. Marcos, *Tableaux for logics of formal inconsistency*, Proceedings of the 2001 International Conference on Artificial Intelligence (IC-AI'2001), held in Las Vegas, USA, June 2001 (H. R. Arabnia, ed.), vol. II, CSREA Press, Athens GA, USA, 2001, pp. 848–852. http://logica.rug.ac.be/~joao/Publications/Congresses/tableauxLFIs.pdf.

[13] W. A. Carnielli, J. Marcos, and S. de Amo, *Formal inconsistency and evolutionary databases*, Logic and Logical Philosophy **8** (2000), 115–152. http://www.cle.unicamp.br/e-prints/abstract_6.htm.

[14] N. C. A. da Costa, *Calculs propositionnels pour les systèmes formels inconsistants*, Comptes Rendus d'Academie des Sciences de Paris **257** (1963), 3790–3792.

[15] N. C. A. da Costa and E. H. Alves, *A semantical analysis of the calculi C_n*, Notre Dame Journal of Formal Logic **18** (1977), 621–630.

194   C. Caleiro, W. Carnielli, M.E. Coniglio and J. Marcos

[16] N. C. A. da Costa, J.-Y. Béziau, and O. A. S. Bueno, *Malinowski and Suszko on many-valued logics: On the reduction of many-valuedness to two-valuedness*, Modern Logic **3** (1996), 272–299.

[17] V. L. Fernández and M. E. Coniglio, *Combining valuations with society semantics*, Journal of Applied Non-Classical Logics **13** (2003), no. 1, 21–46. http://www.cle.unicamp.br/e-prints/abstract_11.html.

[18] G. Malinowski, *Many-Valued Logics*, Oxford Logic Guides 25, Clarendon Press, Oxford, 1993.

[19] J. Marcos, *Possible-Translations Semantics (in Portuguese)*, Master's thesis, State University of Campinas, Brazil, 1999. http://www.cle.unicamp.br/students/J.Marcos/

[20] D. Scott, *Background to formalisation*, Truth, Syntax and Modality (H. Leblanc, ed.), North-Holland, Amsterdam, 1973, pp. 244–273.

[21] D. Scott, *Completeness and axiomatizability in many-valued logic*, Proceedings of Tarski Symposium (L. Henkin *et. al.*, ed.), Proceedings of Symposia in Pure Mathematics, vol.25, Berkeley 1971, 1974, pp. 411–436.

[22] A. M. Sette, *On the propositional calculus P^1*, Mathematica Japonicae **18** (1973), 173–180.

[23] R. Suszko, *Abolition of the Fregean Axiom*, Logic Colloquium: Symposium on Logic held at Boston, 1972–73 (R. Parikh, ed.), Lecture Notes in Mathematics, vol. 453, Springer-Verlag, 1972, pp. 169–239.

[24] R. Suszko, *Remarks on Łukasiewicz's three-valued logic*, Bulletin of the Section of Logic **4** (1975), 87–90.

[25] R. Suszko, *The Fregean axiom and Polish mathematical logic in the 1920's*, Studia Logica **36** (1977), 373–380.

[26] M. Tsuji, *Many-valued logics and Suszko's Thesis revisited*, Studia Logica **60** (1998), no. 2, 299–309.

[27] R. Wójcicki, *Logical matrices strongly adequate for structural sentential calculi*, Bulletin de l'Academie Polonaise des Sciences, *Série des Sciences Mathématiques, Astronomiques et Physiques* **17** (1969), 333–335.

Carlos Caleiro
SQIG / IT / IST, Av. Rovisco Pais, 1049-001 Lisbon, Portugal
URL: http://slc.math.ist.utl.pt/ccal.html
e-mail: ccal@math.ist.utl.pt

Walter Carnielli and Marcelo E. Coniglio
CLE and IFCH / Unicamp, CP 6133 – Barão Geraldo, 13083-970 Campinas, SP, Brazil
and SQIG / IT / IST, Av. Rovisco Pais, 1049-001 Lisbon, Portugal
URL: http://www.cle.unicamp.br/prof/carnielli/
URL: http://www.cle.unicamp.br/prof/coniglio/
e-mail: carniell@cle.unicamp.br, coniglio@cle.unicamp.br

João Marcos
DIMAp / UFRN, Brazil
and SQIG / IT / IST, Av. Rovisco Pais, 1049-001 Lisbon, Portugal
URL: http://geocities.com/jm_logica/
e-mail: jmarcos@dimap.ufrn.br

J.-Y. Beziau (Ed.), *Logica Universalis*, *2nd edition*, 195–224
© 2007 Birkhäuser Verlag Basel/Switzerland

Friendliness and Sympathy in Logic[†]

David Makinson

Abstract. We define and examine a notion of *logical friendliness*, which is a broadening of the familiar notion of classical consequence. The concept is studied first in its simplest form, and then in a syntax-independent version, which we call *sympathy*. We also draw attention to the surprising number of familiar notions and operations with which it makes contact, providing a new light in which they may be seen.

1. Friendliness

1.1. Rationale, Definition, Notation

Recall the definition of classical consequence in propositional logic. Let A be any set of formulae, and x any individual formula. Then x is said to be a classical consequence of A, written $A \vdash x$, iff for every valuation v on all letters of the language, if $v(A) = 1$ then $v(x) = 1$.

Trivially, the only letters that count here are those occurring in A or in x. So the definition may be rephrased as: $A \vdash x$ iff for every partial valuation v on $E(A,x)$, if $v(A) = 1$ then $v(x) = 1$. Equivalently again, $A \vdash x$ iff for every partial valuation v on $E(A)$, if $v(A) = 1$ then $v^+(x) = 1$ for every extension v^+ to $E(A, x)$.

Expressed in this last way, classical consequence is a $\forall\forall$ concept. It is natural to ask: what does the corresponding $\forall\exists$ concept look like, and how does it behave? This simple question, born of no more than curiosity, is the starting point of our investigation.

The definition is straightforward:

[†] This paper revises and extends the version that appeared in the first edition of *Logica Universalis*. Specifically, it adds several new sections (1.8-1.10, 3.5-3.6, and all of part 2) as well as additional material in other sections (notably the axiomatization of friendliness in 1.5, a much stronger version of compactness in 1.6, more information about interpolant formulae in 1.7 and 3.5, and counterexamples to proof by exhaustion and to compactness for sympathy in 3.2). The present version also appeared as part of a festschrift for Dov Gabbay, see [Makinson 2005a].

- We say that A is *friendly* to x and write $A \approx x$, iff every partial valuation v on $E(A)$ with $v(A) = 1$ may be extended to a partial valuation v^+ on $E(A, x)$ with $v^+(x) = 1$.

- Equivalently: iff for every partial valuation v on $E(A)$ with $v(A) = 1$ there is a partial valuation w on $E(x)$ agreeing with v on letters in $E(A) \cap E(x)$, with $w(x) = 1$.

- Equivalently: iff for every valuation v on the set E of all elementary letters of the language with $v(A) = 1$ there is a valuation w (on all letters) agreeing with v on letters in $E(A)$, with $w(x) = 1$.

The notation used in these definitions is fairly straightforward, but we state it explicitly for reference. We use lower case $a, b, \ldots, x, y, \ldots$, to range over *formulae* of classical propositional logic. It will be convenient to include the zero-ary falsum $\perp$ among the primitive connectives. *Sets of formulae* are denoted by upper case letters $A, B, \ldots, X, Y, \ldots$, reserving L for the set of all formulae, E for the set of all elementary letters, and $F, G, \ldots$ for subsets of the elementary letters. For any formula a, we write $E(a)$ to mean the set of all *elementary letters* occurring in a. Similarly for sets A of formulae. For any set A of formulae, L_A stands for the sub-language generated by $E(A)$, i.e., the set of all formulae y with $E(y) \subseteq E(A)$. Thus $L_A = L_{E(A)}$.

Classical consequence is written as $\vdash$ when treated as a relation, Cn when viewed as an operation. The relation of *classical equivalence* is written $\dashv\vdash$. When we speak of a *valuation*, we always mean a Boolean valuation, i.e., a function into $\{0,1\}$ defined on the entire set E of elementary letters of the language and extended to cover all formulae in the usual way. A *partial valuation* is a restriction of a valuation to a subset of E.

To lighten notation, we follow the common convention of usually writing A, x for $A \cup \{x\}$. $A \vdash B$ is short for '$A \vdash b$ for all $b \in B$'. Also, $v(A) = 1$ is short for '$v(a) = 1$ for all $a \in A$', while $v(A) = 0$ is short for '$v(a) = 0$ for some $a \in A$'.

1.2. Remarks on the Definition

Of the three equivalent ways of defining friendliness, we will usually be working with the first. Thus throughout the paper (except for the appendix) we will be talking about *partial* valuations rather than full ones. In this context, it is essential to keep in mind some fine distinctions, which are easy to overlook because they are without much significance for classical consequence.

- $E(a)$ is the set of all elementary letters *actually occurring* in a, rather than the least set of letters needed to get a formula classically equivalent to a. For example, if $a = p \wedge (q \vee \neg q)$ then $E(a)$ is $\{p, q\}$, not $\{p\}$. We will look at least letter-sets and a corresponding notion of *sympathy* later, in section 3.

- When we speak of a partial valuation v on a set F of elementary letters, we mean one with *exactly* F as domain. Any valuation on a proper superset F^+ of F, agreeing with v over F, will be called an *extension* of v.

It will sometimes shorten formulations to apply the notion of friendliness to partial valuations themselves. Let F be any set of elementary letters, and let v be any partial valuation on F. Let x be any formula. We say that v is *friendly* to x iff it may be extended to a partial valuation v^+ on $F \cup E(x)$ with $v^+(x) = 1$. Clearly, whenever a partial valuation is friendly to a formula then so too are all its restrictions. In other words, whenever a partial valuation is not friendly to a formula, none of its extensions are friendly to it. The first definition of $A \mathrel{|\!\approx} x$ may thus be expressed concisely as follows:

- $A \mathrel{|\!\approx} x$ iff every partial valuation v on $E(A)$ with $v(A) = 1$ is friendly to x.

Similar definitions of friendliness may be made for first-order logic, speaking of (partial) models rather than partial valuations. It should be noted, however, that in the first-order case there are several ways of understanding the notion of an extension of a model, which give rise to variant concepts of friendliness. On the one hand, we could require that when we extend a partial model the domain of discourse must remain fixed, as well as the interpretations into it of the already given predicate letters; in the literature this is usually called an 'expansion'. On the other hand, we may allow the domain to increase. In this case we have sub-options to choose from, according to whether we keep the interpretations of the already given predicate letters fixed, or allow them to flow out into the enlarged domain in some way.

But for simplicity, in this paper *we will remain within the propositional context*. We will not discuss the question of what would be the most interesting way of generalizing the definition of friendliness to the first-order context. Nor, apart from some passing negative observations, will we tabulate which among our results for the propositional context carry over to which among the first-order notions.

In section 2 we discuss links between the notion of friendliness and several other operations and concepts in the literature. Readers of a historical bent may prefer to start there and return, but we begin by clarifying the behaviour of the friendliness relation itself.

1.3. Properties that Fail

At first sight, the relation of friendliness seems to be hopelessly ill behaved. It fails many familiar features of classical consequence. In particular:

- It is not closed under substitution for elementary letters. Example: $p \mathrel{|\!\approx} p \wedge q$ where p, q are (here and always) distinct elementary letters, but $p \mathrel{|\!\not\approx} p \wedge \neg p$.
- It fails monotony and left strengthening. Example: $p \mathrel{|\!\approx} p \wedge q$, but $\{p, \neg q\} \mathrel{|\!\not\approx} p \wedge q$ and similarly $p \wedge \neg q \mathrel{|\!\not\approx} p \wedge q$.
- It fails cautious monotony and cautious left strengthening. Example: $p \mathrel{|\!\approx} q$ and $p \mathrel{|\!\approx} \neg q$, but $\{p, q\} \mathrel{|\!\not\approx} \neg q$ and likewise $p \wedge q \mathrel{|\!\not\approx} \neg q$.
- It fails left classical equivalence. Example: $p \mathrel{|\!\approx} p \wedge q$ but $p \wedge (q \vee \neg q) \mathrel{|\!\not\approx} p \wedge q$.
- It fails conjunction in the conclusion. Example: $p \mathrel{|\!\approx} q$, $p \mathrel{|\!\approx} \neg q$, but $p \mathrel{|\!\not\approx} q \wedge \neg q$.
- For essentially the same reason, it fails a general form of cumulative transitivity. Example: $p \mathrel{|\!\approx} q$, $p \mathrel{|\!\approx} \neg q$, and $p \wedge q \wedge \neg q \mathrel{|\!\approx} \neg p$, but $p \mathrel{|\!\not\approx} \neg p$.

- It fails plain transitivity. Example: $p \approx\!\!\!\mid q, q \approx\!\!\!\mid \neg p$, but $p \not\approx\!\!\!\mid \neg p$.
- It fails disjunction in the premises. Example: $p \approx\!\!\!\mid p \leftrightarrow q, q \approx\!\!\!\mid p \leftrightarrow q$, but $p \vee q \not\approx\!\!\!\mid p \leftrightarrow q$.

Nevertheless, friendliness does have positive properties including 'local' versions of some of the above, which we now describe.

1.4. Relationship of Friendliness to Classical Consequence

We begin by clarifying the relation of friendliness to classical consequence.

Supraclassicality. Whenever $A \vdash x$ then $A \approx\!\!\!\mid x$. Briefly: $\vdash\, \subseteq\, \approx\!\!\!\mid$

Verification. Immediate from the definition of $\approx\!\!\!\mid$.   □

The inclusion is proper; for example, when p, q are distinct elementary letters then $p \approx\!\!\!\mid q$ but not $p \vdash q$. Friendliness is not the trivial relation over the language; for example, when a is a tautology and x a contradiction, $a \not\approx\!\!\!\mid x$. For a less extreme example, $p \vee q \not\approx\!\!\!\mid p \wedge q$ where p, q are distinct elementary letters.

However, there are special cases where friendliness collapses into classical consequence, and others where it collapses into non-consequence of the negation.

First Reduction Case. Whenever $E(x) \subseteq E(A)$ then $A \approx\!\!\!\mid x$ iff $A \vdash x$.

Verification. Right to left is given unconditionally by supraclassicality, so we need only show left to right. Suppose $E(x) \subseteq E(A)$ and $A \approx\!\!\!\mid x$. Let v be any partial valuation on $E(A)$ with $v(A) = 1$. We need to show that $v^+(x) = 1$ for every extension v^+ of v to $E(A, x)$. Since $A \approx\!\!\!\mid x, v^+(x) = 1$ for some extension v^+ of v to $E(A, x)$. But since $E(x) \subseteq E(A), E(A, x) = E(A)$, so the unique extension of v to $E(A, x)$ is v itself. Thus $v(x) = 1$ and indeed $v^+(x) = 1$ for every extension v^+ of v to $E(A, x)$.   □

Second Reduction Case. Suppose A is consistent and for each elementary letter $p \in E(A)$, either $A \vdash p$ or $A \vdash \neg p$. Then $A \approx\!\!\!\mid x$ iff $A \nvdash \neg x$.

Verification. Under the hypotheses, suppose first that $A \approx\!\!\!\mid x$. Since A is consistent, there is some partial valuation v on $E(A)$ with $v(A) = 1$. Choose any one such v. Since $A \approx\!\!\!\mid x$, we have $v^+(x) = 1$ for some extension v^+ of v to $E(A, x)$. Thus $v^+(\neg x) = 0$ while $v^+(A) = 1$, so $A \nvdash \neg x$.

For the converse, suppose $A \nvdash \neg x$. Then there a partial valuation v on $E(A)$ with $v(A) = 1$ that can be extended to a partial valuation v^+ on $E(A, x)$ with $v^+(x) = 1$. Since either $A \vdash p$ or $A \nvdash \neg p$, for each elementary letter $p \in E(A)$, v is the only partial valuation on $E(A)$ with $v(A) = 1$. Hence every partial valuation w on $E(A)$ with $w(A) = 1$ can be extended to a partial valuation w^+ on $E(A, x)$ with $w^+(x) = 1$.   □

We also have the following important characterization of friendliness in terms of classical consistency.

Characterization in Terms of Consistency. $A \mathrel{\vert\approx} x$ iff every set B of formulae in L_A that is consistent with A, is consistent with x.

Verification. Suppose first that $A \mathrel{\vert\approx} x$. Let B be any set of formulae in L_A that is consistent with A. Then there is a partial valuation v on $E(A)$ with $v(A) = 1, v(B) = 1$. From the supposition, v may be extended to a partial valuation v^+ on $E(A, x)$ with $v^+(x) = 1$. Since v^+ extends v and $v(B) = 1$ we have $v^+(B) = 1$. Hence B is consistent with x, as desired.

For the converse, suppose that $A \mathrel{\not\vert\approx} x$. Then there is a partial valuation v on $E(A)$ with $v(A) = 1$, such that $v^+(x) = 0$ for every extension v^+ of v to $E(A, x)$. Put B to be the state-description (set of literals) in L_A that corresponds to v; in the limiting case that $E(A) = \emptyset$ put $B = \{\top\}$.

We complete the verification by showing that B is consistent with A but not consistent with x. The former is immediate from the fact that $v(A) = 1$ and by construction also $v(B) = 1$. For the latter, we observe that by construction, v is the only partial valuation on $E(B) = E(A)$ with $v(B) = 1$, and by hypothesis $v^+(x) = 0$ for every extension v^+ of v to $E(A, x)$. Thus there is no partial valuation w on $E(B, x) = E(A, x)$ with $w(B) = 1$ and $w(x) = 1$. In other words, B is inconsistent with x. □

This characterization can be refined. Our first refinement says, in effect, that in the characterization individual formulae c can do all the work of sets B of formulae.

First Refinement. $A \mathrel{\vert\approx} x$ iff $A \vdash c$ for every $c \in L_A$ with $x \vdash c$.

Verification. Suppose first $A \mathrel{\vert\approx} x$. Applying the characterization from left to right, we have that every formula in L_A that is consistent with A, is consistent with x. Contrapositively, whenever $c \in L_A$ and $x \vdash c$ then $A \vdash c$.

In the other direction, suppose $A \mathrel{\not\vert\approx} x$. Applying the characterization from right to left, there is a set B of formulae in L_A that is consistent with A, but is not consistent with x. Since B is not consistent with x, compactness tells us that is has a finite subset C that is not consistent with x. Then $x \vdash c$, where $c = \neg \wedge C$. But $A \nvdash c$, since A is consistent with B and so with its subset C. □

A second refinement will be useful for proving compactness for friendliness. In effect, in the characterization it suffices to consider only formulae $c \in L_A \cap L_x$, i.e., with $E(c) \subseteq E(A) \cap E(x)$.

Second Refinement. $A \mathrel{\vert\approx} x$ iff $A \vdash c$ for every $c \in L_A \cap L_x$ with $x \vdash c$.

Verification. Left to right is immediate from the first corollary. For the converse, suppose $A \mathrel{\not\vert\approx} x$. Then by the first corollary, there is a $d \in L_A$ with $x \vdash d$ but $A \nvdash d$. Since $x \vdash d$, classical interpolation tells us that there is a $c \in L_d \cap L_x \subseteq L_A \cap L_x$ with $x \vdash c \vdash d$. Since $c \vdash d$ and $A \nvdash d$ we have $A \nvdash c$ as desired. □

We note in passing that in the first-order context, if we define friendliness in terms of expansions (see section 1.2), then the second reduction case, the characterization in terms of consistency, and its two refinements, all fail in their right-to-left part. A single example serves for the three. Consider the language L with just one unary predicate letter P (no equality symbol, no individual constants), and put $\Gamma = Cn(\forall x(Px))$ to be the complete and consistent theory in that language. Let φ be the formula $\exists x \exists y (Rxy \wedge \neg Ryx)$, containing the additional letter R not available in L. On the one hand $\Gamma \nvdash \neg\varphi$; also every set Δ of formulae in L that is consistent with Γ, is consistent with φ. On the other hand, there is a model that satisfies Γ which has no expansion satisfying φ. Take any model with a singleton domain interpreting P as the whole domain. This satisfies Γ, but it cannot be expanded to a model satisfying Γ,φ, which would require two elements in the domain.

1.5. Closure Properties of Friendliness

We now see which among the familiar properties of classical consequence remain for friendliness. We begin with two that carry over without restriction.

Right Weakening. Whenever $A \mathrel{\approx\mkern-9mu\mid} x \vdash y$ then $A \mathrel{\approx\mkern-9mu\mid} y$.

Verification. Immediate from the definition of $\mathrel{\approx\mkern-9mu\mid}$.   □

It follows from this, of course, that the relation is syntax-independent in its right argument, i.e., satisfies right classical equivalence: whenever $x \dashv\vdash y$ then $A \mathrel{\approx\mkern-9mu\mid} x$ iff $A \mathrel{\approx\mkern-9mu\mid} y$. This contrasts with the already noted syntax-dependence on the left.

Singleton Cumulative Transitivity. Whenever $A \mathrel{\approx\mkern-9mu\mid} x$ and $A, x \mathrel{\approx\mkern-9mu\mid} y$ then $A \mathrel{\approx\mkern-9mu\mid} y$.

Verification. Suppose $A \mathrel{\approx\mkern-9mu\mid} x$ and $A,x \mathrel{\approx\mkern-9mu\mid} y$. Let v be any partial valuation on $E(A)$ with $v(A) = 1$. By the first hypothesis, v may be extended to a partial valuation v^+ on $E(A,x)$ with $v^+(x) = 1$, so also $v^+(A,x) = 1$. By the second hypothesis, v^+ may be extended to a partial valuation v^{++} on $E(A,x,y)$ with $v^{++}(y) = 1$. Restrict v^{++} to $E(A,y)$, call it v^{++-}. Then v^{++-} is still an extension of v with domain $E(A)$, and $v^{++-}(y) = 1$.   □

We now formulate some properties that carry over in a restricted form only. The following are straightforward; compactness and interpolation are subtler and will be discussed in the following sections.

Local Left Strengthening. Suppose $E(B) \subseteq E(A)$. Then $B \vdash A \mathrel{\approx\mkern-9mu\mid} x$ implies $B \mathrel{\approx\mkern-9mu\mid} x$.

Verification. Suppose $B \vdash A \mathrel{\approx\mkern-9mu\mid} x$. Consider any partial valuation v on $E(B)$ with $v(B) = 1$; we need to show that v is friendly to x. Extend v to any partial valuation v^+ on $E(A) \supseteq E(B)$. Then $v^+(B) = v(B) = 1$, and so since $B \vdash A$ we have $v^+(A) = 1$. Since $A \mathrel{\approx\mkern-9mu\mid} x$, there is an extension v^{++} of v^+ to $E(A,x)$ with $v^{++}(x) = 1$. Restrict v^{++} to $E(B,x)$, call it v^{++-}. Then clearly $v^{++-}(x) = v^{++}(x) = 1$. But v^{++-} is still an extension of v with domain $E(B)$. Hence v is friendly to x, as desired.   □

Local Left Equivalence. Suppose $E(B) \subseteq E(A)$. Then $A \mathrel{\approx\!\!\!\mid} x$ and $A \dashv\vdash B$ together imply $B \mathrel{\approx\!\!\!\mid} x$.

Verification. When $A \dashv\vdash B$ then $B \vdash A$ so we can apply local left strengthening.
□

Local Monotony. Suppose $E(B) \subseteq E(A)$. If $A \mathrel{\approx\!\!\!\mid} x$ and $A \subseteq B$ then $B \mathrel{\approx\!\!\!\mid} x$.

Verification. When $A \subseteq B$ then $B \vdash A$; apply local left strengthening.   □

Local Disjunction in the Premisses. Suppose $E(b_2) \subseteq E(A,b_1)$ and $E(b_1) \subseteq E(A,b_2)$. Then $A, b_1 \mathrel{\approx\!\!\!\mid} x$ and $A, b_2 \mathrel{\approx\!\!\!\mid} x$ together imply $A, b_1 \vee b_2 \mathrel{\approx\!\!\!\mid} x$.

Verification. Suppose $A, b_1 \vee b_2 \mathrel{\not\approx\!\!\!\mid} x$. Then there is a partial valuation v on $E(A, b_1 \vee b_2)$ with $v(A, b_1 \vee b_2) = 1$ that is not friendly to x. By the hypotheses, $E(A, b_1 \vee b_2) = E(A, b_1) = E(A, b_2)$. Since $v(A, b_1 \vee b_2) = 1$ either $v(A, b_1) = 1$ or $v(A, b_2) = 1$. Hence either v is a partial valuation on $E(A, b_1)$ with $v(A, b_1) = 1$ but not friendly to x, or similarly with b_2. That is, either $A, b_1 \mathrel{\not\approx\!\!\!\mid} x$ or $A, b_2 \mathrel{\not\approx\!\!\!\mid} x$.   □

Proof by Exhaustion. $A, b \mathrel{\approx\!\!\!\mid} x$ and $A, \neg b \mathrel{\approx\!\!\!\mid} x$ together imply $A \mathrel{\approx\!\!\!\mid} x$.

Verification. Clearly $E(\neg b) = E(b) \subseteq E(A, b)$ and conversely $E(b) = E(\neg b) \subseteq E(A, \neg b)$ so we may apply local disjunction in the premisses to get $A, b \vee \neg b \mathrel{\approx\!\!\!\mid} x$. Clearly also $E(A) \subseteq E(A, b \vee \neg b)$ and also $A \vdash (A, b \vee \neg b) \mathrel{\approx\!\!\!\mid} x$, so we may apply local left strengthening to get $A \mathrel{\approx\!\!\!\mid} x$ as desired.   □

The properties obtained so far lead to another characterization. In a broad sense of the term, it can be seen as an axiomatization of the relation of friendliness, modulo classical consequence. 'A broad sense', since the right-hand side of the third condition is not closed under substitution.

Observation. Friendliness is the least relation R between sets of formulae and individual formulae that satisfies the following three conditions:

1. $\vdash \subseteq R$,
2. $\langle A, x \rangle \in R$ whenever $\langle A \cup \{b\}, x \rangle \in R$ and $\langle A \cup \{\neg b\}, x \rangle \in R$,
3. $\langle A, x \rangle \in R$ whenever $A \not\vdash \neg x$ and for each elementary letter $p \in E(A)$, either $A \vdash p$ or $A \vdash \neg p$.

Verification. First observe that the total relation between sets of formulae and individual formulae satisfies these three conditions, and so there is at least one such relation. Further, the intersection of any non-empty set of such relations is itself such a relation (despite the negative term $A \not\vdash \neg x$ in the third condition, which negates classical consequence rather than the relation R). Thus there is a unique least such relation R, call it R_0.

We already know that $\mathrel{\approx\!\!\!\mid}$ satisfies all three conditions (supraclassicality, proof by exhaustion, second reduction case). Thus $R_0 \subseteq \mathrel{\approx\!\!\!\mid}$.

For the converse, suppose $\langle A, x \rangle \notin R_0$; we need to show that $A \mathrel{\not\approx\!\!\!\mid} x$. Let $p_1, \ldots, p_n$ be all the elementary letters in $E(A)$. Define sets $A_0, \ldots, A_n$ by setting

$A_0 = A$ and putting $A_{i+1} = A_i \cup \{p_{i+1}\}$ if $\langle A_i \cup \{p_{i+1}\}, x\rangle \notin R_0$ and otherwise $A_{i+1} = A_i \cup \{\neg p_{i+1}\}$. By hypothesis, $\langle A_0, x\rangle \notin R_0$ and an easy induction using condition (2) gives us $\langle A_n, x\rangle \notin R_0$. But for each elementary letter $p \in E(A)$, either $A_n \vdash p$ or $A_n \vdash \neg p$, so condition (3) tells us that $A_n \vdash \neg x$. Also, since $\langle A_n, x\rangle \notin R_0$, condition (1) tells us that A_n is consistent, so there is at least one partial valuation v on $E(A_n) = E(A)$ with $v(A_n) = 1$. Since $A_n \vdash \neg x$, we have $v^+(x) = 0$ for every extension v^+ of v to $E(A, x)$, so $A \not\approx x$ as desired. $\square$

1.6. Compactness

In the context of friendliness, some care must be taken with the formulation of compactness. When the property is formulated in exactly the same way as in classical logic, it tells us very little. For suppose $A \approx x$. Then:

- On the one hand, in the limiting case that x is inconsistent the definition of $\approx$ implies that A must also be inconsistent, so by classical compactness there is a finite inconsistent subset $B \subseteq A$, so that by the definition of $\approx$ again, $B \approx x$.
- On the other hand, in the principal case that x is consistent, we have immediately that $\emptyset \approx x$. This leaves us hungry, for while the empty set is certainly finite we would like something more substantial.

This motivates the following strengthened formulation. Bearing in mind that friendliness does not satisfy monotony, it is quite strong.

Compactness. Let A be a non-empty set with $A \approx x$. Then there is a finite subset $B \subseteq A$ such that $C \approx x$ for every C with $B \subseteq C \subseteq A$.

Proof. Suppose $A \approx x$. By the second refinement of the characterization of friendliness in terms of consistency, whenever $c \in L_A \cap L_x$ and $x \vdash c$ then $A \vdash c$. Hence by compactness for classical consequence, for every $c \in L_A \cap L_x$ with $x \vdash c$ there is a finite subset $B_c \subseteq A$ with $B_c \vdash c$. Since x is an individual formula, there are only finitely many $c \in L_A \cap L_x \subseteq L_x$ up to classical equivalence. Taking the finite union of the corresponding sets B_c, we conclude that there is a finite subset $B \subseteq A$ such that $B \vdash c$ for every $c \in L_A \cap L_x$ with $x \vdash c$.

Now let C be any set with $B \subseteq C \subseteq A$. We need to show that $C \approx x$. Since $B \subseteq C$, monotony for classical consequence gives us $C \vdash c$ for every $c \in L_A \cap L_x$ with $x \vdash c$. Also, since $C \subseteq A$, we have $L_C \subseteq L_A$ and so $C \vdash c$ for every $c \in L_C \cap L_x$ with $x \vdash c$. Applying again the second refinement of the characterization of friendliness, we have $C \approx x$ as desired. $\square$

1.7. Interpolation

As in the case of compactness, interpolation for friendliness is trivial when formulated in the way customary in classical logic. For suppose $A \approx x$; we want to show that there is a formula b with $E(b) \subseteq E(A) \cap E(x)$ such that both $A \approx b$ and $b \approx x$. On the one hand, if A is inconsistent, we can put $b = \bot$ giving us $A \vdash b \vdash x$ so $A \approx b \approx x$. On the other hand, if A is consistent then since $A \approx x, x$ must also be consistent, so we can put $b = \top$, so that $A \vdash b$ and thus $A \mathrel{\vertbar\kern-0.4em\approx} b$, and also $b \mathrel{\vertbar\kern-0.4em\approx} x$ using the consistency of x.

The following formulation strengthens the property by guaranteeing that in suitable conditions, b can be chosen more informatively.

Interpolation. Whenever $A \mathrel{\vertbar\kern-0.4em\approx} x$ there is a finite set $F \subseteq E(A) \cap E(x)$ of elementary letters such that for every finite set G of elementary letters with $F \subseteq G \subseteq E(A)$ there is a formula b with the following properties:

1. $E(b) = G$
2. $A \mathrel{\vertbar\kern-0.4em\approx} b$ (indeed $A \vdash b$)
3. $b \mathrel{\vertbar\kern-0.4em\approx} x$
4. b is consistent, provided A is consistent
5. b is not a tautology, provided there is a non-tautology $y \in L_A \cap L_x$ with $A \vdash y$.

Remark. Before giving the proof, we note that the rather odd proviso in property (5) cannot be weakened to, say: A and x are not tautologous. Example: $A = p \vee q$, $x = q \vee r$. Then $A \mathrel{\vertbar\kern-0.4em\approx} x$, but the only formulae b with $E(b) \subseteq E(A) \cap E(x) = \{q\}$ and both $A \mathrel{\vertbar\kern-0.4em\approx} b$ and $b \mathrel{\vertbar\kern-0.4em\approx} x$ are the tautologies containing at most the letter q.

Proof. Suppose $A \mathrel{\vertbar\kern-0.4em\approx} x$. Since x is a single formula, $E(x)$ is finite, and thus so too is $E(A) \cap E(x)$. Hence, up to classical equivalence, there is a strongest formula a with $E(a) \subseteq E(A) \cap E(x)$ and $A \vdash a$. Take any such a and put $F = E(a)$, which is clearly finite. Let G be any finite set of letters with $F \subseteq G \subseteq E(A)$. Form b by conjoining with a the disjunctions $q \vee \neg q$ for the finitely many letters q in $G \setminus F$. We claim that b fulfils all requirements.

Property (1) is immediate by construction. Also by construction $A \vdash a \dashv\vdash b$ and so by supraclassicality, $A \mathrel{\vertbar\kern-0.4em\approx} b$, giving (2). For property (4), if A is consistent then since $A \vdash b$, b is also consistent. For (5), suppose there is a non-tautology $y \in L_A \cap L_x$ with $A \vdash y$. Then by its construction, a is not a tautology, and so since $a \dashv\vdash b$, b is not a tautology.

It remains to show (3). Suppose $b \mathrel{\not\vertbar\kern-0.4em\approx} x$; we derive a contradiction. Since $b \mathrel{\not\vertbar\kern-0.4em\approx} x$ there is a partial valuation v on $E(b) = G \subseteq E(A)$ with $v(b) = 1$, which is not friendly to x, i.e., such that $v^+(x) = 0$ for every extension v^+ of v to $E(b,x)$. Fix such a v for the remainder of the proof.

Write k for the state-description formula in L_b that corresponds to v. Then clearly $v(k) = 1$ and also $k \vdash \neg x$. Put $b^* = b \wedge \neg k$. We show that b^* is a formula in L_A with $A \vdash b^*$ and $b \nvdash b^*$, thus contradicting the construction of b.

For $b^* \in L_A$: This is immediate since both $b, \neg k \in L_A$.

For $b \nvdash b^*$: It suffices to show $b \nvdash \neg k$, i.e., that $k \nvdash \neg b$. We have by its construction that $k \vdash b$; and since $v(k) = 1$, k is satisfiable, so $b \nvdash \neg k$ as desired.

For $A \vdash b^*$: Since $A \vdash b$ it suffices to show $A \vdash \neg k$. As a preliminary observation, we show that there is no extension w of v to $E(A)$ with $w(A) = 1$. For let w be such an extension. Since by hypothesis $A \mathrel{\vertbar\kern-0.4em\approx} x$, there is an extension w^+ of w to $E(A,x)$ with $w^+(x) = 1$. Clearly, w^+ is also an extension of v to

$E(A, x)$. Now restrict w^+ to $E(b, x)$, which is possible since $E(b) \subseteq E(A)$ so that $E(b, x) \subseteq E(A, x)$, and call it w^{+-}. Clearly $w^{+-}(x) = 1$ and also w^{+-} is still an extension of v, which has domain $E(b)$. But this contradicts the fact that v is not friendly to x. This completes the preliminary step of showing that there is no extension w of v to $E(A)$ with $w(A) = 1$.

Now let w be any partial valuation on $E(A) \supseteq E(b) = E(k)$ with $w(\neg k) = 0$, i.e., $w(k) = 1$. It remains to show that $w(A) = 0$. Restrict w to $E(k) = E(b)$ = domain(v), call it w^-. Clearly $w^-(k) = 1$. Hence by the construction of k as a state-description in L_b corresponding to v, $w^- = v$. Thus w is an extension of v to $E(A)$. So by the preliminary observation, $w(A) = 0$ as desired. □

1.8. Friendliness as an Operation

Up to now, we have treated friendliness as a relation between formulae (or sets of formulae) on the left and formulae on the right. But just as in the case of classical consequence and well-known nonmonotonic consequences, we can consider it as an operation, taking sets of formulae to sets of formulae, by defining $Fr(A) = \{x : A \approx x\}$.

However, this may not be a very useful perspective for friendliness, in contrast to the situation for the usual nonmonotonic consequence relations. The reason is that friendliness is much further from being a closure relation. It fails monotony but also, as we have seen in section 1.3, it fails both conjunction in the conclusion and general cumulative transitivity. Expressed as an operation, it also fails idempotence (the same counterexample can be used as for cumulative transitivity). These properties are all satisfied by the usual nonmonotonic consequence relations (see e.g., Makinson 2005), and their absence makes the operational notation much less convenient to use.

So, in this section we examine just one question regarding the operational version: when do we have $Fr(A) = Fr(B)$ for sets A, B of formulae?

Observation. $Fr(A) = Fr(B)$ iff either $A \dashv\vdash B$ and $E(A) = E(B)$ or else A, B are both contradictions.

Verification. In one direction, suppose RHS. We want to show $Fr(A) = Fr(B)$. In the limiting case that A,B are both contradictions, we have $Fr(A) = L = Fr(B)$ vacuously from the definition of friendliness. So consider the principal case that $A \dashv\vdash B$ and $E(A) = E(B)$. Then $Fr(A) = Fr(B)$ by two applications of local left equivalence (section 1.5).

For the other direction, suppose $Fr(A) = Fr(B)$. Suppose that A, B are not both contradictions. We need to show that $E(A) = E(B)$ and $A \dashv\vdash B$.

First, we observe that neither of A,B is a contradiction. For suppose A, say, is a contradiction. Then $A \vdash \bot$ and so by supraclassicality of friendliness, $A \approx \bot$ and so since $Fr(A) = Fr(B)$ we have $B \approx \bot$, so B is a contradiction.

Next, we show $E(A) = E(B)$. It suffices to show $E(A) \subseteq E(B)$; the converse is similar. Suppose $p \in E(A)$ but $p \notin E(B)$; we derive a contradiction. Since $p \notin E(B)$ clearly $B \not\approx p$ and also $B \not\approx \neg p$. Since $Fr(A) = Fr(B)$, this gives us

$A \kern-0.5em\not\kern-0.1em\approx p$ and also $A \kern-0.5em\not\kern-0.1em\approx \neg p$. Since $p \in E(A)$, the first reduction case for friendliness tells us that $A \vdash p$ and also $A \vdash \neg p$ so that A is inconsistent, contradicting what has been shown.

Finally, we show $A \dashv\vdash B$. It suffices to show $A \vdash B$; the converse is similar. Take any $b \in B$. we need to show $A \vdash b$. Now $B \vdash b$ so by supraclassicality $B \kern-0.5em\not\kern-0.1em\approx b$ so since $Fr(A) = Fr(B)$ we have $A \kern-0.5em\not\kern-0.1em\approx b$. Since $E(A) = E(B)$ and $b \in B$ we have $E(b) \subseteq E(A)$ so by the first reduction case for friendliness, $A \vdash b$ as desired, and the proof is complete. $\qquad\square$

1.9. Joint Friendliness: Two Notions

For classical consequence, we have followed the common convention of writing $A \vdash B$ to mean that $A \vdash b$ for all $b \in B$. For friendliness, it is tempting to write $A \kern-0.5em\not\kern-0.1em\approx B$ analogously. But care is needed, for there is an important distinction that does not arise in the classical case. We must distinguish between two relationships:

- $A \kern-0.5em\not\kern-0.1em\approx_{\forall\forall\exists} B$: for every partial valuation v on $E(A)$ with $v(A) = 1$ and every $b \in B$, there is an extension v^+ of v to $E(A, b)$ with $v^+(b) = 1$.
- $A \kern-0.5em\not\kern-0.1em\approx_{\forall\exists\forall} B$: For every partial valuation v on $E(A)$ with $v(A) = 1$ there is an extension v^+ of v to $E(A, B)$ with $v^+(B) = 1$, i.e., with $v^+(b) = 1$ for every $b \in B$.

The former says the same as $A \kern-0.5em\not\kern-0.1em\approx b$ for all $b \in B$. But the latter says more. For classical consequence, where conjunction in the conclusion is satisfied, no such distinction arose. We call $\kern-0.5em\not\kern-0.1em\approx_{\forall\forall\exists}$ *weak* joint friendliness, $\kern-0.5em\not\kern-0.1em\approx_{\forall\exists\forall}$ *strong*. When we refer to joint friendliness (sections 2.2 and 3.4), we will specify clearly which is intended.

1.10. Internalizing the Relation

It is natural to ask whether we can internalize the relation of friendliness as a conditional connective of the object language.

It can be done quite trivially by adding an iterable two-place connective $\rightsquigarrow$ to the object language and adding to the familiar Boolean rules the following one. To bring the formulation as close as possible to standard ones for propositional connectives, we state it with v, w, u understood as full valuations, i.e., defined on the set E of all elementary letters.

$v(a \rightsquigarrow x) = 1$ iff for every full valuation w with $w(a) = 1$ there is a full valuation u that agrees with w on all elementary letters in $E(a)$ and such that $u(x) = 1$.

The same effect can be achieved by means of indexed unary modal operators. Consider a language with operators $\square_a$ and $\lozenge_a$ for all formulae a. This is a little unusual, as the set of connectives is not fixed in advance, but is defined inductively along with the formulae in which they occur; but that is not a problem. We read these connectives by the following rules:

$v(\square_a x) = 1$ iff for every valuation w that agrees with v on all elementary letters in $E(a)$, we have $w(x) = 1$.

$v(\Diamond_a x) = 1$ iff for some valuation w that agrees with v on all elementary letters in $E(a)$, we have $w(x) = 1$.

We may then identify plain $\Box$ and $\Diamond$ as $\Box_\top$ and $\Diamond_\top$ (or equivalently $\Box_\bot$ and $\Diamond_\bot$), giving us the familiar evaluation rules:

$v(\Box x) = v(\Box_\top x) = v(\Box_\bot x) = 1$ iff $w(x) = 1$ for every valuation w.

$v(\Diamond x) = v(\Diamond_\top x) = v(\Diamond_\bot x) = 1$ iff $w(x) = 1$ for some valuation w.

With this equipment, we may represent $a \not\approx x$ in the object language by the formula $\Box(a \to \Diamond_a x)$. Given the rules given above for evaluating indexed modal operators, this formula will satisfy the same evaluation condition that we gave for the trivial internalization. It will come out as true under one valuation iff it does so under all valuations, and that iff the relation $a \not\approx x$ holds.

However, it should be understood that when we internalize the relation of friendliness (whether directly or via indexed modal operators) the resulting system is rather unusual. The set of all valid formulae (defined as those formulae that are true under every valuation) is not closed under substitution, for the very same reason as the relation of friendliness was not so closed. The same example can be used to illustrate the failure. On the one hand, the formula $(p \to \Diamond_p(p \land q)$ is valid, while its substitution instance $(p \to \Diamond_p(p \land \neg p)$ is not.

Thus while internalization is perfectly possible, the propositional system that it gives us is unlike most modal and other non-classical propositional logics, for which the set of valid formulae is closed under substitution. In the author's view, this difference is not a disqualification — see e.g the discussion in Makinson (2005). But it is not clear that internalization provides any insights that are not already available when friendliness is treated as a relation between formulae.

2. Links with Familiar Notions

Friendliness has many friends: several other notions familiar from the literature are connected with it. Roughly speaking, the links are of two main kinds.

- Certain well-known operations from the history of logic, distant and recent, can be seen as *instances* of friendliness.
- There are also more general *conceptual links*, notably with Ramsey eliminability and related notions that have been studied in the context of first-order logic.

We begin with some instances of friendliness.

2.1. Forgetting Letters from Formulae

Consider any formula a and any subset F of its elementary letters, i.e., $F \subseteq E(a)$. Let $\sigma_1, \ldots, \sigma_k$ be the $k = 2^n$ substitutions of $\bot, \top$ for the n letters in F. Following Weber (1987) and later papers such as Lin and Reiter (1994) and Lang, Liberatore, Marquis (2003), we may define $f_F(a)$, the result of *forgetting* the letters in F from a, as $\sigma_1(a) \lor \ldots \lor \sigma_k(a)$. Equivalently, in recursive form, $f_\emptyset(a) = a$, and

$f_{F,q}(a) = \sigma_\perp(f_F(a)) \vee \sigma_\top(f_F(a))$, where the functions $\sigma_\perp$ and $\sigma_\top$ substitute $\perp, \top$ for the letter q.

As is well known, $a \vdash f_F(a)$. The converse fails, i.e., $f_F(a) \nvdash a$; for example $f_F(p) = \perp \vee \top \nvdash p$. However, $f_F(a)$ is easily shown to be the strongest formula b in the language generated by $E(a) \setminus F$ such that $a \vdash b$.

In fact, the notion goes back to Boole, whose focus was however rather different. From his point of view, the central logical relation was equality, coresponding to classical equivalence. Accordingly, the most important fact for him about what we now call forgetting was the equality that he introduced under the name of 'development' in Boole (1847): $a \dashv\vdash (\neg p \wedge \sigma_\perp(a)) \vee (p \wedge \sigma_\top(a))$. The consequence $a \vdash \sigma_\perp(a) \vee \sigma_\top(a) = f_a(a)$ is however implicit (in dual form) in the discussion of the 'elimination' of a term in an equation, in Boole (1854).

Observation. $f_F(a) \approx a$.

Verification. Let v be any partial valuation on $E(f_F(a)) = E(a) \setminus F$ and suppose $v(f_F(a)) = 1$. Then $v(\sigma_i(a)) = 1$ for some $i \leq k$. Extend v to v^+ on $E(a)$ by putting $v^+(q) = 0,1$ according as $\sigma_i(q) = \perp, \top$ for each $q \in F$. Then clearly by induction on length of formulae, $v^+(a) = v(\sigma_i(a)) = 1$ and we are done. □

2.2. Ejective Substitution

It is natural to ask whether this observation can be extended to a more general result linking friendliness and substitution. It cannot cover all substitutions, for we do not always have $\sigma(a) \approx a$, even when σ is a one-one correspondence on letters. Consider for example the formula $a = p \wedge \neg q$ and the substitution σ that simply interchanges the two letters, putting $\sigma(p) = q$ and $\sigma(q) = p$ so that $\sigma(a) = q \wedge \neg p \napprox a = p \wedge \neg q$ (witness the only partial valuation that makes the premiss true).

Nevertheless, we do have a positive result for a certain class of substitutions. Let σ be any substitution on the set E of all elementary letters, and let A be any set of formulae. We call σ *ejective for* A iff for every letter $p \in E(A)$, either $\sigma(p) = p$ or $p \notin E(\sigma(A))$.

Observation. Let a be any formula, and let σ be any substitution that is ejective for a. Then $\sigma(a) \approx a$. More generally, when A is a set of formulae and σ is ejective for A then $\sigma(A) \approx_{\forall\exists\forall} A$.

Verification. The notation $\approx_{\forall\exists\forall}$ for strong joint friendliness is explained in section 1.9. Consider any partial valuation v on $E(\sigma(A))$ with $v(\sigma(A)) = 1$. We extend v to v^+ on $E(\sigma(A), A)$ by putting $v^+(q) = v(\sigma(q))$ for each letter q in $E(A) \setminus E(\sigma(A))$. We want to show that $v^+(A) = v(\sigma(A)) = 1$. It suffices to show by induction that for every subformula b of any formula in A, $v^+(b) = v(\sigma(b))$.

For the basis, if b is a letter p then either $\sigma(p) = p$ or $p \notin E(\sigma(A))$. In the former case $p \in E(\sigma(A))$, so $v(p)$ is defined, so since v^+ extends v we have $v^+(p) = v(p) = v(\sigma(p))$ as desired. In the latter case, $p \in E(A) \setminus E(\sigma(A))$, so that $v^+(p) = v(\sigma(p))$ by definition.

The induction step is then routine using the definitions of a substitution and of a Boolean valuation. $\qquad\square$

This observation covers the 'friendly forgetfulness' property $f_F(a) \mathrel{\vbox{\hbox{$\approx$}}} a$ as a special case. For when a function σ substitutes $\bot, \top$ for some of the elementary letters in a (and is the identity on all other letters) then it is ejective for a. Indeed, it is ejective *tout court*, in the stronger sense that for every letter p, either $\sigma(p) = p$ or $p \notin E(\sigma(L)) = E(\sigma(E))$. Thus we have $f_F(a) = \sigma_1(a) \vee \ldots \vee \sigma_k(a)$ where each substitution σ_i is ejective, so that each $\sigma_i(a) \mathrel{\vbox{\hbox{$\approx$}}} a$. But $E(\sigma_i(a)) = E(a) \setminus F = E(\sigma_j(a))$ for all $i, j \leq k$ and so we may apply local disjunction in the premisses (section 1.5) putting $A = \emptyset$ to conclude that $\sigma_1(a) \vee \ldots \vee \sigma_k(a) \mathrel{\vbox{\hbox{$\approx$}}} a$ as desired.

2.3. Identifying Letters

The above observation has a further corollary. By an *identification of letters* we mean a substitution σ on E into E such that for every letter p, either $\sigma(p) = p$ or $p \neq \sigma(q)$ for all letters q. Equivalently: such that whenever $p = \sigma(q)$ for some letter q then $\sigma(p) = p$. Equivalently: such that for some partition of E and some choice function γ on that partition, $\sigma(p) = \gamma(|p|)$.

Corollary. $\sigma(A) \mathrel{\vbox{\hbox{$\approx$}}}_{\forall\exists\forall} A$ for any identification σ of letters. In particular, when a is an individual formula and σ is an identification of letters, then $\sigma(a) \mathrel{\vbox{\hbox{$\approx$}}} a$.

Verification. By the observation in section 2.2, it suffices to observe that every identification of letters is ejective *tout court*, and so ejective for A. Let σ be any identification of letters. Suppose $p \in E(A)$ and $\sigma(p) \neq p$. Since σ is an identification of letters, this gives us $p \neq \sigma(q)$ for all letters q. Since σ takes E into E this implies that $p \notin E(\sigma(E)) = E(\sigma(L))$. $\qquad\square$

2.4. Existential Quantification

The concept of forgetting can also be expressed in the language of quantified Boolean formulae. Put $g_F(a) = \exists p_1 \ldots \exists p_n(a)$ where $F = \{p_1, \ldots, p_n\}$. Then under the standard semantics for quantified Boolean formulae, $g_F(a)$ has exactly the same truth conditions as $f_F(a)$. So, with the notion of friendliness suitably enlarged to cover such formulae (rather than just unquantified Boolean formulae, as in this paper), we can say that $g_F(a)$ is friendly to a.

More generally, it is clear that in any language admitting existential quantifiers over a syntactic category of items, the existential quantification $\exists i_1 \ldots \exists i_n(a)$ over selected variables from that category will, under a natural enlargement of the notion, be friendly to a.

However, it should also be observed that the forgetting function $f_F(a)$, its quantified Boolean analogue $g_F(a)$, and existentialization $\exists i_1 \ldots \exists i_n(a)$ all have a more intimate relation to their argument a than mere friendliness. For we have not only $f_F(a) \mathrel{\vbox{\hbox{$\approx$}}} a, g_F(a) \mathrel{\vbox{\hbox{$\approx$}}} a, \exists i_1 \ldots \exists i_n(a) \mathrel{\vbox{\hbox{$\approx$}}} a$ but also the classical consequences in the reverse direction: $a \vdash f_F(a), a \vdash g_F(a), a \vdash \exists i_1 \ldots \exists i_n(a)(a)$. This contrasts with the fact that for friendliness in general we may have $b \mathrel{\vbox{\hbox{$\approx$}}} a$ without $a \vdash b$: witness the example $p \mathrel{\vbox{\hbox{$\approx$}}} q$ but $q \nvdash p$ where p, q are distinct elementary letters.

2.5. Skolemization

The process of Skolemization of a formula of first-order logic manifests friendliness in a very special way. Taking for example the formula $\alpha = \forall x \exists y (Rxy)$, we can introduce a function letter f and consider both the formula $sk(\alpha) = \forall x(Rxf(x))$ and its existential quantification $\exists f(sk(\alpha)) = \exists f \forall x (Rxf(x))$. These formulae belong respectively to first-order logic with function letters, and second-order logic.

As Skolem observed, we have $sk(\alpha) \vdash \alpha$ in first-order logic, and also $\alpha \dashv\vdash \exists f(sk(\alpha))$ in second-order logic (assuming the axiom of choice in our metalanguage). The equivalence between α and $\exists f(sk(\alpha))$ means that the relation between these two is much tighter than for plain existentialization.

While $sk(\alpha) \vdash \alpha$, the converse fails: $\alpha \nvdash sk(\alpha)$. But we do have $\alpha \mathrel{|\!\approx} sk(\alpha)$ where $\mathrel{|\!\approx}$ is the friendliness in the first-order context, understood in terms of expansions (section 1.2). For every (partial) model interpreting the predicate letter R in a domain, if that model satisfies α then it has an expansion also interpreting the function letter f in the same domain that satisfies $sk(\alpha)$.

Here again there is an especially close relationship. As is well known, a and $sk(\alpha)$ are equivalent for logical truth, i.e., a is true in all first-order models iff $sk(\alpha)$ is. This does not hold for friendliness in general. In our base territory of classical propositional logic, $p \vee \neg p \mathrel{|\!\approx} q$ but the left is a tautology while the right is not.

As is well known, the passage from α to $sk(\alpha)$ also contrasts with existentialization in this regard. For example $\exists x(\exists x(Px) \to Px)$ is friendly to $\exists x(Px) \to Px$, but the left is a logical truth while the right is not.

2.6. Ramsey Eliminability

As well as the above particular instances of friendliness, there are also more general connections with concepts that have arisen elsewhere. Of these, the closest is with Ramsey eliminability of a predicate or other term in a theory.

This notion takes its origin in the philosophy of science, and more specifically in discussions concerning the relation between the observational and theoretical components of empirical scientific theories. It was first sketched in rough terms by F. P. Ramsey in notes of 1929, published in the posthumous collection Ramsey (1931, chapter 'Theories'). It was taken up and given its name by Sneed (1971, chapter 3); and subsequently discussed in a number of books and papers including van Benthem (1978) and Rantala (1991). All of these are expressed in the context of first-order languages.

Formulations differ in subtle but significant respects. What they all have in common is that every model of one set Γ of (first-order) formulae should be capable of expansion to a model of a larger set Δ that possibly contains further letters (individual constants, predicates, or function signs). We recall that by an *expansion* of a model is meant another model with the same domain, same interpretations of the letters that were interpreted in the first model, plus interpretations of whatever new letters are concerned.

D. Makinson

Where the formulations differ is in what Γ and Δ are taken to be; which of them is taken to be an arbitrary set of formulae while the other is taken as a function of it. The story is as follows.

- For Rantala (1991, pages 150–151): Γ is taken to be an arbitrary set of first-order formulae, and Δ is put as $\Gamma \cup \{\varphi\}$ where φ is a (likewise first-order) formula. Rantala focusses on the case that this formula has just one new letter beyond those occurring in Γ, thought of as a candidate for reduction; however the definition is meaningful without that restriction. The concept is envisaged as expressing a property of the new letter(s) in φ modulo the set $\Gamma \cup \{\varphi\}$, rather than a relation between Γ and $\Delta = \Gamma \cup \{\varphi\}$.
- By contrast, for van Benthem (1978, page 325), it is Δ that is is taken to be an arbitrary set of first-order formulae, while Γ is taken to be $Cn(\Delta) \cap L_0$, where L_0 is an arbitrarily chosen sublanguage of the language L of Δ. Again, the concept is envisaged as expressing a property of the omitted letter set in $L \setminus L_0$ modulo the formula set Δ.

Typically, L_0 will be made up of all the letters in L except for one, which is thought of as a candidate for reduction. In that case, we have exactly a notion introduced by de Bouvère (1959, chapter II.2). He used the failure of this property of the omitted letter (say, a predicate P) modulo a theory Δ, as a method for showing that P is not explicitly definable in Δ. This contrasts with the better-known technique going back to Padoa (1901), which proceeds by showing that some model of Γ can be expanded in two distinct ways to a model of Δ. Unlike de Bouvère's method, that of Padoa is complete for the task, as shown in a celebrated theorem of Beth (1956).

As is well known, the formulations of Rantala and van Benthem are not equivalent. On the one hand, when $\Delta = \Gamma \cup \{\varphi\}$ and L_0 is the language of Γ, then $\Gamma \subseteq Cn(\Delta) \cap L_0$. Hence, if every model of Γ can be expanded to a model of Δ, then every model of $Cn(\Delta) \cap L_0$ can too. In other words, Ramsey eliminability in the sense of Rantala implies the same in the sense of van Benthem. But in general, Γ may be a proper subset of $Cn(\Delta) \cap L_0$. So it may happen that whilst every model of $Cn(\Delta) \cap L_0$ can be expanded to one of Δ, there is some model of Γ (but not satisfying $Cn(\Delta) \cap L_0$) that cannot be so expanded. Thus Ramsey eliminability in the sense of van Benthem does not imply the same in the sense of Rantala. Specific examples have been given in the literature.

To compare these two concepts with friendliness as studied in this paper, we extract the purely propositional content, and write it in the notation that we have been using. We write $L_{E(B) \setminus F}$ for the language generated by the letters that are in $E(B) \setminus F$.

- From Rantala: The letters in $E(x) \setminus E(A)$ are Ramsey eliminable from a set A, x of formulae iff every partial valuation v on $E(A)$ with $v(A) = 1$ can be extended to a partial valuation v^+ on $E(A, x)$ with $v^+(A, x) = 1$.
- From van Benthem: Consider any set B of formulae and any set F of elementary letters with $F \subseteq E(B)$. The letters in F are Ramsey eliminable from B

iff every partial valuation v on $E(B) \setminus F$ with $v(Cn(B) \cap L_{E(B)\setminus F}) = 1$ can be extended to a partial valuation v^+ on $E(B)$ with $v^+(B) = 1$.

Of these, the Rantala-style concept is equivalent to friendliness of A to x, as defined and studied in this paper.

Observation. Let A be any set of propositional formulae and x a propositional formula. Then A is friendly to x iff the letters in $E(x) \setminus E(A)$ are Ramsey eliminable from A, x in the sense of Rantala.

Verification. The only difference between the definition of friendliness and the propositional reduction of Rantala's version of Ramsey eliminability is that whereas the former requires the extension v^+ to satisfy x, the latter requires it to satisfy A, x. But these are equivalent when v^+ extends v and $v(A) = 1$. □

We have already remarked that even in the first-order context, the formulation of van Benthem is weaker than that of Rantala. Indeed, it is very much weaker since, as is well-known, every *finite* model of $Cn(\Delta) \cap L_0$ can be expanded to a model of Δ. In the purely propositional context, it becomes so much weaker that it always holds, as we now show.

Observation. Let B be any set of propositional formulae and $F \subseteq E(B)$ any subset of its elementary letters. Then the letters in F are Ramsey eliminable from B in the sense of van Benthem.

Proof. We need to show that every partial valuation v on $E(B) \setminus F$ with $v(Cn(B) \cap L_{E(B)\setminus F}) = 1$ can be extended to a partial valuation v^+ on $E(B)$ with $v^+(B) = 1$.

Let v be a partial valuation on $E(B) \setminus F$ with $v(Cn(B) \cap L_{E(B)\setminus F}) = 1$. Suppose for reductio ad absurdum that v cannot be extended to a partial valuation v^+ on $E(B)$ with $v^+(B) = 1$. Let S be the state-description corresponding to v, i.e., the set of all literals in $L_{E(B)\setminus F}$ that are true under v. In the limiting case that $F = E(B)$ so that $E(B) \setminus F = \emptyset$, put $S = \{\top\}$.

We note first that $S \cup B$ is inconsistent. Reason: For any partial valuation w on $E(S \cup B) = E(B)$ with $w(S \cup B) = 1$ we have $w(S) = 1$ so w must must agree with v over F, so w is an extension of v to $E(B)$. Also $w(B) = 1$, contrary to the supposition.

Since $S \cup B$ is inconsistent, compactness tells us that there is a formula s that is the conjunction of finitely many elements of S, such that $\neg s \in Cn(B)$. But also by construction, $\neg s \in L_{E(B)\setminus F}$. Hence $\neg s \in Cn(B) \cap L_{E(B)\setminus F}$ and so by hypothesis $v(\neg s) = 1$, contradicting the fact that by the construction of S we have $v(s) = 1$. □

This argument is along much the same lines as that for the characterization of friendliness in terms of consistency in section 1.4. Like that characterization, it does not carry over to first-order contexts; indeed, the counterexample given in section 1.4 also serves here.

Corollary. De Bouvère's method can never be used in purely propositional logic as a way of showing that an elementary letter is not explicitly definable given a set A of propositional formulae.

Verification. Apply the observation with F chosen to be a singleton subset of $E(B)$. $\square$

2.7. Leśniewski's Criterion of Conservativity

Friendliness is also closely related to the criterion of conservativity (alias noncreativity) in the theory of definition.

In lectures of the early 1920s, Leśniewski articulated two criteria that we usually want definitions to satisfy: eliminability and conservativity. A published account was given in Leśniewski (1931), with an easily accessible exposition in Suppes (1957, chapter 8). It is conservativity that connects with friendliness. The concept is usually formulated in the context of first-order logic. To clarify the link with friendliness, we again extract the purely propositional content.

Let A be any set of propositional formulae and let x be a formula. A,x is said to be a *conservative extension* of A iff $Cn(A, x) \cap L_A \subseteq Cn(A)$, i.e., iff $A \vdash c$ for every $c \in L_A$ such that $A, x \vdash c$.

Observation. In the propositional context: $A \mathrel{|\!\approx} x$ iff A,x is a conservative extension of A.

Proof. We already know from the first refinement of the characterization of friendliness in terms of consistency, in section 1.4, that $A \mathrel{|\!\approx} x$ iff (1) $A \vdash c$ for every $c \in L_A$ with $x \vdash c$. So we need only show the equivalence of this with (2) $A \vdash c$ for every $c \in L_A$ with $A, x \vdash c$.

One direction is immediate: by the monotony of classical consequence, (2) clearly implies (1). For the converse, suppose (1). Suppose $c \in L_A$ and $A, x \vdash c$; we need to show $A \vdash c$. Since $A, x \vdash c$ compactness tells us that $a, x \vdash c$ where a is the conjunction of some finite subset of A, and so also $x \vdash a \to c$. Clearly since $c \in L_A$ we also have $a \to c \in L_A$ So we may apply (1) to get $A \vdash a \to c$, and so since $A \vdash a$ we have $A \vdash c$ as desired. $\square$

Corollary. On the level of propositional logic: A, x is a conservative extension of A iff the letters in $E(x) \setminus E(A)$ are Ramsey eliminable from A, x in the sense of Rantala.

Verification. By the observation just established, A, x is a conservative extension of A iff $A \mathrel{|\!\approx} x$. By the first observation of section 2.6, $A \mathrel{|\!\approx} x$ iff the letters in $E(x) \setminus E(A)$ are Ramsey eliminable from A, x in the sense of Rantala. $\square$

Again this corollary is known to fail in the first-order context, where only the right-to-left half holds. An equivalence does hold, but it is between the left and a weaker version of the right: Γ, φ is a conservative extension of Γ iff every model of Γ is elementary equivalent to (i.e., satisfies the same first-order formulae as) some model of Γ that can be expanded to a model of Γ, φ.

2.8. Information-Preserving Paraconsistent Consequence

A less intimate connection with friendliness can be found in the construction of a certain paraconsistent consequence relation, effected in Pietruszczak (2004). This relation, which is a subrelation of classical consequence, is defined by Pietruszczak using a notion of preservation of information. But he also gives it an alternative characterization (his theorem 6.1) that makes contact with friendliness, or more precisely, with its syntax-independent counterpart sympathy, which we will define below in section 3.1.

Specifically, Pietruszczak's relation of information-preserving consequence holds between a formula a and a formula x iff four conditions hold: a classically entails x; a is classically consistent; x is not a tautology; and a further condition, formulated in terms of valuations, also holds. This further condition is not given a name, but is exactly the relation of sympathy, holding in the reverse direction from x to a.

Thus, roughly speaking, the syntax-independent version of friendliness has been used as one of the ingredients to construct a certain kind of paraconsistent subrelation of classical consequence. We have, in other words, an application of the relation.

The present author would comment, however, that the paraconsistent consequence so defined has a rather mixed bag of properties. As well as failing certain consequences that the paraconsistent logician desperately seeks to avoid (e.g., implication from $a \wedge \neg a$ to any proposition whatsoever, and from any proposition to $x \vee \neg x$), and failing others that some are willing to lose in order to achieve this (e.g., from a to $a \vee x$ for any x) the relation fails certain other properties that few paraconsistent logicians would be happy to see depart.

One of these is closure of the consequence relation under uniform substitution (of arbitrary formulae for elementary letters). Others are implication from $p \wedge q$ to any of $p \vee q, p \leftrightarrow q, p \to q, q \to p$, and likewise from $p \leftrightarrow q$ to either of $p \to q, q \to p$. Verification of all these failures is straightforward: none of the right formulae is friendly to the left one.

2.9. Coupled Semantic Decomposition Trees

Finally, we mention a connection with the theory of semantic decomposition trees (alias semantic tableaux) in classical logic. Developed by Beth, Hintikka and others, these trees entered the arena of textbooks with Jeffrey [1967]. Designed to test formulae for satisfiablility, the trees can of course be used to test an inference for invalidity by checking the satisfiability of the set (or conjunction) consisting of the premises and negation of the conclusion. But Jeffrey also suggested another technique for the purpose, which he called 'coupled trees'.

Roughly speaking, he constructed a (signed) tree for the premises, and another one for the conclusion. If every open branch of the former tree contains all the signed elementary letters (alias literals) that occur on some open branch of the latter one, then the inference is valid. However, as Jeffrey noted, the converse is not true without qualification. This is due to the possible absence of elementary letters in branches of the first tree, as for the inference from p to $q \vee \neg q$, likewise from p to $(p \wedge q) \vee (p \wedge \neg q)$. For this reason, he introduced an additional rule allowing the introduction of new elementary letters (by branching to an arbitrary formula and to its negation) when constructing a tree.

In the revised version of the textbook, published in 1981, Jeffrey omitted the technique of 'coupled trees' altogether, presumably because of the inelegance of the additional rule. In the meantime, Dunn [1976] showed that it could be adapted neatly to the so-called first-degree entailments of relevance logics. One simply requires that every branch (even closed) of the former tree contains all the signed elementary letters that occur on some branch (even closed) of the latter tree. This characterizes first-degree entailment without the need for any additional rules.

We remark that the technique of 'coupled trees' is even more naturally suited to determining whether a set A of formulae is friendly to another formula x. Construct the two (signed) trees as before. Call two branches *compatible* iff they do not contain any elementary letter with opposite signs. To test whether A is friendly to x, we simply check whether every open branch of the tree for A is compatible with some open branch of the tree for x. This characterizes friendliness without additional rules. We omit the straightforward verification.

3. From Friendliness to Sympathy

3.1. Definitions

We now consider a normalized version of friendliness that is syntax-independent on the left as well as on the right.

It is well known that for any finite set A of Boolean formulae, there is a unique least set F of elementary letters such that A is classically equivalent to some set of formulae in the language generated by F.

Although this is usually stated and proven for finite sets A only, it also holds for infinite ones. More specifically, let A be any set of formulae:

- Put $E!(A)$ to be the set of all letters p that are *essential for* A, in the sense that there are two valuations v, w, on the set E of all elementary letters of the language, that agree on all letters other than p but disagree in the value they give to A. Clearly $E!(A) \subseteq E(A)$.
- Put A^* to be the set of all formulae x with both $A \vdash x$ and $E(x) \subseteq E!(A)$. Clearly $E(A^*) = E!(A)$.

Clearly, whenever $A \dashv\vdash B$ then $E!(A) = E!(B)$ and also $A^* = B^*$. Moreover, as we show in the Appendix:

Least letter-set theorem. $A \dashv\vdash A^*$, and for every set B of formulae with $A \dashv\vdash B, E(A^*) \subseteq E(B)$.

We say that a set A of formulae is *sympathetic* to x and write $A \mathrel{|\!\!\sim} x$, iff $A^* \mathrel{\approx\!\!\!|} x$. This notion can be seen as a normalized version of friendliness, making it syntax-independent in the left argument.

Unrestricted Left Classical Equivalence for $\mathrel{|\!\!\sim}$. Whenever $A \mathrel{+\!\!+} B$, then $A \mathrel{|\!\!\sim} x$ iff $B \mathrel{|\!\!\sim} x$.

Verification. Whenever $A \mathrel{+\!\!+} B$ then as noted $A^* = B^*$, so $A^* \mathrel{\approx\!\!\!|} x$ iff $B^* \mathrel{\approx\!\!\!|} x$, i.e., $A \mathrel{|\!\!\sim} x$ iff $B \mathrel{|\!\!\sim} x$. $\square$

From the least letter-set theorem we have immediately the following useful criterion for membership in $E!(A)$.

Criterion for Membership in $E!(A)$. Let p be any elementary letter. Then $p \in E!(A)$ iff $p \in E(B)$ for every set B of formulae with $B \mathrel{+\!\!+} A$.

We also have the following four criteria for sympathy.

Criteria for Sympathy. Each of the following is equivalent to $A \mathrel{|\!\!\sim} x$:

(a) $B \mathrel{\approx\!\!\!|} x$ for every B with $A \mathrel{+\!\!+} B$ and $E(B) = E!(A)$
(b) $A^* \mathrel{\approx\!\!\!|} x$
(c) $B \mathrel{\approx\!\!\!|} x$ for some B with $A \mathrel{+\!\!+} B$ and $E(B) = E!(A)$
(d) $B \mathrel{\approx\!\!\!|} x$ for some B with $A \mathrel{+\!\!+} B$.

Verification. $A \mathrel{|\!\!\sim} x$ is defined as (b), and immediately (a) $\Rightarrow$ (b) $\Rightarrow$ (c) $\Rightarrow$ (d). So we need only show (d) $\Rightarrow$ (a). Suppose $B \mathrel{\approx\!\!\!|} x$ for some B with $A \mathrel{+\!\!+} B$. Let $A \mathrel{+\!\!+} C$ and $E(C) = E!(A)$. We need to show $C \mathrel{\approx\!\!\!|} x$. Let v be any partial valuation on $E(C)$ with $v(C) = 1$. We need to find an extension v^+ of v to $E(C, x)$ with $v^+(x) = 1$. Since $E(C) = E!(A) = E(A^*) \subseteq E(B)$ by the least letter-set theorem, we may fix an arbitrary extension w of v to $E(B)$. Since $C \mathrel{+\!\!+} A \mathrel{+\!\!+} B$, we have $w(B) = 1$. Since $B \mathrel{\approx\!\!\!|} x$ there is an extension w^+ of w to $E(B, x)$ with $w^+(x) = 1$. Then w^+ is an extension of v to $E(B, x)$. Since $E(C) \subseteq E(B)$ we also have $E(C, x) \subseteq E(B, x)$, so we may restrict w^+ to $E(C, x)$, call it w^{+-}. Clearly w^{+-} is still an extension of v and also $w^{+-}(x) = w^+(x) = 1$, so we may put $v^+ = w^{+-}$ and it has the desired properties. $\square$

Corollary: Broadening. Whenever $A \mathrel{\approx\!\!\!|} x$ then $A \mathrel{|\!\!\sim} x$.

Verification. By criterion (d). $\square$

Evidently, the inclusion converse to broadening fails. Example: $p \wedge (q \vee \neg q) \mathrel{\not\approx\!\!\!|} p \wedge q$ but $(p \wedge (q \vee \neg q)) \mathrel{|\!\!\sim} p \wedge q$ since $(p \wedge (q \vee \neg q)) \mathrel{+\!\!+} p \mathrel{\approx\!\!\!|} p \wedge q$.

3.2. Property Failures for Sympathy: Inherited and New

All of the property failures that we bulleted for $\mathrel{\approx\!\!\!|}$ in section 1.3 are also failures for $\mathrel{|\!\!\sim}$. We can take the same counterexamples and observe that for each premiss a, $E!(a) = E(a)$. On the other hand and perhaps surprisingly, there are two important properties that succeeded for $\mathrel{\approx\!\!\!|}$ but fail for $\mathrel{|\!\!\sim}$: local disjunction in the premisses and compactness.

The following example, due to Pavlos Peppas (personal communication) illustrates the failure of local disjunction in the premises.

Counterexample to Local Disjunction in the Premises. Put $a = p \vee r$, $b_1 = p \wedge q$, $b_2 = \neg q$, and $x = \neg q \vee \neg r$. Then $E(b_2) \subseteq E(a, b_1)$; $E(b_1) \subseteq E(a, b_2)$; $a, b_1 \mid\!\sim x$; $a, b_2 \mid\!\sim x$; but $a, b_1 \vee b_2 \not\mid\!\sim x$.

Verification. Clearly $E(b_2) \subseteq E(a, b_1)$ and indeed $E!(b_2) \subseteq E!(a, b_1)$. Also $E(b_1) \subseteq E(a, b_2)$ and indeed $E!(b_1) \subseteq E!(a, b_2)$. Also $a, b_1 \mid\!\sim x$ since $\{a, b_1\} \dashv\vdash b_1 \approx x$, applying criterion (d) for sympathy. Also $a \wedge b_2 \vdash x$ so that $a \wedge b_2 \approx x$ and thus $a, b_2 \mid\!\sim x$. But $a, (b_1 \vee b_2) \not\mid\!\sim x$.

To check the last, note that $a \wedge (b_1 \vee b_2) = (p \vee r) \wedge ((p \wedge q) \vee \neg q) \dashv\vdash p \vee (r \wedge \neg q)$ so that $E!(a, (b_1 \vee b_2)) = \{p, q, r\}$. So by criterion (a) for sympathy, it suffices to check that $p \vee (r \wedge \neg q) \not\approx \neg q \vee \neg r$. Since every letter on the right already occurs on the left, it suffices to show $p \vee (r \wedge \neg q) \not\vdash \neg q \vee \neg r$ by the reduction case for friendliness (section 1.4). But this is clear putting $v(p) = v(q) = v(r) = 1$. □

By suitably tweaking this example, we can turn it into one that illustrates the failure, for sympathy, of the closely related rule of proof by exhaustion.

Counterexample to Proof by Exhaustion. Put $a = p \vee \neg q \vee r$, $b = p \wedge q$, $x = \neg q \vee \neg r$. Then $a, b \mid\!\sim x$; $a, \neg b \mid\!\sim x$; but $a \not\mid\!\sim x$.

Verification. Similar to that of the preceding example, but we give the details. Again we have $a, b \mid\!\sim x$ since $\{a, b\} \dashv\vdash b \approx x$, applying criterion (d) for sympathy. Also $a, \neg b \mid\!\sim x$ since $a, \neg b \vdash x$. But $a \not\mid\!\sim x$ since $E!(a) = \{p, q, r\}$, so by criterion (a) for sympathy, it suffices to check that $p \vee \neg q \vee r \not\approx \neg q \vee \neg r$. Since every letter on the right already occurs on the left, it suffices to show $p \vee \neg q \vee r \not\vdash \neg q \vee \neg r$ by the reduction case for friendliness. But this is clear putting $v(p) = v(q) = v(r) = 1$. □

The next example illustrates the failure of compactness for sympathy. Consider a language with countably many elementary letters $q, p_1, p_2, \ldots$.

Counterexample to Compactness. Put A to be the set of all formulae a_n that are of the form $(p_1 \wedge \ldots \wedge p_n) \vee q$ for odd $n \geq 1$, or of the form $(p_1 \wedge \ldots \wedge p_n) \vee \neg q$ for even $n \geq 1$. Then $A \mid\!\sim q$ but $B \not\mid\!\sim q$ for every finite non-empty subset $B \subseteq A$.

Verification. To show $A \mid\!\sim q$ it suffices, by criterion (d) for sympathy, to find an $X \dashv\vdash A$ with $X \not\approx q$. Putting $X = \{p_i : i \geq 0\}$ we clearly have the former, and since q does not occur in any formula in X we also have the latter.

Now let B be any finite non-empty subset of A. To complete the verification of the example, we need to show that $B \not\mid\!\sim q$, i.e., that $B^* \not\approx q$.

First, we show that q is essential to B. Consider the largest n such that $a_n \in B$; this exists because B is finite and non-empty. We examine the case that n is odd, so that $a_n = (p_1 \wedge \ldots \wedge p_n) \vee q$; the case for even n is similar. Put $v(p_i) = w(p_i) = 1$ for all $i < n$, $v(p_n) = w(p_n) = 0$, and $v(q) = 1$ while $w(q) = 0$. Then $w(a_n) = 0$ so that $w(B) = 0$. On the other hand, $v(a_n) = 1$ (since $v(q) = 1$) and also $v(a_i) = 1$ for all $i < n$ (since p_n does not occur in any such a_i) so that

$v(B) = 1$. Since v, w agree on all p_i for all $i \leq n$ while disagreeing on B, this shows that q is essential to B, as desired.

We can now show that $B^* \not\approx q$. Put $u(p_i) = 1$ for all $i \leq n$ and $u(q) = 0$. Then $u(a_i) = 1$ for all $i \leq n$ so that $u(B) = 1$ and hence $B \not\vdash q$; so since $B \dashv\vdash B^*$ we have $B^* \not\vdash q$. But since q is essential to B, q occurs in B^*. So by the first reduction case of section 1.4, since $B^* \not\vdash q$ we have finally $B^* \not\approx q$ completing the verification of the example.   □

3.3. Property Successes for Sympathy

Apart from disjunction in the premises and compactness, all of the other properties that we noted as satisfied by friendliness also hold for sympathy. We consider them one by one. Whenever possible, we derive the property for $\mathrel{\mid\!\sim}$ from the one for $\approx$, rather than argue from scratch. Most of the verifications are straightforward; only singleton cumulative transitivity is rather tricky, needing some lemmas on least letter-sets.

Supraclassicality for $\mathrel{\mid\!\sim}$. Whenever $A \vdash x$ then $A \mathrel{\mid\!\sim} x$.

Verification. Suppose $A \vdash x$. Then $A \approx x$ by supraclassicality for $\approx$, so $A \mathrel{\mid\!\sim} x$ by broadening.   □

Reduction Case for $\mathrel{\mid\!\sim}$. Whenever $E(x) \subseteq E!(A)$ then $A \mathrel{\mid\!\sim} x$ iff $A \vdash x$.

Verification. Right to left is given by supraclassicality. For the converse, suppose $E(x) \subseteq E!(A)$. Suppose $A \mathrel{\mid\!\sim} x$. By definition, $A^* \approx x$. Recalling that $E!(A) = E(A^*)$ so that $E(x) \subseteq E(A^*)$, the reduction case for friendliness tells us $A^* \vdash x$. Since $A \dashv\vdash A^*$ we have $A \vdash x$ as desired.   □

Characterization of $\mathrel{\mid\!\sim}$ in Terms of Consistency. $A \mathrel{\mid\!\sim} x$ iff every set of formulae in $L_{E!(A)}$ that is consistent with A, is consistent with x.

Verification. By definition, $A \mathrel{\mid\!\sim} x$ iff $A^* \approx x$. Applying the corresponding consistency characterization of $\approx$ and the fact that $A^* \dashv\vdash A$, the desired equivalence follows.   □

Right Weakening for $\mathrel{\mid\!\sim}$. Whenever $A \mathrel{\mid\!\sim} x \vdash y$ then $A \mathrel{\mid\!\sim} y$

Verification. From the definition of $\mathrel{\mid\!\sim}$ and right weakening for $\approx$.   □

This implies right classical equivalence for sympathy: whenever $x \dashv\vdash y$ then $A \mathrel{\mid\!\sim} x$ iff $A \mathrel{\mid\!\sim} y$. The relation $\mathrel{\mid\!\sim}$ is thus syntax-independent on both left and right.

Local Left Strengthening for $\mathrel{\mid\!\sim}$. Suppose $E!(B) \subseteq E!(A)$. If $B \vdash A \mathrel{\mid\!\sim} x$ then $B \mathrel{\mid\!\sim} x$.

Verification. Immediate from the corresponding property of $\approx$, the definition of $\mathrel{\mid\!\sim}$, and the fact that $A^* \dashv\vdash A$.   □

Local Monotony for $\mathrel{|\!\sim}$. Suppose $E!(B) \subseteq E!(A)$. If $A \mathrel{|\!\sim} x$ and $A \subseteq B$ then $B \mathrel{|\!\sim} x$.

Verification. If $A \subseteq B$ then $B \vdash A$.   □

Note that in these two 'local' properties, the locality condition concerns $E!(A), E!(B)$ rather than $E(A), E(B)$.

3.4. Singleton Cumulative Transitivity for Sympathy

We have postponed consideration of singleton cumulative transitivity because its proof requires two lemmas about least letter-sets.

Lemma. $E!(A, B) \subseteq E!(A) \cup E!(B) \subseteq E!(A) \cup E(B)$.

Verification. The right inclusion is immediate from $E!(B) \subseteq E(B)$. For the left inclusion, suppose $p \in E!(A, B)$. Then there are partial valuations v_0, v_1 on $E(A, B)$ that agree on all letters in this domain other than p, with $v_0(A, B) = 0$ and $v_1(A, B) = 1$. Since $v_0(A, B) = 0$, either $v_0(A) = 0$ or $v_0(B) = 0$.

Suppose the former; the argument for the latter is similar. Restrict v_0, v_1 to $E(A)$, call them v_0^-, v_1^-. Then $v_0^-(A) = 0$ whilst $v_1^-(A) = 1$, but v_0^-, v_1^- agree on all letters in their common domain other than p. Hence $p \in E!(A) \subseteq E!(A) \cup E!(B)$ as desired.   □

Lemma. If $A \mathrel{|\!\approx} x$ then $E!(A) \subseteq E!(A, x)$. Indeed, more generally: If $A \mathrel{|\!\approx}_{\forall\exists\forall} B$ then $E!(A) \subseteq E!(A, B)$.

Verification. Suppose $A \mathrel{|\!\approx}_{\forall\exists\forall} B$ (defined in section 1.9) and $p \in E!(A)$. From the latter, there are partial valuations v_0, v_1 on $E(A)$ that agree on all letters in this domain other than p, with $v_0(A) = 0$ and $v_1(A) = 1$. Since $A \mathrel{|\!\approx}_{\forall\exists\forall} B$, v_1 can be extended to a valuation v_1^+ on $E(A, B)$ with $v_1^+(B) = 1$, so $v_1^+(A, B) = 1$. Now extend v_0 to $E(A, B)$ by putting $v_0^+(q) = v_1^+(q)$ for every letter $q \in E(A, B) \backslash E(A)$. Then clearly v_0^+, v_1^+ agree on all letters in their common domain except p, and disagree on A, B since $v_1^+(A, B) = 1$ while $v_0^+(A, B) = 0$ since $v_0(A) = 0$. Hence $p \in E!(A, B)$ as desired.   □

Singleton Cumulative Transitivity for $\mathrel{|\!\sim}$. Whenever $A \mathrel{|\!\sim} x$ and $A, x \mathrel{|\!\sim} y$ then $A \mathrel{|\!\sim} y$.

Proof. Suppose $A \mathrel{|\!\sim} x$ and $A, x \mathrel{|\!\sim} y$. From the hypotheses we have $A^* \mathrel{|\!\approx} x$ and $(A, x)^* \mathrel{|\!\approx} y$. We need to show $A^* \mathrel{|\!\approx} y$.

Let v be any partial valuation on $E(A^*) = E!(A)$ with $v(A^*) = 1$. We need to find an extension w of v to $E(A^*, y) = E!(A) \cup E(y)$ with $w(y) = 1$.

Since $A^* \mathrel{|\!\approx} x$ and $v(A^*) = 1$, v can be extended to a v^+ on $E(A^*, x) = E!(A) \cup E(x)$ with $v^+(x) = 1$. By the first lemma, we may restrict v^+ to the subset $E!(A, x)$ of its domain, call it v^{+-}. By the second lemma, since $A^* \mathrel{|\!\approx} x$ we have $E(A^*) = E!(A) \subseteq E!(A, x)$, so v^{+-} is an extension of v. Also, $v^{+-}((A, x)^*) = v^+((A, x)^*) = v^+(A^*, x)$. Also $v^+(A^*) = v(A^*) = 1$ and $v^+(x) = 1$. Putting this together, $v^+(A^*, x) = 1$ so $v^{+-}((A, x)^*) = 1$.

Hence, since $(A,x)^* \approx y$, v^{+-} may be extended from $E!(A,x)$ to a valuation v^{+-+} on $E!(A,x) \cup E(y)$ with $v^{+-+}(y) = 1$. Since v^{+-} is an extension of v it follows that v^{+-+} is also an extension of v. Finally, restrict v^{+-+} to $E!(A) \cup E(y)$, which by the second lemma again is a subset of $E!(A,x) \cup E(y)$; call it v^{+-+-}. This is still an extension of v, defined on $E!(A)$, and also $v^{+-+-}(y)$ is well defined with $v^{+-+-}(y) = v^{+-+}(y) = 1$. Put $w = v^{+-+-}$ and the proof is complete.   □

3.5. Interpolation for Sympathy

An interpolation property for sympathy follows readily from its counterpart for friendliness. We need to be careful, however, about where we can write A, versus A^*, in the formulation.

Interpolation for $\vdash\!\sim$. Whenever $A \vdash\!\sim x$ there is a finite set $F \subseteq E(A^*) \cap E(x) \subseteq E(A) \cap E(x)$ of elementary letters, such that for every finite set G of elementary letters with $F \subseteq G \subseteq E(A^*)$ there is a formula b with the following properties:

1. $E(b) = G$
2. $A \vdash\!\sim b$ (indeed $A \vdash b$)
3. $b \vdash\!\sim x$
4. b is consistent, provided A is consistent
5. b is not a tautology, provided there is a non-tautology $y \in L_A \cap L_x$ with $A \vdash y$.

Proof. Suppose $A \vdash\!\sim x$. By definition, $A^* \approx x$. So by interpolation for friendliness, we have the above but with A^* in place of A in properties (2), (4), (5). Since $A \dashv\vdash A^*$ we also have (2), (4) for A. It remains to check condition (5).

Suppose there is a non-tautology $y \in L_A \cap L_x$ with $A \vdash y$. We need to find a non-tautology $z \in L_{A^*} \cap L_x$ with $A^* \vdash z$. Consider the 2^k formulae that can be obtained from y by substituting $\top, \bot$ for the k letters ($k \geq 0$) in $E(y)$ that are not in $E(A^*)$. Since y is not a tautology, at least one of these 2^k formulae is not a tautology; choose one as z. Clearly $z \in L_{A*} \cap L_x$. Also, since $A \vdash y$ and $A \dashv\vdash A^*$ we have $A^* \vdash y$ and so since the substitution producing z is the identity on A^* we have $A^* \vdash z$ and the verification is complete.   □

3.6. Further Remarks on the Concept of an Essential Letter

Karl Schlechta (personal communication) has observed that it is possible to generalize the notion of an essential letter, making it relative to an arbitrary *set of valuations* rather than to a *set of formulae*. In detail: let W be an arbitrary set of valuations. We say that a letter p is *essential to W* iff there are two valuations that agree on all letters other than p, but one in and the other outside W.

As is often the case when we pass to arbitrary sets of valuations in place of sets of formulae (which correspond to definable sets of valuations), we get an equivalent notion in the finite case, but a more general one in the infinite case with loss of some properties. Without following this through systematically, we give one example. When dealing with sets of formulae, we have the following:

Observation. Let A be any set of formulae. Then A is contingent (neither a tautology nor a contradiction) iff at least one of its elementary letters is essential to it.

Verification. Right to left is immediate from the definition of an essential letter. For the converse, suppose that A contingent. Then there are two partial valuations v, w on $E(A)$, with $v(A) = 1$ and $w(A) = 0$. From the latter, there is a formula $a \in A$ with $w(a) = 0$. Let v_w be the partial valuation on $E(A)$ defined by putting $v_w(p) = w(p)$ for all letters in $E(a)$, and $v_w(p) = v(p)$ for all other letters. Then v, v_w disagree on only finitely many letters, and we have $v(A) = 1$ while $v_w(a) = 0$ so that $v_w(A) = 0$.

Since v, w_v disagree on only finitely many letters, there is a finite chain $v_1, \ldots, v_n$ of partial valuations on $E(a)$ beginning with $v_1 = v$ and ending with $v_n = v_w$, each disagreeing with its predecessor on just one letter. Take the last v_k in the chain with $v_k(A) = 1$. Then $k < n$ and $v_{k+1}(A) = 0$. Thus v_k, v_{k+1} are partial valuations on $E(A)$ that agree on all letters except one, but give A different values, so that letter is essential to A. $\qquad\square$

This argument goes through no matter what the cardinality of the set of the elementary letters, and independently of whether they can be well ordered. But the observation fails for its counterpart in terms of sets of valuations, even for a countable language.

The counterpart says: Let W be any subset of the set of all valuations; then W is proper and non-empty iff at least elementary letter is essential to it. Right to left does hold: if at least one letter is essential to W, then immediately from the definition W is neither empty nor the set of all valuations. But left to right fails. Example: put W to be the set of all valuations that make only finitely many elementary letters true. This is neither empty nor the set of all valuations. But when a valuation is in W, so is every valuation that differs from it at exactly one letter.

4. Open Questions

4.1. Specific Problems

- Can we give an axiomatic characterization of friendliness (or for sympathy) that is more traditional in style than the one at the end of section 1.5?
- What is the most interesting way of defining friendliness in a first-order context, and which of its properties carry over?
- Which properties of the notion of an essential letter carry over when that notion is understood modulo an arbitrary set of valuations, as in section 3.6, rather than modulo a set of formulae?

4.2. Open-Ended Questions

- How much of the theory of friendliness remains if we generalize from the classical two-valued context to a many-valued one?
- Is it helpful to characterize friendliness and sympathy using appropriate three-valued possible worlds structures, with a relation between possible worlds representing the extension of one partial valuation by another?
- Are there any interesting connections between the theory of friendliness and possible-worlds semantics for intuitionistic logic?

5. Appendix

5.1. Proof of Least Letter-Set Theorem

As remarked in the text, proofs of the least letter-set theorem usually cover only the finite case. Perhaps the most elegant such proof, given for example by Parikh (1999), uses interpolation for classical logic. We recall it briefly.

Let A be any finite set of Boolean formulae. Since A is finite, $E(A)$ is also finite, so there is at least one minimal subset $F \subseteq E(A)$ with the property that A is classically equivalent to some set of formulae in the language generated by F. So we need only show that F is unique. Let G be any other such minimal set of letters. Then there are sets B, C of formulae in L_F, L_G respectively with $B \dashv\vdash A \dashv\vdash C$ so $B \vdash C$ so by interpolation for classical logic there is a set X of formulae in $L_{F \cap G}$ with $B \vdash X \vdash C$ so $A \vdash B \vdash X \vdash C \vdash A$ so $A \dashv\vdash X$. But since F, G were both minimal, it follows that $F = F \cap G = G$ and we are done.

Unfortunately, this elegant argument is not available in the infinite case, as we cannot assume that there is a minimal F with the property. We give a different proof covering the infinite as well as the finite case. We have not been able to ascertain whether such a proof already occurs in the literature.

We recall from section 3.1 the definitions that will be needed.

- $E!(A)$ is the set of all letters p that are *essential for* A, in the sense that there are two valuations v, w, on the set E of all elementary letters of the language, that agree on all letters other than p but disagree in the value they give to A. Clearly $E!(A) \subseteq E(A)$, and whenever $A \dashv\vdash B$ then $E!(A) = E!(B)$.
- A^* is the set of all formulae x with both $A \vdash x$ and $E(x) \subseteq E!(A)$. Clearly $E(A^*) = E!(A)$. Clearly, whenever $A \dashv\vdash B$ then $A^* = B^*$.

Clearly, it would be equivalent to formulate the definition of $E!(A)$ in terms of partial valuations on $E(A)$ rather than full valuations on the entire set E of elementary letters, but working with full valuations here streamlines the argument.

We proceed via a lemma. Roughly speaking, it says that letters that are individually inessential to a set of formulae, are also jointly so.

Lemma. Let v, w be any two valuations on E that agree on $E!(A)$. Then $v(A) = 1$ iff $w(A) = 1$.

Proof. First we use induction to show that the lemma holds whenever v, w disagree on only finitely many letters. Then we use this to show that it holds when they disagree on infinitely many letters.

For the basis of the induction put $n = 0$, i.e., suppose that v, w disagree on no letters. Then $v = w$ and we are done. For the induction step, suppose that the lemma holds whenever two valuations disagree on just n letters. Suppose v, w disagree on just $n+1$ letters $p_1, \ldots, p_n, p_{n+1}$. Let w' be a valuation that is just like w except that $w'(p_{n+1}) = v(p_{n+1})$. Then w' disagrees with v on just n letters, and so by the induction hypothesis $v(A) = 1$ iff $w'(A) = 1$. But also w' disagrees with w on just the one letter p_{n+1}. Since v, w agree on $E!(A)$ while disagreeing on p_{n+1} we know that $p_{n+1} \notin E!(A)$, i.e., p_{n+1} is not essential for A. Hence since w, w' agree on every letter other than p_{n+1} we have by the definition of essential letters that $w(A) = 1$ iff $w'(A) = 1$. Putting these together, $v(A) = 1$ iff $w(A) = 1$ as desired. This completes the induction.

Now suppose that v, w are any two valuations on L that agree on $E!(A)$ but differ on infinitely many letters. We want to show that $v(A) = 1$ iff $w(A) = 1$. Suppose otherwise; we obtain a contradiction. Then either $v(A) = 1$ while $w(A) = 0$, or $w(A) = 1$ while $v(A) = 0$. Consider the former; the latter case is similar.

Since $w(A) = 0$, we have $w(a) = 0$ for some $a \in A$. Let v_w be the valuation like v except for the letters in a, where it is like w. Then v_w disagrees with v on just finitely many letters. Moreover, none of those letters are in $E!(A)$. For suppose $v_w(p) \neq v(p)$. Then the letter p occurs in a, so $v_w(p) = w(p)$ so $w(p) \neq v(p)$ and thus $p \notin E!(A)$ by the supposition that v, w agree on $E!(A)$. Hence the finite part of the lemma gives us $v(A) = 1$ iff $v_w(A) = 1$. By supposition, $v(A) = 1$ so we have $v_w(A) = 1$. Since $a \in A$ this gives $v_w(a) = 1$. But $w(a) = 0$ and by the construction of v_w we have $v_w(a) = w(a)$. Hence $v_w(a) = 0$ giving us the desired contradiction. $\square$

Least Letter-set Theorem. $A \dashv\vdash A^*$, and for every set B of formulae with $A \dashv\vdash B$, $E(A^*) \subseteq E(B)$.

Proof. We need to show (1) $E!(A) \subseteq E(B)$ for every B with $A \dashv\vdash B$, and (2) $A \dashv\vdash A^*$.

For (1), suppose $A \dashv\vdash B$, $p \in E!(A)$, but $p \notin E(B)$; we obtain a contradiction. The diagram illustrates the argument that follows.

$v(A)$	$\neq$	$w(A)$
$=$		$=$
$v(B)$	$=$	$w(B)$

Since $p \in E!(A)$ there are valuations v, w on L with $v(q) = w(q)$ for all letters q with $q \neq p$, but $v(A) \neq w(A)$ (top row). Since $p \notin E(B)$ this implies

$v(B) = w(B)$ (bottom row). But since $A \dashv\vdash B$ we have both $v(A) = v(B)$ and $w(A) = w(B)$ (side columns), giving a contradiction.

For (2), by construction, we have $A \vdash A^*$. Suppose $A^* \nvdash A$; we derive a contradiction. Since $A^* \nvdash A$ there is a valuation v with $v(A^*) = 1$ and $v(A) = 0$, i.e., $v(a) = 0$ for some $a \in A$. Let S be the set of all literals $\pm q$ with $q \in E(A^*)$ such that $v(\pm q) = 1$. Then clearly $S \vdash A^*$. We break the argument into two cases, deriving a contradiction in each.

Case 1. Suppose S is inconsistent with A. Then by classical compactness, some finite subset $S_f \subseteq S$ is inconsistent with A. Hence $A \vdash \neg \wedge S_f$. Since all letters in $\neg \wedge S_f$ are in $E(A^*)$ it follows that $\neg \wedge S_f \in A^*$, so since $v(A^*) = 1$ we have $v(\neg \wedge S_f) = 1$. But by the construction of S we also have $v(\wedge S_f) = 1$, giving us the desired contradiction.

Case 2. Suppose S is consistent with A. Then there is a valuation w with $w(S) = w(A) = 1$. Since $w(S) = 1$ it follows that w agrees with v on all letters in $E(A^*)$. So the lemma tells us that $v(A) = 1$ iff $w(A) = 1$. So since $w(A) = 1$ we have $v(A) = 1$. Since $a \in A$, this gives $v(a) = 1$, contradicting $v(a) = 0$ and completing the proof of (2). □

References

[Beth,1953] Beth, Evert W. 1953. On Padoa's method in the theory of definition, *Nederl. Akad. Wetensch. Proc. Ser. A* 56: 330-339; also *Indagationes Mathematicae* 15: 330-339.

[Boole,1847] Boole, George. 1847. *The Mathematical Analysis of Logic*. Cambridge: Macmillan.

[Boole,1854] Boole, George. 1854. *An Investigation into the Laws of Thought*. London: Walton.

[de Bouvère,1959] de Bouvère, K.L. 1959. *A Method in Proofs of Undefinability*. Amsterdam: North Holland.

[Dunn,1976] Dunn, J.M. 1976. Intuitive semantics for first-degree entailments and coupled trees, *Philosophical Studies*, **29**, 149–168.

[Jeffrey,1967] Jeffrey, R.C. 1967. *Formal Logic: Its Scope and Limits* (second edition 1981). New York: McGraw-Hill.

[Lang *et al.*,2003] Lang, J., P. Liberatore, P. Marquis. 2003. Propositional independence: formula-variable independence and forgetting, *Journal of Artificial Intelligence Reseach*, 18: 391–443.

[Leśniewski,1931] Leśniewski, S. 1931. Über definitionen in der sogennanten Theorie der Deduktion. *Comptes Rendus des Séances de la Société des Sciences et des Lettres de Varsovie*, Classe 3, XXIV: 300-302.

[Lin and Reiter,1994] Lin, F. and R. Reiter (1994). Forget it!. In R. Greiner and D. Subramanian, eds. *Working Notes on AAAI Fall Symposium on Relevance*. Menlo Park: AAAI Press.

[Makinson,2005] Makinson, David. 2005. *Bridges from Classical to Nonmonotonic Logic*. London: College Publications. Series: Texts in Computing, vol 5.

[Makinson,2005a] Makinson, David. 2005. Friendliness for logicians. In Sergei N. Artemov, Howard Barringer, Artur S. d'Avila Garcez, Luis C. Lamb, and John Woods,
editors, *We Will Show Them! Essays in Honour of Dov Gabbay, Volume Two*, pages
259-292. College Publications, 2005.

[Padoa,1901] Padoa, A. 1901. Essai d'une théorie algébrique des nombres entiers, précédé
d'une introduction logique à une théorie deductive quelconque. *Bibliothèque du
Congrès International de Philosophie, Paris 1900*, vol 3: 309-365. Paris: Armand
Colin.

[Parikh,1999] Parikh, R. 1999. Beliefs, belief revision, and splitting languages. Pages 266–
278 of L. Moss et al eds, *Logic, Language and Computation*, vol 2. CSLI Lecture Notes
n° 96: 266-278. California: CSLI Publications.

[Pietruszczak,2004] Pietruszczak, A. 2004. The consequence relation preserving logical
information, *Logic and Logical Philosophy* 13: 89-120.

[Ramsey,1931] Ramsey, F.P. 1931. *The Foundations of Mathematics and Other Logical
Essays* ed. R.B. Braithwaite. London: Kegan Paul, Trench, Trubner.

[Rantala,1991] Rantala, V. 1991. Definitions and definability, pages 135-159 of James
H. Fetzer et al *Definitions and Definability: Philosophical Perspectives*. Dordrecht :
Kluwer.

[Sneed,1971] Sneed, J.D. 1971. *The Logical Structure of Mathematical Physics*. Dordrecht:
Reidel.

[Suppes,1957] Suppes, P. 1957. *Introduction to Logic*. Princeton: Van Nostrand.

[van Benthem,1978] Van Benthem, J.F.A.K. 1978. Ramsey eliminability, *Studia Logica*
37: 321-336.

[Weber,1987] Weber, A. 1987. Updating propositional formulae, pages 487–500 in L. Ker-
schberg, ed. *Proceedings of the First Conference on Expert Data Systems*. Benjamin
Cummings.

Acknowledgments

Many friendly logicians helped in various ways. In particular, thanks to Pavlos
Peppas for the counterexample to disjunction in the premiss for the relation of
sympathy in section 3.2, Lloyd Humberstone for discussions on links in part 2,
and Karl Schlechta for the concept of a letter essential to a set of valuations in
section 3.6. Anatoli Degtyarev, Kurt Engesser, Maribel Fernández, Dov Gabbay,
Jamie Gabbay, George Kourousias and Odinaldo Rodrigues also commented on
various versions.

David Makinson
Dept. of Philosophy, Logic & Scientific Method, London School of Economics
Houghton Street, London WC2A 2AE
United Kingdom
e-mail: david.makinson@gmail.com

J.-Y. Beziau (Ed.), *Logica Universalis*, *2nd edition*, 225–246

Logical Discrimination

Lloyd Humberstone

Abstract. We discuss conditions under which the following 'truism' does indeed express a truth: the weaker a logic is in terms of what it proves, the stronger it is as a tool for registering distinctions amongst the formulas in its language.

Mathematics Subject Classification (2000). Primary 03B20; Secondary 03B60.

Keywords. logics, consequence relations, synonymous formulas.

1. Introduction

Our topic is the idea that deductive strength varies inversely with discriminatory strength: the more a logic proves, the fewer distinctions (or discriminations) it registers. This is a thought often voiced, either in general terms, or with reference to a specific case. Here, for example, is what David Nelson had to say about the relationship between classical and intuitionistic logic:

> As we have suggested earlier, an argument favouring intuitionistic logic over the classical is the fact that the intuitionistic logic allows the classical distinctions in meaning and further ones besides. Classical logic is open to possible objection in that it identifies certain constructively distinct entities. Since we are speaking here of formal systems, we are interested in the general question of finding when one formal system allows distinctions among concepts which are not possible in another ([27], p. 215).

In a similar vein, Anderson and Belnap [1] write as follows when comparing the implicational fragments $\mathbf{T}_\rightarrow$, $\mathbf{E}_\rightarrow$, and $\mathbf{R}_\rightarrow$ (cited here in order of increasing deductive strength) of their logics of ticket entailment, entailment, and relevant implication; the initially mentioned "two systems" are the first and third just listed:

> These two systems, both intensional, exhibit two quite different ways of demolishing the theory of necessity enshrined in $\mathbf{E}_\rightarrow$: $\mathbf{R}_\rightarrow$ by making

stronger assumptions about identity or intersubstitutability (and hence having fewer propositional entities), and $\mathbf{T}_\rightarrow$ by making weaker assumptions (and hence having more distinct propositional entities). Modal systems, generally being weaker than their cousins, tend to make more distinctions; in $\mathbf{E}_\rightarrow$ we can distinguish A from $\Box A$, since, though the latter entails the former, the converse is neither true nor provable. As we saw in §5, adding $A \rightarrow .A \rightarrow A \rightarrow A$, i.e., $A \rightarrow \Box A$, to $\mathbf{E}_\rightarrow$ ruins this distinction and produces $\mathbf{R}_\rightarrow$: a stronger assumption produces fewer propositional entities ([1], p. 47).[1]

Sometimes the additional distinctions made available by passage to a weaker logic are thought of as making for an *embarras de richesses* when that logic is applied as the logic of a particular (typically, mathematical) theory. Troelstra and van Dalen [37] devotes a Section (3.7 of Chapter 1, entitled "Splitting of Notions") to replying to this objection – as it arises specifically in the passage from classical to intuitionistic logic – mainly by suggesting that in fact far fewer than the in-principle available distinct versions of what would in the classical case be alternative equivalent definitions of the same notion are of practical significance. In what follows we shall be concerned neither to sing the praises nor to lament the consequences of weakening a logic and thereby increasing the number of distinctions that have to be made as a result, contenting ourselves with an examination of the question of what background assumptions need to be in place in the general case for this "thereby" to be justified. We shall be concerned to see what these assumptions are, as well as to illustrate how, in cases in which they do not hold, a weaker logic may yet fail to support a greater number of distinctions. (See the discussion following Proposition 2.5 in this regard.) Alternatively put, strengthening a logic deductively need not, in such cases, result in collapsing any distinctions. We also consider the possibility, conversely, that a decrease (or increase) in discriminatory power need not signal a corresponding increase (or decrease, respectively) in deductive strength. While the particular distinctions that arise, to return to the previous example, in intuitionistic as opposed to classical mathematics – nonempty vs. inhabited, apartness vs. inequality (non-identity), etc. – might call for a logical discussion at the level of predicate logic, the general issue about discriminatory and deductive strength varying inversely can be illustrated without going beyond purely propositional logic. In the interest of simplicity, then, our general discussion as well as our illustrations are drawn from amongst propositional logics. The discussion presents a few elementary observations and examples, which might provide a stimulus for a general and systematic study of the topic, without itself pretending to constitute such a study.

---

[1]The passage continues with: "And of course further strengthening in the direction of the two-valued calculus produces a system which cannot tell the difference between Bizet's being French and Verdi's being Italian", rather lowering the tone since no formalization of "Bizet is French" and "Verdi is Italian" would render these two *logically* equivalent by the lights of classical logic.

Although we shall find the dictum that deductive and discriminatory strength vary inversely is not universally correct, it does hold up over a wide range of logics, so it is interesting to see the opposite presumption expressed in print. This is what we find in J. R. Lucas' discussion of a past-tense version of A. N. Prior's argument for the logical possibility of time without change (a version of which, purged of errors in an earlier formulation, appears in Prior [30]). Lucas [23], p. 10, writes: "Even within the austere framework of Lemmon's minimal tense logic K_t, we can distinguish between a dawn of creation in which the stars started in their courses the moment time began and a more leisurely inauguration in which they spent part of the morning doing nothing in unison". That is – presumably – we can distinguish the hypothesis that time had a beginning from the hypothesis that change had a beginning. Although what Lucas says may not seem especially clearly to amount to this, the present objection is different. It is to the confusion underlying any claim of the form "Even within the austere framework of Lemmon's minimal tense logic K_t, we can distinguish between X and Y." The *weaker* the logic, the *more* distinctions it allows, according to the by-and-large correct dictum enunciated above: so there is no "even" about it.[2]

We close this introduction with remarks on three related issues we shall not be further attending to. The first concerns the general theme of discrimination in logic, one aspect of which is the issue of more and less discriminating accounts of what a logic is. In Section 2 and 3, for example, we shall be concerned with logics as sets of formulas and logics as consequence relations.[3] It is well known that many distinct consequence relations on a given language induce (by taking the consequences of the empty set) the same logic-as-set-of-formulas,[4] and in this sense we may say that the 'consequence relations' account of what a logic is counts as *more discriminating* than the set-of-formulas account. Similarly, the use of generalized consequence relations in the style of Scott [32] (or more generally – see the preceding footnote – logics as sets of multiple-succedent sequents) is more discriminating still.[5] Another dimension of variation consists in how much attention is paid to rules: taking, for example, single-succedent sequents, we could say that two proof systems which render provable the same set of such sequents count as

---

[2]Setting aside the issue specifically about distinctions supported, teaching experience attests to the difficulty that students have with talk of one logic's being stronger than another, invariably intended by logicians, when no further qualification is added, to mean *deductively stronger*, but often suggesting the reverse to students, the stronger logic being taken to be the one making the more stringent demands in respect of what is provable. (Many examples of the customary usage alluded to here may be found in Mortensen and Burgess [26] and authors there quoted. The issue under discussion is whether for this or that purpose a stronger logic is better or worse than a weaker logic.)

[3]The latter could themselves be viewed as a special case of 'logics as sets of (single-succedent) sequents' – see the discussion after Proposition 3.2 below.

[4]In the terminology of Section 3 below, these are consequence relations which, though distinct, '1-agree'.

[5]See Gabbay [11], Theorem 13 on p. 8, Theorem 4 on p. 28, for example.

two systematizations of the same logic, or we could be more demanding and require for this that not only the same sequents should be provable but the same sequent-to-sequent rules should be derivable (= primitive or derived). Interesting as these issues are in their own right, they are not what we are talking about here. The discriminations we are concerned with are those made by a logic – however conceived – between formulas, not discriminations in respect of the individuation of logics themselves. This allows us derivatively to speak, for instance, of one logic making a finer discrimination between connectives than another, in the sense in which substructural logics support a distinction between, e.g., multiplicative and additive conjunction, which is collapsed in classical or intuitionistic logic – because this amounts to saying that the former logics discriminate, and the latter do not, between the *formulas* formed from two distinct propositional variables by compounding them on the one hand with the one connective and on the other with the other. (See pp. 15–17 of Paoli [28] for some discussion of the "suppression of distinctions" objection to the structural rules.)

Secondly, we are not directly concerned with semantically based measures of relative expressive power, such as closeness to functional completeness in logics determined by reducts of some single matrix, or the ability to distinguish between more frames (validating the logic) amongst normal modal logics interpreted by the Kripke semantics. ([14] gives one example of this kind of enterprise; note the title.) This is not to say that there are no connections between such issues and the more straightforwardly syntactical matter of discriminatory strength as understood here: just that we are not addressing any such connections here. Note that an 'inverse proportionality' between deductive strength and some such measures of expressive power is often remarked on – for example in Tennant [36] *à propos* of expressive power as the power to discriminate between non-isomorphic structures. There is also the matter of discrimination between elements within an individual structure, stylishly explored in Quine [31]. An algebraic incarnation of the latter theme arises with the (ternary) discriminator function t satisfying for arbitrary elements a, b, c: $t(a, b, c) = c$ if $a = b$, and $t(a, b, c) = a$ otherwise. Discriminator varieties – varieties generated by a class of algebras in which this function is a (fixed) term function – have turned out to have striking applications outside the realm of universal algebra: Burris ([6], esp. Section 5) shows how to 'reduce', in one reasonable sense of that word, an arbitrary first-order theory to an equational theory, in the context of such varieties.[6] No doubt there are further things that could go under the name of discriminatory strength from the point of view of interpretations of formal languages (and thus outside our present purview) but that should suffice by way of example.

Finally, a remark is in order on the measure of deductive strength we are employing, according to which one logic, S_1, is at least as strong as another, S_0, (resp., strictly stronger than S_0) when $S_0 \subseteq S_1$ (resp., $S_0 \subsetneq S_1$). Those formulations are suited to the logics-as-sets-of-formulas of the following section, while for Section

---

[6]See Bignall and Spinks [3] for some further developments and references.

3 the corresponding inclusions are between consequence relations. Though in the main cases we consider below the language does not change, the definition does not rule out the possibility that the language of S_0 should be properly included in that of S_1, in which case the issue of relative deductive strength becomes clouded by the possibility of a translation from the larger language to the smaller which allows for a faithful embedding of S_1 into S_0. Thus Łukasiewicz [24] argued that intuitionistic propositional logic should be regarded as a extension rather than – as is customarily maintained – a sublogic of classical propositional logic, because if the connectives of classical logic were taken as defined in terms of conjunction and negation, the similarly notated but now to be distinguished connectives of intuitionistic logic could be regarded as new non-classical primitives (somewhat in the style of modal logic). This alternative point of view was available because of Gödel's observation that the conjunction–negation fragments of classical and intuitionistic logic coincided, which is of course so on the 'set-of-formulas' conception of logics (favoured by Łukasiewicz), though not on the 'consequence relation' conception; however, other examples can be given of apparent reversals in comparative deductive strength attendant upon judicious definitional manoeuvres which do work equally well at the level of consequence relations. One such example is given a particularly crisp presentation in Béziau [2], where the puzzling nature of the general phenomenon is also emphasized. (Further discussion of the phenomenon, as well as of Béziau's specific example, appears in [18].) Here we simply set such matters to one side, taking relative deductive strength as given quite literal-mindedly by set-theoretic inclusion – no "re-notation" permitted.

2. Discrimination In Logics as Sets of Formulas

The idea that the stronger the logic (deductively), the more distinctions it collapses, voiced by the authors quoted in the preceding section, is conveniently formulated in general terms with the aid of the notion of synonymy in the sense of Smiley [34].[7] As also mentioned in that section, we confine ourselves to two (from amongst many possible) conceptions as to what constitutes a logic: the conception of logics as (certain) sets of formulas of a formal language, and the somewhat richer conception of logics as consequence relations on such a language. Working with the former conception, we say that formulas A and B are *synonymous* according to (or "in") a logic S (considered as a set of formulas from some language to which A and B belong) when for any formula $C(A)$ in which A occurs zero or more times as a subformula, and any formula $C(B)$ resulting from replacing zero or more such occurrences by B, we have $C(A) \in S$ if and only if $C(B) \in S$. (We can regard the 'context' $C(\cdot)$ as a formula $C(q)$ in which amongst others there occurs the propositional variable q, with $C(A)$, $C(B)$ the results of uniformly substituting A,

---

[7]In fact Smiley writes "synonymity", as do the authors of [10], explaining at p. 34 there the relation of this concept to the main concepts of abstract algebraic logic in the tradition alluded to in note 9 below.

B, respectively, for that variable.[8] Compare the notation with Δ below, in which the exhibited variables are the only ones allowed to occur.) In some formulations to follow, we render "$C(A) \in S$" in words by saying that $C(A)$ is provable in – or is a theorem of – the logic S. (Though we are confining ourselves to sentential logic for illustrative purposes here, a similar notion of synonymy could be given in an obvious way for expressions of arbitrary syntactic categories.) If we are thinking, as on the second conception of logic mentioned above, of a logic as a consequence relation, $\vdash$, then we say that A and B are *synonymous according to* $\vdash$ when, with the $C(\cdot)$ notation understood as above, $C_1(A), \ldots, C_n(A) \vdash C_{n+1}(A)$ if and only if $C_1(B), \ldots, C_n(B) \vdash C_{n+1}(B)$. Of course, as remarked in Section 1, many still richer conceptions of what should constitute a logic are possible, but our purposes will be served by considering only these two. In the present section, we stick with the first 'logics-as-sets-of-formulas' conception. In such a setting, the general idea that increasing deductive strength goes with reducing discriminatory power is embodied in (1) below, in which we denote, for logics S_0 and S_1, for simplicity presumed to have the same language, the relation of synonymy according to S_i by $\equiv_i$ (to avoid a proliferation of subscripts).

$$S_0 \subseteq S_1 \text{ if and only if } \equiv_0 \subseteq \equiv_1 \tag{1}$$

Since the original idea is that increasing discriminatory power goes with decreasing logical strength, a formulation in terms of strictly increasing and decreasing discrimination and strength, respectively, may be found attractive:

$$S_0 \subsetneq S_1 \text{ if and only if } \equiv_0 \subsetneq \equiv_1 \tag{2}$$

It is the "only if" direction of (2) that most directly encapsulates the dictum that the weaker a logic is deductively, the more discriminating it is amongst formulas, since it says that whenever one logic, here S_0, is strictly weaker (deductively) than another, S_1, then the former logic collapses strictly fewer distinctions between pairs of formulas than the latter, thus making finer discriminations between formulas. Arguably, in adding the converse, the biconditional formulation of (2) captures the idea that deductive and discriminatory strength vary inversely. (2) is a consequence of (1), but we shall concentrate on (1) itself, considering separately the possibility of counterexamples to its "if" and "only if" directions, and begin with some simple conditions which suffice to rule out such counterexamples. We follow a similar pattern in Section 3, except that there we take logics to be consequence relations rather than collections of formulas. In either case, we take the languages concerned to be based on a countable supply of propositional variables (sentence letters) amongst which are p, q, and r, with formulas generated from these by application of sentence connectives in the usual way. To avoid complications, when two logics are considered in the same breath (as with the S_0, S_1 of (1) and (2) above) we assume for the most part that they are logics in the same language.

---

[8]The substitution of B for A, or better, *replacement* of A by B in the transition from $C(A)$ to $C(B)$ is of course itself required to be uniform.

Let Δ be a set of formulas in which the only propositional variables to appear are p and q, to emphasize which we write Δ as $\Delta(p, q)$, with $\Delta(A, B)$ as the result of substituting the formula A for every occurrence of p and B for every occurrence of q in the formulas in $\Delta(p, q)$. Adapting a usage of T. Prucnal and A. Wroński (see Czelakowski [8], [9]), we call logic S (in the set-of-formulas sense) *equivalential* if there is a set $\Delta(p, q)$ of formulas in the language of S with the property that A and B are synonymous according to S if and only if $\Delta(A, B) \subseteq S$. (*Cf.* also Porte [29], where the terminology of formula-definable congruences is used instead.)[9] The simplest example of such a $\Delta(p, q)$ would be $\{p \leftrightarrow q\}$, which shows, amongst many others, classical logic to be equivalential. In a purely implicational logic, such as **BCI** logic, which comprises all the consequences under the rule Modus Ponens of instances of the three schemas **B**, **C**, and **I** below, we obtain a similar effect by taking $\Delta(p, q)$ to be $\{p \rightarrow q, q \rightarrow p\}$.

B $\qquad (B \rightarrow C) \rightarrow ((A \rightarrow B) \rightarrow (A \rightarrow C))$

C $\qquad (A \rightarrow (B \rightarrow C)) \rightarrow (B \rightarrow (A \rightarrow C))$

I $\qquad A \rightarrow A$

We shall return to this logic and some of its close relatives below. (These logics were intensively investigated by C. A. Meredith, to whom the combinator-derived labelling – "**BCI**" etc. – is also due. Discussion and extensive bibliographical references may be found in Hindley [12].) For the moment, we need to consider the following variation on this theme. Call logics, S_0 and S_1, presumed for simplicity to be in the same language, *similarly equivalential* if there is a set $\Delta(p, q)$ of formulas of that language with, for $i = 0, 1$, A and B are synonymous according to S_i if and only if $\Delta(A, B) \subseteq S_i$. Thus S_0 and S_1 are not just equivalential in that there is some set of formulas licensing the interreplaceability of arbitrary formulas A and B – i.e., the provability of appropriate substitution instances of which is necessary and sufficient for the synonymy of A and B, but it must be the *same* set for both logics. This relationship between S_0 and S_1 provides a simple and obvious sufficient condition for the "only if" direction of (1) above:

---

[9]In the original usage, it is logics as consequence relations rather than as sets of formulas, that are said to be equivalential. That usage has considerable currency in the literature on contemporary 'abstract algebraic logic' – [4], [8], [10], *q.v.* for the definition of "equivalential" as applied to consequence relations (or 'deductive systems' as this literature would have it). We have chosen to write "Δ" here to echo the choice made in Blok and Pigozzi [4] – but without their infix notation – for what they call a set of 'equivalence formulas' (though arguably 'congruence formulas' would be a more appropriate description). Though we make little direct contact with this tradition, there are some connections, especially as suggested in the following remark from Font *et al.* [10], p. 24: "One of the reasons why classical logic has its distinctive algebraic character lies precisely in the fact that logical equivalence and logical truth are reciprocally definable." (*Cf.* the proof of Proposition 2.3 below.) The remark just quoted could convey the misleading impression that the relation of logical equivalence – or more to the point, synonymy – associated with classical propositional logic is not thus associated with any other logic. If we take S as the set of classical tautologies in the language with, say, negation and implication as primitive connectives and S' as the set of formulas in this same language whose negations like in S, then S-synonymy and S'-synonymy coincide, even though $S \neq S'$.

Proposition 2.1. *For any similarly equivalential logics S_0 and S_1, $S_0 \subseteq S_1$ implies $\equiv_0 \subseteq \equiv_1$.*

Proof. Suppose S_0 and S_1 are similarly equivalential, with replacement-licensing formulas $\Delta(p,q)$, that $S_0 \subseteq S_1$, and that $A \equiv_0 B$. Since $A \equiv_0 B$, we have $\Delta(A,B) \subseteq S_0$, so since $S_0 \subseteq S_1$, $\Delta(A,B) \subseteq S_1$, and thus, finally $A \equiv_1 B$. $\qquad\square$

Remark 2.2. As this proof shows, the requirement of being similarly equivalential is stronger than is actually called for (and was employed for the sake of a succinct formulation). If we let Δ_i be the set of replacement-licensing formulas for S_i ($i = 1, 2$), then all we have is that $\Delta_1 \subseteq \Delta_0$ – and not also the converse inclusion.

For the other direction of (1), we are also able to find a fairly simple sufficient condition, frequently satisfied in practice. Again some terminology is needed for its formulation. A logic S is *monothetic* if all its theorems are synonymous, i.e, if for all $A, B \in S$, we have $A \equiv_S B$. (The terminology is motivated by the consideration that for such logics there is, to within synonymy, only one theorem or 'thesis'.) Note that if the language of S has a binary connective $\rightarrow$ for which $\{p \rightarrow q, q \rightarrow p\}$ licenses replacements, and S is closed under Modus Ponens for this connective, then as long as every instance of the schema $\boldsymbol{K}$ is provable:

$$\boldsymbol{K} \qquad A \rightarrow (B \rightarrow A),$$

S is monothetic. This applies to all the intermediate logics, intuitionistic and classical logic included, as well as to $\boldsymbol{BCK}$ logic, a pure implicational logic axiomatized as $\boldsymbol{BCI}$ logic was above, except putting $\boldsymbol{K}$ in place of $\boldsymbol{I}$ (all instances of which are now derivable). $\boldsymbol{BCI}$ logic itself, as well as $\boldsymbol{BCIW}$ logic, for which we add the contraction schema

$$\boldsymbol{W} \qquad (A \rightarrow (A \rightarrow B)) \rightarrow (A \rightarrow B)$$

are well known non-monothetic logics. (These last two are the implicational fragments, respectively, of Girard's linear logic and of the the relevant logic $\mathbf{R}$, according to neither of which are the provable formulas $p \rightarrow p$ and $q \rightarrow q$ synonymous. See the discussion following Proposition 2.5 below.)

Proposition 2.3. *Let S_0 and S_1 be monothetic logics with $S_0 \cap S_1 \neq \varnothing$. Then $\equiv_0 \subseteq \equiv_1$ implies $S_0 \subseteq S_1$.*

Proof. Assuming S_0 and S_1 as described, choose $B \in S_0 \cap S_1$. Suppose that $\equiv_0 \subseteq \equiv_1$ and that $A \in S_0$, with a view to showing that $A \in S_1$. Since $A \in S_0$ and S_0 is monothetic, $A \equiv_0 B$, and so $A \equiv_1 B$. Since S_1 is also monothetic and $B \in S_1$, $A \in S_1$. $\qquad\square$

We turn to the negative business for this section, with a counterexample – or family of counterexamples – to the "if" direction of (1) above. To describe the examples, we need to mention another schema, all instances of which are provable in $\boldsymbol{BCI}$ logic:

$$\boldsymbol{B'} \qquad (A \rightarrow B) \rightarrow ((B \rightarrow C) \rightarrow (A \rightarrow C)).$$

As with the other Meredith-style labelling, $BB'I$ logic comprises the Modus Ponens consequences of all instances of the schemata named in the label. We use this convention without further comment for other cases as they arise, and further, write such things as "$BB'I \subseteq BCI$" to abbreviate the claim – just made – that $BB'I$ logic is a sublogic of BCI logic. For a proof of the following, see Theorem 5.1 in Martin and Meyer [25], as well as the discussion in their introductory section:

Lemma 2.4. (E. Martin) *If for formulas A, B, we have $A \to B$ and $B \to A$ both provable in $BB'I$ logic then A is the same formula as B.*

Proposition 2.5. *Let S be any logic with $I \subseteq S \subseteq BB'I$. Then the relation $\equiv_S$ is the relation of identity between formulas.*

Proof. Since $\equiv_S$ is reflexive for any S, we have only to show that for S between I and $BB'I$, if $A \equiv_S B$ then $A = B$. Since $A \to A \in S$ for any $S \supseteq I$, if $A \equiv_S B$ then $A \to B \in S$ and $B \to A \in S$. But we are also supposing that $S \subseteq BB'I$, so each of $A \to B$ and $B \to A$ is also $BB'I$-provable, implying by Lemma 2.4 that $A = B$.   □

As a corollary to Proposition 2.5, then, we have that all logics between I and $BB'I$ have the same synonymy relation, giving rise to a range of counterexamples to the "if" half of (1):

Example. (A range of examples, really.) If we take S_0 as $BB'I$ logic and S_1 as any one of $I, BI, B'I$, we have $\equiv_0 \subseteq \equiv_1$ while $S_0 \not\subseteq S_1$. (Alternatively, we can see these as counterexamples to the "only if" half of (2).)

Proposition 2.3 gave sufficient conditions which together ruled out this situation, namely (i) that each of S_0 and S_1 was monothetic, and (ii) that $S_0 \cap S_1 \neq \varnothing$. Clearly in the present instance condition (ii) is satisfied – indeed for the cases just listed, we have $S_1 \subseteq S_0$ – so it is condition (i) that fails. Like BCI logic, all of the logics here fail to be monothetic. (We can see that for all these logics, BCI included, $p \to p$ and $q \to q$ are both provable though the result of replacing the first occurrence of the former by the latter in the equally provable $(p \to p) \to (p \to p)$ is unprovable – an often-made observation with many interesting repercussions not germane to the present study.[10]) The example of S and S' at the end of note 9 also gave a counterexample to the "if" direction of (1), taking these as S_0 and S_1 respectively, or indeed vice versa. In this case, condition (i), the monotheticity condition, is satisfied and it is condition (ii) that fails: S and S' are disjoint.

Can we with equal ease illustrate how a failure of the sufficient condition in Proposition 2.1 can give rise to a counterexample to the "only if" half of (1)? The simplest cases in which strengthening a logic results in a loss of synonymies arise with a change of language, and so are not directly pertinent to the present enterprise since we have agreed to concentrate on comparisons amongst logics in

---

[10] *Cf.* Kabziński [21] and Section 4 of Humberstone [20].

the same language. For example, if we take the smallest modal logic,[11] or any of various non-normal modal logics such as Lemmon's S0.5, we have a proper extension of non-modal classical propositional logic in which classically equivalent formulas, synonymous in that logic, are no longer synonymous – indeed in which, as for the logics treated in Proposition 2.5, no two distinct formulas are synonymous. (See Porte [29].) Another well-known example is that of intuitionistic logic with 'strong negation', which we shall consider at the end of Section 3. Abiding by our 'same language' restriction on S_0 and S_1, one simple, if artificial, type of case arises as follows.

Example. Take again the language of (non-modal) classical propositional logic and S_0 as the empty set (certainly a subset of the set of formulas of this language, and answering to the most commonly proposed additional conditions on logics as sets of formulas – such as closure under Uniform Substitution[12]), with S_1 as classical logic. Although $S_0 \subseteq S_1$ we do not have $\equiv_0 \subseteq \equiv_1$, because every pair of formulas stand in the former relation while only formulas which are classically equivalent stand in the latter.

The above example is not very appealing because the empty set may not be regarded as a logic on the 'set-of-formulas' conception of logics (which does not say that *any* old set of formulas constitutes a logic), or is perhaps regarded only as an extreme and degenerate case of a logic. If one is interested in some 'atheorematic' logic such as the classical logic of conjunction and disjunction, one would normally pass to something like the consequence relation conception, noting that the set of pairs $\langle \Gamma, A \rangle$ standing in this relation is far from empty, even though the set of such pairs for which Γ is empty is itself empty.[13] Let us accordingly give a counterexample to the "only if" half of (1) not requiring $\varnothing$ to be acknowledged as a logic.

Example. Let the language have two connectives $\to$ and $\star$, say, of arities 2 and 1 respectively, and let S_0 consist all formulas of the form $\star A$, and S_1 of all all such formulas together with all formulas of the form $A \to A$. Then for any formulas A and B, $\star A \equiv_0 \star B$, though this is not so in the case of $\equiv_1$; for example $\star p$ is not synonymous with $\star q$ in S_1, because $\star p \to \star p \in S_1$ while $\star p \to \star q \notin S_1$.

---

[11] We understand a modal logic here to be a set of formulas in the language of classical propositional logic with some functionally complete set of boolean primitives and one additional 1-ary connective $\square$, containing all classical tautologies and closed under Modus Ponens and Uniform Substitution.

[12] All logics-as-sets-of-formulas we consider satisfy this condition, with the corresponding condition also satisfied for all logics-as-consequence relations in the following section.

[13] This is what we mean by an *atheorematic* consequence relation. Such consequence relations are called 'purely inferential' in Wójcicki [39] – except that Wójcicki tends to prefer formulations in terms of consequence operations rather than consequence relations.

3. Discrimination in Logics as Consequence Relations

We defined synonymy according to a consequence relation in Section 2, before putting this notion to one side in order to compare discriminatory and deductive strength in the simpler setting of logics as sets of formulas. We to take it up again here, to which end the following notation will convenient. For a consequence relation $\vdash$ ($\vdash_i$) we denote by $\equiv_\vdash$ ($\equiv_{\vdash_i}$, or for short $\equiv_i$) the relation of synonymy according to $\vdash$ ($\vdash_i$), and write $A \dashv\vdash B$ to mean "$A \vdash B$ and $B \vdash A$" ($A \dashv\vdash_i B$ to mean "$A \vdash_i B$ and $B \vdash_i A$"). As in Segerberg [33], we call a consequence relation $\vdash$ *congruential* when for all formulas A, B (in the language of $\vdash$) $A \dashv\vdash B$ implies $A \equiv_\vdash B$. (The converse implication holds for any $\vdash$. Thus a congruential consequence relation is one for which logical equivalence – the relation $\dashv\vdash$, that is – and synonymy coincide. Here we rely on the fact that the synonymy of A, B according to a consequence relation $\vdash$, as defined in Section 2, is equivalent to its being the case that for all contexts $C(\cdot)$, we have $C(A) \dashv\vdash C(B)$. Wójcicki [39] uses "self-extensional" for "congruential".)

Conceiving of logics as consequence relations rather than sets of formulas makes for the following modifications to (1) and (2):

$$\vdash_0 \subseteq \vdash_1 \text{ if and only if } \equiv_0 \subseteq \equiv_1 \tag{3}$$

$$\vdash_0 \subsetneq \vdash_1 \text{ if and only if } \equiv_0 \subsetneq \equiv_1 \tag{4}$$

Again, we concentrate on the first of these, and on the case in which $\vdash_0$ and $\vdash_1$ are consequence relations on the same language. Here is a very simple sufficient condition for the "only if" direction of (3):

Proposition 3.1. *If $\vdash_1$ is congruential and $\vdash_0 \subseteq \vdash_1$, then $\equiv_0 \subseteq \equiv_1$.*

Proof. Suppose that $\vdash_0 \subseteq \vdash_1$ for congruential $\vdash_1$, and that $A \equiv_0 B$. Since $A \equiv_0 B$, we have $A \dashv\vdash_0 B$, so since $\vdash_0 \subseteq \vdash_1$, $A \dashv\vdash_1 B$, whence by the congruentiality of $\vdash_1$, we get $A \equiv_1 B$.   □

Proposition 3.1 is (nearly) a special case of the analogue for consequence relations of Proposition 2.1. Although the notion of an equivalential (set-of-formulas) logic was abstracted from the notion of an equivalential consequence relation, the latter turns out not to be the pertinent concept, and what we want instead is the concept of a consequence relations $\vdash$ with *sequent-definable synonymy*, by which we mean (*cf.* [29]) that there is a set $\Sigma(p,q)$ of pairs $\langle \Gamma, C \rangle$, all formulas occurring in which are constructed from only the variables p, q with the property that for all formulas A, B (in the language of $\vdash$) we have $\Sigma(A, B) \subseteq \vdash$ if and only if $A \equiv_\vdash B$. As in Section 2, we immediately pass to a relational version of this concept, saying that consequence relations $\vdash_0$ and $\vdash_1$ have *similarly sequent-definable congruences* if the same set $\Sigma(p,q)$ witnesses the sequent-definability of synonymy for $\vdash_0$ and $\vdash_1$. Then by a simple argument which replaces S_i in the proof of Proposition 2.1 by $\vdash_i$ and substitutions in the set of formulas $\Delta(p,q)$ by substitutions in the set of sequents $\Sigma(p,q)$, we obtain a proof of:

Proposition 3.2. *For any consequence relations* $\vdash_0$ *and* $\vdash_1$ *with similarly sequent-definable synonymies,* $\vdash_0 \,\subseteq\, \vdash_1$ *implies* $\equiv_0 \,\subseteq\, \equiv_1$.

The analogue of Remark 2.2 applies here too.

It may seem stretching things to use the term *sequents* for the ordered pairs $\langle \Gamma, C \rangle$, certain sets of which are consequence relations, since the the 'antecedent' of a sequent might typically be required to be a finite set, whereas these Γ will not all be finite. Indeed on many versions of what a sequent should be (e.g., for the sake of a convenient sequent-calculus), Γ wouldn't be a set (of formulas) at all but a multiset or a sequence. Nevertheless, the terminology is convenient and we ignore those objections to its use here. Let us further follow Blamey [5] in using $\succ$ as our sequent-separator – that is, we notate the sequent $\langle \Gamma, C \rangle$ more suggestively as $\Gamma \succ C$. We are now in a position to see Proposition 3.1 as close to being a special case of Proposition 3.2: a congruential consequence relation is one for which synonymy is defined by the set of sequents $\Sigma(p, q) = \{p \succ q, q \succ p\}$. "Close to being" a special case but not quite there, since Proposition 3.1 demands only that $\vdash_1$ be congruential, whereas the application just envisaged of Proposition 3.2 would appear to require the condition that both $\vdash_0$ and $\vdash_1$ be congruential (since they need to have similarly sequent-definable synonymies).[14]

We turn our attention to the provision of two counterexamples to the "if" direction of (3), each of which features a pair of consequence relations which, though distinct, yield the same synonymy relation. These examples, especially the second (appearing after Remark 3.12), are of some theoretical interest in their own right, and all four logics (playing the $\vdash_0$ and $\vdash_1$ roles in the two examples) are congruential, though that fact does not need to be exploited. The first example (immediately following Coro. 3.6 below) draw attention to a relation we shall call "1-agreement" between consequence relations, isolating which will assist in presenting the second example. After that discussion, we conclude with a counterexample (or two) to the "only if" direction of (3).

For the first of these examples, the language we use has only one connective, the 0-place connective (sentential constant) $\top$; we define $\vdash_0^\top$ to be the least consequence relation $\vdash$ on this language satisfying (5) for $\Gamma \neq \varnothing$, and $\vdash_1^\top$ to be the least consequence relation $\vdash$ on the language satisfying (5) for arbitrary Γ (equivalently, satisfying (5) for $\Gamma = \varnothing$):

$$\Gamma \vdash \top. \tag{5}$$

$\vdash_0^\top$ is a simplified version of idea of Roman Suszko's, described in note 7 of Smiley [34], and it is not hard to check that $\vdash_0^\top \,\subsetneq\, \vdash_1^\top$. Indeed, we will verify this twice over, the second proof following its statement below as Corollary 3.4. It is clear from the definitions that $\vdash_0^\top \,\subseteq\, \vdash_1^\top$; that the converse inclusion does not hold follows from the fact that $\varnothing \vdash_1 \top$ (again from the definition of $\vdash_1^\top$) while $\varnothing \nvdash_0^\top \top$. We can verify this latter fact syntactically by thinking of the above definition of $\vdash_0^\top$ as an inductive definition ("from below") of the class of pairs $\langle \Gamma, A \rangle$ standing in

---

[14]The author has the strong impression of missing an insight here.

this relation, which allows for a proof by induction on the length of a construction which would place $\langle \Gamma, A \rangle$ in $\vdash_0^\top$ only when $\Gamma \neq \varnothing$. (See Scott [32] for this type of argument; the characterization below in terms of valuations is also much inspired by Scott's work.)

An alternative to the above (quasi-)proof-theoretic argument, we can obtain the same conclusion by semantic reasoning, couched in terms of the notion of a consequence relation $\vdash$'s being *determined* by a class V of valuations (bivalent truth-value assignments to the formulas of the language of $\vdash$), a relation defined to hold between $\vdash$ and V just in case for all sets Γ of formulas of the language and all formulas A thereof: $\Gamma \vdash A$ if and only if for each $v \in V$, whenever $v(C) = \text{T}$ for all $C \in \Gamma$, then $v(A) = \text{T}$. (We use "T", "F", to denote the two truth-values; if $\vdash$ has been specified by means of a proof system, the "only if" and the "if" parts of this definition amount to the soundness and the completeness, respectively, of this system, with respect to V.) The easy proof of the following is left to the reader; the reference to valuations in both cases is to valuations for the (common) language of $\vdash_0^\top$ and $\vdash_1^\top$.

Proposition 3.3. *Let v_F be the unique valuation (for the language of $\vdash_0^\top$ and $\vdash_1^\top$) assigning the value F to every formula, and V be the class of all valuations (for this language) satisfying $v(\top) = \text{T}$. Then*

(i) $\vdash_0^\top$ *is determined by* $V \cup \{v_\text{F}\}$
(ii) $\vdash_1^\top$ *is determined by* V.

We now repeat the earlier syntactically argued assertion with its new semantic justification:

Corollary 3.4. $\vdash_0^\top \subsetneq \vdash_1^\top$.

Proof. That $\vdash_0^\top \subseteq \vdash_1^\top$ follows from Prop. 3.3 by a familiar Galois duality between consequence relations and classes of valuations, since $V \subseteq V \cup \{v_\text{F}\}$; the failure of the converse inclusion (between the $\vdash_i^\top$) is illustrated by the fact that $\varnothing \vdash_1^\top \top$ while $\varnothing \nvdash_0^\top \top$. $\qquad\qquad\square$

Remark 3.5. The formula $\top$, as it behaves according to $\vdash_0^\top$, is what is called in Humberstone [16], p. 59, a "mere follower": it follows from every formula and thus from every non-empty set of formulas – but not from the empty set of formulas. Note that so defined, only an atheorematic consequence relation can have a mere follower, and that any two mere followers are logically equivalent (each being a consequence of the other).

From Proposition 3.3 we may also infer (by an argument we leave to the reader) the following:

Corollary 3.6. *For all non-empty Γ and all formulas B, we have $\Gamma \vdash_0^\top B$ if and only if $\Gamma \vdash_1^\top B$.*

In particular, then, we have:

Example. For all formulas A, B, we have $A \dashv\vdash_0^\top B$ if and only if $A \dashv\vdash_1^\top B$. Since $\vdash_0^\top$ and $\vdash_1^\top$ are congruential, the induced synonymy relations $\equiv_0^\top$ and $\equiv_1^\top$ coincide, the fact that $\vdash_0^\top \subsetneqq \vdash_1^\top$ notwithstanding, providing a counterexample to the "if" direction of (3), taking $\vdash_0$ and $\equiv_0$ (resp. $\vdash_1$ and $\equiv_1$) in (3) as $\vdash_1^\top$ and $\equiv_1^\top$ (resp. $\vdash_0^\top$ and $\equiv_0^\top$). Alternatively, we can see this as a counterexample to the "only if" direction of (4) – keeping the subscripts the same, this time. (In fact, the reference to congruentiality is not needed. See Remark 3.7 below.)

There is one aspect of the situation just reviewed we shall isolate for our second example. Say that consequence relations $\vdash$ and $\vdash'$ on the same language *n-agree* when for all formulas A and all sets of formulas Γ of cardinality $n \in \mathbb{N}$, we have $\Gamma \vdash A$ if and only if $\Gamma \vdash' A$. In this terminology Corollary 3.6 says that $\vdash_0$ and $\vdash_1$ *n*-agree for all $n \geq 1$. What actually matters for the above example though, is specifically that these consequence relations 1-agree:

Remark 3.7. Even if $\vdash$ and $\vdash'$ are not congruential, if $\vdash$ and $\vdash'$ 1-agree, then $\equiv_\vdash = \equiv_{\vdash'}$, since, adapting the characterization of congruentiality at the end of the opening paragraph of this section, A and B are synonymous according to a consequence relation just in case for all C, $C(A)$ and $C(B)$ are equivalent. But any 1-agreeing consequence relations also agree in respect of which formulas are synonymous – that is, have the same synonymy relation.

For our second example, included for its intrinsic interest, there is again only one connective in the language, and this time it is binary, and will be written – for reasons to become clear immediately – as "∧". Let $\vdash_0^\wedge$ and $\vdash_1^\wedge$ be the least consequence relations $\vdash$ on this language satisfying, for all formulas A and B and in the case of $\vdash_0^\wedge$, for all Γ of the form $\{C\}$ while in the case of $\vdash_1^\wedge$, for arbitrary Γ, the condition (6):

$$\Gamma \vdash A \wedge B \text{ if and only if } \Gamma \vdash A \text{ and } \Gamma \vdash B. \tag{6}$$

The consequence relations $\vdash_0^\wedge$ and $\vdash_1^\wedge$, or similarly related consequence relations with additional connectives present answering to their own conditions, are distinguished in Koslow [22] and Cleave [7]. $\vdash_1^\wedge$, is the restriction to the language with ∧ of the consequence relations of intuitionistic or classical logic; it is called the logic of 'parametric' conjunction in [22], where essentially reasoning of Prop. 3.8 and Coro. 3.9 may be found (p. 129*f*.).[15]

Proposition 3.8. *Whenever $\Gamma \vdash_0^\wedge A$, we have $C \vdash_0^\wedge A$ for some $C \in \Gamma$.*

Corollary 3.9. $\vdash_0^\wedge \subsetneqq \Gamma \vdash_1^\wedge$.

Proof. Clearly we have $p, q \vdash_1^\wedge p \wedge q$, since we may take Γ in (6) as $\{p, q\}$; but by Proposition 3.8 $p, q \nvdash_0^\wedge p \wedge q$, as otherwise we should have $p \vdash_0^\wedge p \wedge q$ or $q \vdash_0^\wedge p \wedge q$ (neither of which is even the case for $\vdash_1^\wedge$, of course).   □

---

[15] In the case of Cleave [7], pp. 121*ff*. should be consulted. There are many problems with Cleave's discussion, and a few with Koslow's; see [15], esp. p. 478*f*. for these.

It was promised that the example involving the $\vdash_i^\wedge$ would be of some theoretical interest in its own right. There are two points of interest. A philosophical moral to be drawn is most easily seen if condition (6) is recast as a collection of sequent-to-sequent rules, in which case the weaker ($\vdash_0^\wedge$-defining) $\Gamma = \{C\}$ version of (6) emerges as follows, with semicolons separating the premiss-sequents from each other and "/" separating them from the conclusion-sequent:

(i) $C \succ A; C \succ B / C \succ A \wedge B$.   (ii) $C \succ A \wedge B / C \succ A$.   (iii) $C \succ A \wedge B / C \succ B$.

The point of interest is that these rules already uniquely characterize $\wedge$ (to within logical equivalence),[16] even though they are weaker than the standard rules (i.e., the rules with the general set-variable "Γ" replacing C throughout – though it is easy to see that the rule (iii) would not be strengthened by this generalization). Thus it is not open to the intuitionist, for example, to complain that what is wrong with the classical rules governing negation is that they are 'stronger than needed' to characterize this connective uniquely, since the intuitionistically acceptable negation rules already suffice for uniqueness. To take such a line without further qualification would be to leave the intuitionist open to an objection to the intuitionistically accepted rules governing conjunction, since as just observed, these are also stronger than needed for unique characterization. In (the paper abstracted as) [13] it is suggested that the 'further qualification' needed will address the issue of rules being *fully general* in respect of side-formulas (so arbitrary Γ, rather than just C or more explicitly $\{C\}$, for instance), though what this comes to will naturally depend on exactly what form the sequents take – e.g., on whether multiple succedents are to be permitted. (Of course such sequents do not arise in the rules embodying conditions on consequence relations, but we are speaking of sequent-to-sequent rules for a notion of sequent that should be thought of as yet to be settled on, when issues of one logic *vs.* another are being aired.)

Philosophy of logic aside, the case of $\vdash_0^\wedge$ presents us with an interesting task in valuational semantics, namely that of informatively specifying a class of valuations which determines this consequence relation. To attack this problem, which will have dividends for our main business as well (see Coro. 3.11), we need some terminology and notation. If $\#$ is an n-ary connective with which some

---

[16]This is pointed out in Example 4.3(i) on p. 121 of [7]. (The parenthetical "to within logical equivalence" is an allusion to the possible contrast with unique characterization *to within synonymy*, on which see Humberstone [19], Sections 3 and 4.) The result of this is that if one party to a dispute about the logical powers of conjunction endorses only the $\vdash_0^\wedge$ conditions, while the other endorses the stronger $\vdash_1^\wedge$ conditions, they cannot agree to bury their differences by agreeing to adopt a logic with two connectives in place of $\wedge$ – $\wedge_0$ and $\wedge_1$, say – governed by rules embodying the respective conditions: because even the weaker rules have the unique characterization property, the resulting combined logic then has the $\wedge_0$-conjunction of two formulas following from those formulas. The situation is just as with intuitionistic and classical negation, alluded to presently, in which the intuitionist would be ill advised indeed to concede the intelligibility of a connective governed by the rules for classical negation *alongside* and notationally distinguished from the favoured intuitionistic negation: the former's distinctive behaviour will then infect the latter, leaving no room, as Humberstone [13] concludes, for any such 'live and let live' attitude.

preassigned n-ary truth function f is associated, then we call a valuation $\#$-*boolean* if for all formulas $A_1, \ldots, A_n$, $v(\#(A_1 \ldots A_n)) = f(v(A_1), \ldots, v(A_n))$. Thus the class V featuring in Proposition 3.3 is the class of $\top$-boolean valuations, while the $\wedge$-boolean valuations (for a given language) are exactly those which assign the value T to formulas (of that language) $A \wedge B$ when they assign the value T to A and to B. A somewhat less frequently encountered notion is the following. (See [16].) For an arbitrary family of valuations, V we denote by $\sum V$ what we call the *disjunctive combination* of the valuations in V, defined to be the unique valuation u for which for all formulas A, $u(A) = $ T if and only if there is some $v \in V$ with $V(A) = $ T. If $V = \{v_1, v_2\}$, we write $v_1 + v_2$ for $\sum V$. The dual – in the sense of poset duality, not Galois duality[17] – operation on valuations, *conjunctive combination* here denoted by $\prod V$ ($v_1 \cdot v_2$ in the binary case) is similarly defined but with "there is some" replaced by "for all"; these are fairly well known, being a bivalentized version of the notion of a supervaluation over V.[18] Their key logical significance is that the consequence relation determined by a class of valuations remains unaffected by adding conjunctive combinations of valuations to the determining class.[19] This, which is not so for generalized consequence relations, is due to the presence of a single formula on the *right* of the "$\vdash$". (The presence of at most one formula on the right, that is, rather than at least one, as in the preceding note.) In view of Proposition 3.8 above, which says that for the case of $\vdash_0^\wedge$ a consequence statement holds in virtue of a single formula from amongst those on the *left*, suggests the semantic characterization given in Proposition 3.10 below. It is well known that $\vdash_1^\wedge$ is easily seen to be determined by the class of all $\wedge$-boolean valuations; what we need for $\vdash_0^\wedge$ is the class of disjunctive combinations of such valuations, so we pause to observe that disjunctively combining $\wedge$-boolean valuations typically results in a valuation that is not $\wedge$-boolean (whereas the class of $\wedge$-boolean valuations is closed under *conjunctive* combination). We illustrate with the binary mode of combination.

Example. Let u and v be $\wedge$-boolean valuations satisfying: $u(p) = v(q) = $ T, $u(q) = v(p) = $ F. For their disjunctive combination $u + v$ we have $u + v(p) = $ T, since $u(p) = $ T, and also $u + v(q) = $ T since $v(q) = $ T. However, $u + v(p \wedge q) = $ F, since neither u nor v verifies this conjunction, so $u + v$ is not $\wedge$-boolean.

---

[17] In the terminology, though not the notation, of [16], $+$ is Galois dual to $\vee$ and $\cdot$ to $\wedge$. (Upward and downward pointing triangles are used in [16] to symbolize conjunctive and disjunctive combinations, large for the case of families of valuations and small in the case of the binary operation.)

[18] Incidentally, the standard 'gappy' version of what later became known as supervaluations appears already at the end of the second paragraph of §4 in Nelson [27].

[19] A special case is that of $V = \varnothing$, for which $\prod V$ is the valuation v_T assigning the value T to every formula. So any $\vdash$ determined by a class U of valuations is also determined by $U \cup \{v_\mathrm{T}\}$. For the same choice of V, $\sum V$ is the valuation v_F of Proposition 3.3, which taken together with Corollary 3.4 shows that, by contrast with the case of v_T, adding v_F to the determining class can change which consequence relation is determined. The explanation for this lies in the mandatory appearance of a formula on the right of the "$\vdash$" (or "$\succ$", at the level of individual sequents), as contrasted with the possible disappearance of all formulas from the left.

Having shown that we obtain a new class of valuations other than just that consisting of $\wedge$-boolean valuations when passing to arbitrary disjunctive combinations of such valuations, we proceed to our semantic characterization of $\vdash_0^\wedge$

Proposition 3.10. *The consequence relation $\vdash_0^\wedge$ is determined by the class of all valuations which are disjunctive combinations of families of $\wedge$-boolean valuations.*

Proof. We must show that $\Gamma \vdash_0^\wedge A$ if and only if every disjunctive combination of $\wedge$-boolean valuations which verifies each formula in Γ also verifies A. The "only if" direction is essentially a soundness proof for the system with, in addition to basic structural rules, the sequent-to-sequent rules (i), (ii), (iii), above, for which purpose it suffices to check that no disjunctive combination of $\wedge$-boolean valuations verifies all the left hand formulas without verifying the right-hand formula of any provable sequent. Since rules (ii) and (iii) are obviously equivalent (given the structural rules encoding the fact that our sequents are the elements of a consequence relation) to the zero-premiss rules $A \wedge B \succ A$ and $A \wedge B \succ B$ it is sufficient in their case to check that there are no countervaluations in the class w.r.t. which soundness is being shown, We consider the former by way of example. Suppose u is $\sum V$ for a family V of $\wedge$-boolean valuations, and $u(A \wedge B) = \mathrm{T}$. We must show that $u(A) = \mathrm{T}$. As $u(A \wedge B) = \mathrm{T}$ and $u = \sum V$, there is $v \in V$ with $v(A \wedge B) = \mathrm{T}$. But all valuations in V, v included, are $\wedge$-boolean, so $v(A) = \mathrm{T}$, and therefore $u(A) = \mathrm{T}$. We now check (i), showing that if there is a countervaluation to the conclusion sequent $C \succ A \wedge B$ of an application of this rule, then there is a countervaluation to one or another of the premiss-sequents $C \succ A$, $C \succ B$. So suppose that $u = \sum V$ for a collection V of $\wedge$-boolean valuations, and $u(C) = \mathrm{T}$ while $u(A \wedge B) = \mathrm{F}$. Then for some $v \in V$, we have $v(C) = \mathrm{T}$, but since $u(A \wedge B) = \mathrm{F}$, $v(A \wedge B) = \mathrm{F}$. As v is $\wedge$-boolean, either $v(A) = \mathrm{F}$ or $v(B) = \mathrm{F}$, so since v $(= \sum\{v\})$ is itself a disjunctive combination of $\wedge$-boolean valuations, it is either a countervaluation to $C \succ A$ or to $C \succ B$.

We turn to the "if" (completeness) direction of the claim. We must show then whenever $\Gamma \not\vdash_0^\wedge A$, we can find a disjunctive combination of $\wedge$-boolean valuations verifying each formula in Γ but not A. For each $C \in \Gamma$ define the valuation v_C by setting $v_C(B) = \mathrm{T}$ iff $C \vdash_0^\wedge B$ for all formulas B. Note that v_C is guaranteed to be $\wedge$-boolean by the way $\vdash_0^\wedge$ was defined. Also observe that for each $C \in \Gamma$, we have $v_C(A) = \mathrm{F}$, since otherwise we should have $C \vdash_0^\wedge A$ and hence, by a defining property (variously called monotonicity, thinning, weakening,...) of consequence relations, $\Gamma \vdash_0^\wedge A$, contradicting our initial assumption. But together these facts imply that for $u = \sum\{v_C | C \in \Gamma\}$, u is a disjunctive combination of $\wedge$-boolean assigning T to every formula in Γ and F to A, as required. □

Corollary 3.11. *The consequence relations $\vdash_0^\wedge$ and $\vdash_1^\wedge$ 1-agree.*

Proof. Since $\vdash_0^\wedge \subseteq \vdash_1^\wedge$, we have only to show that for all formulas C, A, if $C \vdash_1^\wedge A$, then $C \vdash_0^\wedge A$. So, arguing contrapositively, suppose that $C \not\vdash_0^\wedge A$. By Prop. 3.10 there is a valuation $u = \sum V$ with all $v \in V$ $\wedge$-boolean, with $u(C) = \mathrm{T}$ and

$u(A) = \mathrm{F}$. Thus for some $v \in V$, $v(A) = \mathrm{T}$ while $v(C) = \mathrm{F}$. But v is an $\wedge$-boolean valuation, so since $\vdash_1^\wedge$ is determined by the class of $\wedge$-boolean valuations, $C \nvdash_1^\wedge A$.

$\square$

Remark 3.12. Notice how this argument would have failed if we had tried to show that, for instance, if $C, D \vdash_0^\wedge A$ then $C, D \vdash_0^\wedge A$. In this case we have $u(C) = u(D) = \mathrm{T}$ while $u(A) = \mathrm{F}$, for $u = \sum V$ as above: but this allows $v \in V$ with $v(C) = \mathrm{T}$ and $v' \in V$ with $v'(D) = \mathrm{T}$, with no guarantee that $v = v'$ and so way to complete the argument – since as we saw in the Example preceding Prop. 3.10, $v + v'$ need not be $\wedge$-boolean. (We could have established Coro. 3.11 purely syntactically, but the semantic characterization seems illuminating.)

We have now assembled all the ingredients for the second of the counterexamples to be presented here to the "if" direction of (3).

Example. Although $\equiv_0^\wedge = \equiv_1^\wedge$, by Coro. 3.11 and Remark 3.7, $\vdash_0^\wedge \subsetneq \Gamma \vdash_1^\wedge$ (by Coro. 3.9).

We pause to notice that the $\vdash_i^\top$ and $\vdash_i^\wedge$ pairs ($i = 0, 1$) with which we have illustrated the failure of the "if" direction of (3), are also convenient indicators of the falsity of a conjecture either to the effect that if consequence relations n-agree then they must m-agree whenever $m \leq n$ or to the effect that n-agreeing consequence relations must m-agree whenever $n \leq m$. Counterexamples to these conjectures are given respectively by the cases of the $\vdash_i^\top$, which 1-agree without 0-agreeing, and of the $\vdash_i^\wedge$, which 1-agree without 2-agreeing.

Remark 3.13. It should be noted, however, that the implication in the case of $m \leq n$: "$\vdash, \vdash'$ n-agree $\Rightarrow$ $\vdash, \vdash'$ m-agree" holds under very weak conditions for all $m \geq 1$. The following additional condition secures the implication, for example: that for every formula A there is some formula $B \neq A$ such that $B \vdash A$, and likewise in the case of $\vdash'$. Alternatively, if one wished, one could secure the above implication, still with the $m \geq 1$ proviso in force, by a change in the definition of n-agreement, defining this relation to hold between $\vdash$ and $\vdash'$ just in case for all formulas $A_1, \ldots, A_n, B$: $A_1, \ldots, A_n \vdash B$ iff $A_1, \ldots, A_n \vdash B$. (This differs from the original definition because we can have $A_i = A_j$ when $i \neq j$.)

Before leaving the subject of 1-agreement altogether, we should take a moment to observe that despite its figuring in our examples distinct consequence relations with the same synonymy relation, 1-agreement is by no means a necessary condition for two consequence relations to coincide thus in respect of synonymy. The small reminder we include to that end, Proposition 3.14, requires the following concept. Let us call consequence relations on the same language $\vdash$ and $\vdash'$ *weakly dual* when for all formulas A, B, of that language, $A \vdash B$ if and only if $B \vdash' A$. Many consequence relations will weakly dual to any given consequence relation $\vdash$, and though they will 1-agree with each other, they will typically not 1-agree with $\vdash$ – the point of current interest.[20]

Proposition 3.14. *If $\vdash$ and $\vdash'$ are weakly dual, then they have the same synonymy relation.*

Proof. Weakly dual $\vdash$ and $\vdash'$, though not in general 1-agreeing, still 'agree' in respect of which formulas are equivalent to each other, and so, by the reasoning given in Remark 3.7, agree as to which pairs of formulas are synonymous.   □

The counterexamples we have provided to the "if" direction of (3) have been of cases of differing consequence relations with the same same synonymy relations. We cannot similarly offer counterexamples to the "only if" direction of (3) – our final topic – in which $\vdash_0$ coincides with $\vdash_1$ while $\equiv_0$ and $\equiv_1$ differ, since $\equiv_i$ is fixed by $\vdash_i$. A somewhat artificial counterexample can be obtained by tinkering minimally with that given at the end of Section 2. We use the same language, with connectives $\rightarrow$ and $\star$, and define $\vdash_0$ and $\vdash_1$ as the least consequence relations $\vdash$ on this language such that (for the former) $\varnothing \vdash \star A$ for all A, and (for the latter) $\varnothing \vdash \star A$ as well as $\varnothing \vdash A \rightarrow A$ for all formulas A. The explanation given at the end of Section 2 as to why this is a counterexample applies here also, *mutatis mutandis*. Our final topic will be a more interesting 'naturally occurring' counterexample to the "only if" direction of (3).

Consider first the consequence relations of intuitionistic (propositional) logic, $\vdash_{IL}$, with any familiar set of primitive connectives, and of intuitionistic logic with strong negation $\vdash_{ILS}$, whose language contains a further 1-ary connective ('strong negation') written as "$-$", governed by principles which may be found in any discussion of the subject, such as Chapter 7, Section 2 of Gabbay [11].[21] (If the account specifies a logic in the set-of-formulas sense, by means of an axiomatization using Modus Ponens as the sole rule, then the consequence relation we are interested in relates Γ to A in the following familiar way: A stands at the end of a sequence of formulas each of which is either an axiom, an element of Γ, or

---

[20]From the definition of weak duality given here it is not hard to deduce the following. The smallest consequence relation weakly dual to a given $\vdash$ is the $\vdash'$ defined by: $\Gamma \vdash' A$ iff for some $B \in \Gamma$, $A \vdash B$. The largest consequence relation weakly dual to $\vdash$ is the $\vdash'$ defined by: $\Gamma \vdash' A$ iff for all B such that $C \vdash B$ for each $C \in \Gamma$, we have $B \vdash A$. This latter is essentially the notion of *the dual of* $\vdash$ offered by Wójcicki [38] (see also [35]) and §9.5 of Koslow [22], though there are slight differences. Koslow is discussing what calls implication relations rather consequence relations, which amounts to treating them as relations between finite but non-empty sets of formulas and individual formulas (subject otherwise to the usual defining conditions for consequence relations), while Wójcicki's definition is like ours except that what ours requires of Γ itself for $\Gamma \vdash' A$ to hold is instead required of some finite subset of Γ. With generalized consequence relations, of course, matters are much more straightforward since one can take the dual of such a relation just to be its converse. (See, e.g., Gabbay [11], p. 16.)

[21]Since we quoted Nelson [27] in our opening section, we should stress that the logic of strong negation presented in [27] is definitely *not* what we have in mind here (though it was earlier work by Nelson that inspired what we do have in mind), since that is not an extension of (even the implicational fragment of) intuitionistic logic.

follows from earlier formulas in the sequence by an application of Modus Ponens.) A well-known feature of $\vdash_{ILS}$ is that it is not congruential, since for example, writing "$\neg$" for (ordinary) intuitionistic negation, $\neg\neg\neg p$ and $\neg p$ are $\vdash_{ILS}$-equivalent (being $\vdash_{IL}$-equivalent), whereas $-\neg\neg\neg p$ and $-\neg p$ are not $\vdash_{ILS}$-equivalent.[22] The latter pair of formulas are $\vdash_{ILS}$-equivalent respectively to $\neg\neg p$ and p, which are not $\vdash_{ILS}$-equivalent, since they do not involve strong negation, are not $\vdash_{IL}$-equivalent and $\vdash_{ILS}$ is a conservative extension of $\vdash_{IL}$. Clearly, however, there is *something* non-conservative going on. We could say that the passage from intuitionistic logic to intuitionistic logic with strong negation fails to conserve synonymy – which should raise eyebrows amongst adherents of intuitionistic logic, the conservativity of the extension notwithstanding[23] – since evidently the 'strong negation'-free formulas $\neg p$ and $\neg\neg\neg p$ synonymous according to $\vdash_{IL}$ but not according to $\vdash_{ILS}$. This gives the counterexample we have in mind (and in fact could have presented in a suitably modified form in Section 2, as involving 'formula' logics):

Example. We have assembled the pieces for a counterexample to the following case of the "if" direction of (3):

$$\vdash_{IL} \subseteq \vdash_{ILS} \Rightarrow \equiv_{IL} \subseteq \equiv_{ILS},$$

since, as just observed, although $\vdash_{IL} \subseteq \vdash_{ILS}$, we have $\neg p \equiv_{IL} \neg\neg\neg p$ without $\neg p \equiv_{ILS} \neg\neg\neg p$.

This is, however, a 'two-language' example since strong negation is not a connective in the language of $\vdash_{IL}$, whereas we undertook to seek counterexamples without a change of language. One might think to get around this by considering in place of the consequence relation $\vdash_{IL}$ a variation which has strong negation in its language but enjoying no special logical behaviour, $\vdash_{IL(S)}$, we could call it, much as with $\Box$ in the smallest modal logic or $\perp$ in Minimal Logic (*Minimalkalkül*). For the counterexample, however, we should need $\neg p \equiv_{IL(S)} \neg\neg\neg p$, which is no longer the case since the two formulas involved here give non-equivalent results (relative to $\vdash_{IL(S)}$) when embedded in the scope of the strong negation connective: the very point we were exploiting concerning $\vdash_{ILS}$ (though with $\vdash_{IL(S)}$, the situation is more serious in that, as with several other logics we have considered, no two formulas are synonynmous). If there is a simple one-language counterexample in this vicinity to the claim that inclusion of consequence relations implies the corresponding inclusion of synonymy relations, we leave it for others to find.

References

[1] A. R. Anderson and N. D. Belnap, *Entailment: the Logic of Relevance and Necessity, Vol. I*, Princeton University Press, Princeton, NJ 1975.

---

[22]Though non-congruential, $\vdash_{ILS}$ has a sequent-definable synonymy relation, taking $\Sigma(p,q)$ as $\{p \succ q, -p \succ -q, q \succ p, -q \succ -p\}$.
[23]Or so it is argued in note 27 of Humberstone [17].

[2] Jean-Yves Béziau, 'Classical Negation Can Be Expressed by One of its Halves', *Logic Journal of the IGPL* **7** (1999), 145–151.

[3] R. J. Bignall and M. Spinks, 'Multiple-Valued Logics for Theorem-Proving in First Order Logic with Equality', Proceedings of the 28th IEEE Symposium on Multiple-Valued Logics 1998, pp. 102–107.

[4] Wim J. Blok and Don L. Pigozzi, 'Algebraizable Logics', *Memoirs of the American Math. Soc.* **77** (1989), #396.

[5] Stephen Blamey, 'Partial Logic', pp. 261–353 in D. Gabbay and F. Guenthner (eds.), *Handbook of Philosophical Logic, Second Edition, Vol. 5*, Kluwer, Dordrecht 2002.

[6] Stanley Burris, 'Discriminator Varieties and Symbolic Computation', *Journal of Symbolic Computation* **13** (1992), 175–207.

[7] J. P. Cleave, *A Study of Logics*, Clarendon Press, Oxford 1991.

[8] J. Czelakowski, 'Equivalential Logics (I)', *Studia Logica* **40** (1981), 227–236.

[9] J. Czelakowski, *Protoalgebraic Logics*, Kluwer, Dordrecht 2001.

[10] J. M. Font, R. Jansana and D. Pigozzi, 'A Survey of Abstract Algebraic Logic', *Studia Logica* **74** (2003), 13–97.

[11] D. M. Gabbay, *Semantical Investigations in Heyting's Intuitionistic Logic*, Reidel, Dordrecht 1981.

[12] J. Roger Hindley, 'BCK and BCI Logics, Condensed Detachment and the 2-Property', *Notre Dame Journal of Formal Logic* **34** (1993), 231–250.

[13] Lloyd Humberstone, 'Unique Characterization of Connectives' (Abstract) *Journal of Symbolic Logic* **49** (1984), 1426–1427.

[14] Lloyd Humberstone, 'A More Discriminating Approach to Modal Logic' (Abstract) *Journal of Symbolic Logic* **51** (1986), 503–504.

[15] Lloyd Humberstone, Review of Cleave [7] and Koslow [22], *Australasian Journal of Philosophy* **73** (1995), 475–481.

[16] Lloyd Humberstone, 'Classes of Valuations Closed Under Operations Galois-Dual to Boolean Sentence Connectives', *Publications of the Research Institute of Mathematical Sciences, Kyoto University* **32** (1996), 9–84.

[17] Lloyd Humberstone, 'Contra-Classical Logics', *Australasian Journal of Philosophy* **78** (2000), 437–474.

[18] Lloyd Humberstone, 'Béziau's Translation Paradox', *Theoria* **71** (2005), 138–181.

[19] Lloyd Humberstone, 'Identical Twins, Deduction Theorems, and Pattern Functions: Exploring the Implicative *BCSK* Fragment of **S5**', *Journal of Philosophical Logic* **35** (2006), 435–487.

[20] Lloyd Humberstone, 'Variations on a Theme of Curry', *Notre Dame Journal of Formal Logic* **47** (2006), 101–131.

[21] J. Kabziński, 'BCI-Algebras From the Point of View of Logic', *Bulletin of the Section of Logic* **12** (1983), 126–129.

[22] Arnold Koslow, *A Structuralist Theory of Logics*, Cambridge University Press, Cambridge 1992.

[23] J. R. Lucas, *A Treatise on Time and Space*, Methuen, London 1973.

[24] Jan Łukasiewicz, 'On the Intuitionistic Theory of Deduction', pp. 325–340 in L. Borkowski (ed., *Jan Łukasiewicz: Selected Works*, North-Holland, Amsterdam 1970. (Article originally published 1952.)

[25] E. P. Martin and R. K. Meyer, 'Solution to the P − W Problem', *Journal of Symbolic Logic* **47** (1982), 869–887.

[26] C. Mortensen and T. Burgess, 'On Logical Strength and Weakness', *History and Philosophy of Logic* **10** (1989), 47–51.

[27] D. Nelson, 'Negation and the Separation of Concepts in Constructive Systems', pp. 208–225 in A. Heyting (ed.), *Contructivity in Mathematics: Proceedings of a conference held at Amsterdam 1957*, North-Holland, Amsterdam 1959.

[28] Francesco Paoli, *Substructural Logics: A Primer*, Kluwer, Dordrecht 2002.

[29] Jean Porte, 'Congruences in Lemmon's S0.5', *Notre Dame Journal of Formal Logic* **21** (1980), 672–678.

[30] A. N. Prior, 'Time and Change', *Ratio* **10** (1968), 173–177.

[31] W. V. Quine, 'Grades of Discriminability', pp. 129–133 in Quine, *Theories and Things*, Harvard University Press, Cambridge, Mass. 1981. (Article originally published 1976.)

[32] Dana Scott, 'Rules and Derived Rules', pp. 147–161 in S. Stenlund (ed.), *Logical Theory and Semantic Analysis*, Reidel, Dordrecht 1974.

[33] K. Segerberg, *Classical Propositional Operators*, Clarendon Press, Oxford 1982.

[34] Timothy Smiley, 'The Independence of Connectives', *Journal of Symbolic Logic* **27** (1962), 426–436.

[35] M. Spasowski, 'Some Properties of the Operation *d*', *Reports on Mathematical Logic* **3** (1974), 53–56.

[36] Neil Tennant, 'Deductive Versus Expressive Power: A Pre-Gödelian Predicament', in *Journal of Philosophy* **97** (2000), 257–277.

[37] A. S. Troelstra and D. van Dalen, *Constructivism in Mathematics: An Introduction*, Vols. I and II, North-Holland, Amsterdam 1988.

[38] R. Wójcicki, 'Dual Counterparts of Consequence Operations', *Bulletin of the Section of Logic* (Polish Academy of Sciences) **2** (1973), 54–57.

[39] R. Wójcicki, *Theory of Logical Calculi*, Kluwer, Dordrecht 1988.

Lloyd Humberstone
Department of Philosophy,
Monash University,
Wellington Rd.,
Clayton, Victoria 3800,
Australia.
e-mail: Lloyd.Humberstone@arts.monash.edu.au

9 783764 383534